27th Summer School and International Symposium on the Physics of Ionized Gases

(SPIG 2014)

AA009066

Journal of Physics: Conference Series Volume 565

Belgrade, Serbia
26 – 29 August 2014

Editors:

D. Maric
A. Milosavljevic
Z. Mijatovic

ISBN: 978-1-63439-751-3
ISSN: 1742-6588

Printed from e-media with permission by:

Curran Associates, Inc.
57 Morehouse Lane
Red Hook, NY 12571

Some format issues inherent in the e-media version may also appear in this print version.

Printed by Curran Associates, Inc. (2015)

For permission requests, please contact the Institute of Physics
at the address below.

Institute of Physics
Dirac House, Temple Back
Bristol BS1 6BE UK

Phone: 44 1 17 929 7481
Fax: 44 1 17 920 0979

techtracking@iop.org

Additional copies of this publication are available from:

Curran Associates, Inc.
57 Morehouse Lane
Red Hook, NY 12571 USA
Phone: 845-758-0400
Fax: 845-758-2634
Email: curran@proceedings.com
Web: www.proceedings.com

TABLE OF CONTENTS

Journal of Physics: Conference Series **565** (2014) 011001 doi:10.1088/1742-6596/565/1/011001

Preface

This volume of Journal of Physics: Conference Series contains a selection of papers presented at the 27th Summer School and International Symposium on the Physics of Ionized Gases – SPIG 2014, as General Invited Lectures, Topical Invited Lectures, Progress Reports and associated Workshop Lectures. The conference was held in Belgrade, Serbia, from 26-29 August 2014 at the Serbian Academy of Sciences and Arts. It was organized by the Institute of Physics Belgrade, University of Belgrade and Serbian Academy of Sciences and Arts, under the auspices of the Ministry of Education, Science and Technological Development, Republic of Serbia.

A rare virtue of a SPIG conference is that it covers a wide range of topics, bringing together leading scientists worldwide to present and discuss state-of-the art research and the most recent applications, thus stimulating a modern approach of interdisciplinary science. The Invited lectures and Contributed papers are related to the following research fields:

1. **Atomic Collision Processes** (Electron and Photon Interactions with Atomic Particles, Heavy Particle Collisions, Swarms and Transport Phenomena),
2. **Particle and Laser Beam Interactions with Solids** (Atomic Collisions in Solids, Sputtering and Deposition, Laser and Plasma Interaction with Surfaces),
3. **Low Temperature Plasmas** (Plasma Spectroscopy and other Diagnostic Methods, Gas Discharges, Plasma Applications and Devices) and
4. **General Plasmas** (Fusion Plasmas, Astrophysical Plasmas and Collective Phenomena).

Additionally, the 27th SPIG encompassed three workshops that are closely related to the scope of the conference:

- **The Workshop on Dissociative Electron Attachment (DEA)** – Chaired by Prof. Nigel J Mason, OBE, The Open University, United Kingdom
- **The Workshop on X-ray Interaction with Biomolecules in Gas Phase (XiBiGP)**, Chaired by Dr. Christophe Nicolas, Synchrotron SOLEIL, France
- **The 3rd International Workshop on Non-Equilibrium Processes (NonEqProc)** – Chaired by Prof. Zoran Lj. Petrović, Institute of Physics Belgrade, University of Belgrade, Serbia

The Editors would like to thank the members of the Scientific and Advisory Committees of SPIG conference for their efforts in proposing the program of the conference and to the referees that have reviewed submitted papers, as well as the chairmen of the associated workshops for their efforts and help in organizing them and a selection of excellent invited talks. We particularly acknowledge the efforts of all the members of the Local Organizing Committee in the organization of the Conference. We are grateful to all sponsors of the conference: SOLEIL synchrotron, RoentDek Handels GmbH, Klett Publishing House Ltd, Springer (EPJD and EPJ TI), IoP Publishing (IoP Conference Series), DEA club, Austrian Cultural Forum Belgrade, Institut français de Serbie and Collegium Hungaricum Belgrade.

Holding on to a long tradition is never easy and the only way to achieve that is to have a large number of people who appreciate the conference, so we would like to thank all the invited speakers and participants for taking part in the 27th SPIG conference.

Editors of the issue:
Dr Dragana Marić (Instutute of Physics Belgrade, University of Belgrade)
Dr Aleksandar R. Milosavljević (Instutute of Physics Belgrade, University of Belgrade)
Prof Zoran Mijatović (Faculty of Sciences, University of Novi Sad)

Peer review statement

All papers published in this volume of *Journal of Physics: Conference Series* have been peer reviewed through processes administered by the proceedings Editors. Reviews were conducted by expert referees to the professional and scientific standards expected of a proceedings journal published by IOP Publishing.

Journal of Physics: Conference Series **565** (2014) 012001 doi:10.1088/1742-6596/565/1/012001

Future studies on electron scattering; a renaissance

Nigel J Mason

Department of Physical Sciences, The Open University, Walton Hall, Milton Keynes MK7 6AA United Kingdom

E-mail: nigel.mason@open.ac.uk

Abstract. 2014 is the centenary of the first announcement of the Franck-Hertz experiment [1], now regarded as one of the pivotal experiments of modern physics. The Franck-Hertz experiment is widely regarded as an experiment that provided validation of the Bohr theory of atomic structure, itself only published in 2013, however it should also be viewed as the first quantitative experiment in electron scattering and the birth of scientific study of atomic and molecular phenomena by collisions. Today we recognize that electron-atom and electron-molecule collisions are prevalent across nature, describing disparate phenomena whilst the exploitation of such collisions underpins many of the technologies upon which modern society relies. The centenary of the Franck-Hertz experiment is thus a suitable opportunity to review both our current knowledge of electron interactions and to consider the directions of future research. In this article I therefore aim to both review our current state of knowledge and look forward, proposing that recent advances are providing something of a renaissance to the field and are vital for emerging technologies as well as answering some of the greatest scientific challenges of the 21st century.

The Franck-Hertz Experiment

The original Franck-Hertz experiment [1], passed a beam of electrons through a vapour of mercury (interestingly described as mercury molecules) with subsequent detection of the kinetic energy of the transmitted electrons. A characteristic series of peaks were observed, separated by 4.9eV (254nm), precisely the energy needed to excite the first excited state of the mercury atom Figure 1. Thus an incident electron was described as scattering from the atom, transferring its kinetic energy to the target in exciting the atom. Hence by measuring the energy lost by scattering electrons it should be possible to determine excitation energies in the target and this is the basis of the now standard analytical technique of Electron Energy Loss Spectroscopy (EELS) capable of not only revealing electronic states but vibrational energy levels and, in some cases, even rotational structure. EELS is applied to gas phase targets providing a comparison with optical techniques [2] and is a standard technique in surface science [3] where it may be used to detect constituents of absorbates and may reveal surface phenomena such as plasmons. Electron spectroscopy in general is also used to explore ionization phenomena e.g. through photoelectron spectroscopy (PES) [4]. Thus the Frank Hertz experiment can be considered as the pioneer of energy loss spectroscopy.

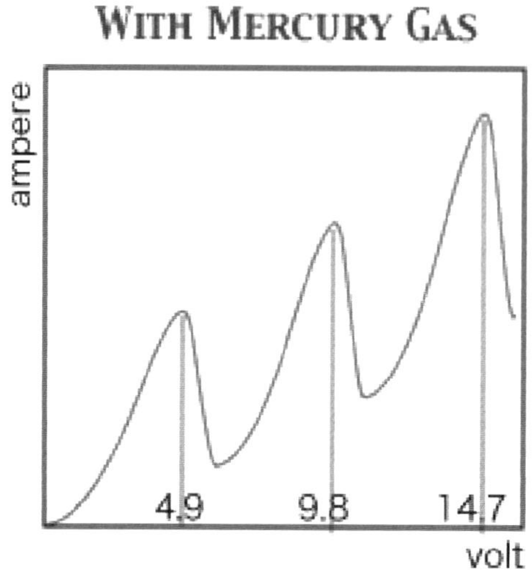

Figure 1. The typical Franck-Hertz electron curve showing the presence of a series of peaks separated by 4.9eV (254nm) characteristic of the excitation energy of the first electronic state of mercury.

Electron collisions and quantum mechanics

The 1920's are regarded as the decade in which quantum mechanics was born with the publication of the Schrodinger equation and the Heisenberg uncertainty principle. However what is often neglected is the role that electron collisions played in the development of the field. Once the true nature of the Franck-Hertz experiment was understood (the authors originally believing it to be an ionization phenomena), and due to both being called on to serve in WW1 this was not until 1919 with their award of the Nobel prize being in 1926, a series of electron transmission experiment were performed. Ramsauer [5] developed a more sophisticated apparatus using magnetic fields to collimate a beam of electrons whose energy could be more accurately defined (Figure 2a). At low energies (less than 1eV) when electrons passed through a chamber containing one of the heavier rare gases (Ar, Kr and Xe) the gas appeared to be 'transparent' with little or no scattering occurring, even for elastic scattering (figure 2b). Such a phenomena could not be explained by any classical theory which, from its 'billiard balls approach', would expect both scattering and at lower energies (longer interactions times) might expect a larger scattering cross section. However by applying the Schrodinger equation and considering the electron projectile as an incident wave composed of partial waves characterised by the momentum of the electron, this Ramsauer-Townsend minimum (Townsend conducting similar experiments at Oxford in the UK) could be explained. Accordingly quantum mechanics could be employed to study electron-atom/molecular interactions and calculate scattering cross sections which could be validated by experiment.

Thus it became possible to quantify electron collisions and use these cross sections to model natural and industrial phenomena. Led by Mott and Massey [6,7] in the UK the era of theoretical collision physics began. The first applications of this theory coincided with the technological development of radar and the discovery of the ionosphere. Ionospheric models and a physical description of that most magnificent of natural phenomena, the aurora, were based on theoretical descriptions and experimental measurement of electron-atom and electron-molecule collisions leading to excitation and ionization of the target species.

Journal of Physics: Conference Series **565** (2014) 012001 doi:10.1088/1742-6596/565/1/012001

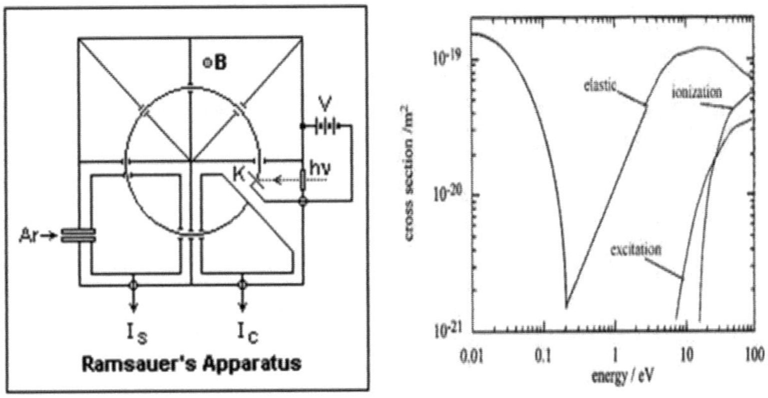

Figure 2. (a) The Ramsauer apparatus used to measure low energy scattering from atoms and molecules revealing (b) the Ramsauer-Townsend minimum in the total and elastic scattering cross section of the rare gases.

The first 'golden age' of electron collisions.

In the 1960's humanity, recovering from the destruction of the second world war, turned to science and in an unparalleled frenzy of activity supported scientific endeavours across the world. Many universities and research institutes were established in this period (hence the current plethora of 50 year anniversaries) and generations of new scientists were able to study the world providing an explosion of scientific data. Atomic and molecular physics was one discipline that enjoyed such a growth, indeed many of the now famous conference series (e.g. ICPEAC) were founded to disseminate this 'data explosion'. With new studies came new results as apparatus were developed to probe collision effects in greater detail. A major breakthrough came with the development of 'monochromated' electron beams, beams with resolution of less than 100meV capable of resolving vibrational structure in energy loss spectra. However such high resolution electron beams also revealed unexpected fine structures in collision cross sections for example a sharp dip in the elastic scattering cross section of electrons from helium at 19.317eV (Figure 3) [8].

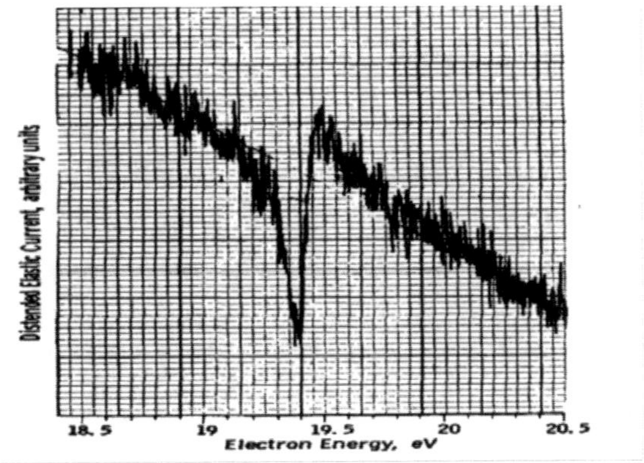

Figure 3. The original trace revealing the presence of resonances in electron scattering from helium [8].

Such a phenomena below the first excited state of helium (19.819eV) appeared to have no explanation but similar effects were seen in other atoms and molecules [9]. Schulz proposed that these 'resonances' could be explained if the electron was temporarily 'captured' by the target to form a 'temporary negative ion (TNI)' which lives for only a fs or ps before decaying by autodetachment hence performing delayed elastic scattering.

$$e + AB \rightarrow AB^- \rightarrow AB + e$$

Subsequent experiments, particularly in nitrogen [9], revealed that this effect may also leave the target excited, indeed such phenomena may dominate electron induced vibrational excitation.

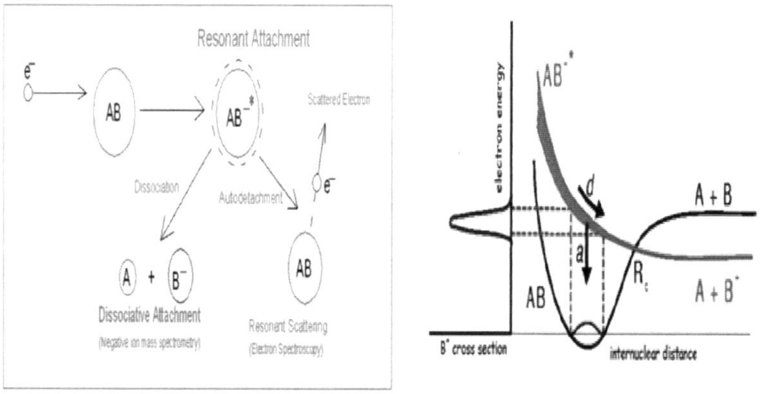

Figure 4. A schematic of the Dissociative Electron Attachment process (DEA) with an illustration of the potential energy surfaces f neutral and temporary negative ionic states [10].

An alternative route of decay leads to the fragmentation of the TNI to form an anionic product

$$AB^- \rightarrow A + B^-$$

a process entitled Dissociative Electron Attachment (DEA). First revealed in the 1960's it is a process common to most molecules though the probability (cross section) is low ($< 10^{-18}$ cm^2) [10]. However since the anionic product is charged it can be readily detected (e.g. by conventional mass spectrometry) and therefore was the subject of extensive studies. DEA cross sections are strongest for targets containing halogen atoms (since they are electronegative) but may be large in compounds containing 'psuedo-halogen' chemical groups such as –CN. DEA occurs naturally in the Earth's ionosphere where O^- anions are formed by DEA to molecular oxygen.

DEA may be readily explained by an understanding of the potential energy surfaces of the anion and the neutral parent (Figure 4). The overlap of the anion curve and the neutral defining both the cross section and the residual kinetic energy of the product anions.

The desire to understand the dynamics of electron scattering led to ever more sophisticated experiments including study of electron-photon coincidences [11] to probe electron induced excitation (mainly of rare gases); electron-ion coincidence experiments [12] to explore electron induced ionization and correlation between incident, scattered and ejected electrons; the study of spin polarized electrons scattering from atoms and molecules [13] and electron scattering in the presence of laser fields[14] to couple electron scattering with photon excitation of the target (including study of 'electron induced de-excitation or 'super-elastic' scattering and simultaneous electron photon excitation). Such detailed experiments provided challenging tests of theoretical representations of electron collisions. Theory developed commensurate with the experimental progress coupling

Journal of Physics: Conference Series **565** (2014) 012001 doi:10.1088/1742-6596/565/1/012001

quantum mechanical descriptions of the scattering with quantum chemistry descriptions of the target structures. Several 'formalisms' were developed and have stood the test of time with the R-matrix techniques [15], translated from nuclear physics perhaps being the most well known, although there are many other methods each with strengths and weaknesses and applicability to one scattering process over another, for a comprehensive review see [16].

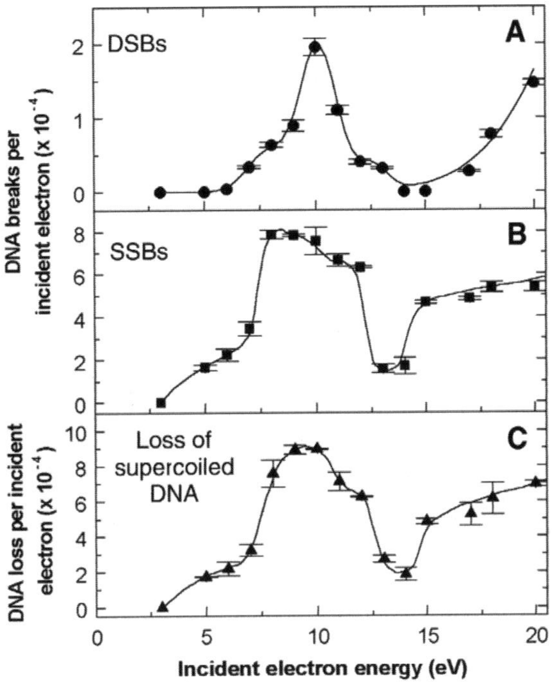

Figure 5. The now famous result of Sanche and co-workers showing resonance structure in electron interactions with DNA [22].

The new age.

In the late 1990's the study of electron collisions seemed to have reached a point where the community had gained an understanding of the collision processes that was sufficient and new challenges in atomic and molecular physics appeared to offer more intellectual and technical challenges (e.g. cold atom physics with the creation of BECs and ultrafast lasers capable of 'imaging' molecular dynamics). The collisions community began to contract and it seemed that the centenary of Franck-Hertz might deem the study of electron collisions as 'mission accomplished'. However, then a series of unrelated events made our need to understand electron interactions at the forefront of technology and the grand challenges of 21st century science.

In 1985 British Antarctic Survey scientists, Joseph Farman , Brian Gardiner, and Jonathan Shanklin reported massive falls in the terrestrial ozone layer in the Antarctic polar spring [17], levels collapsing by upwards of 90%. These were found to be due to a catalytic chemical reaction involving halogens (mainly chlorine) released by solar UV from chloroflourocarbons (CFCs) [18]. Many CFCs are purely manufactured and have no natural sources, indeed the first CFCs were created in 1928 as non-toxic, non-flamable refrigerants, and were first produced commercially in the 1930's by DuPont. The unfettered release of CFCs had led to a build up in their concentrations in the terrestrial atmosphere to such an extent that when they are photolysed the liberated free radicals can destroy many hundreds of ozone molecules before being destroyed themselves. This has led to the creation if the 'ozone hole'.

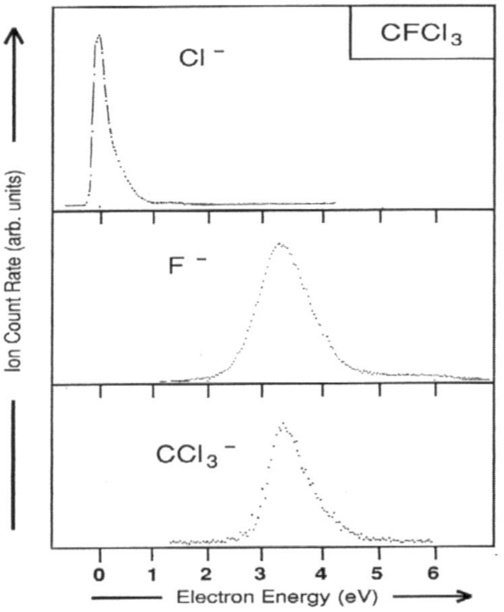

Figure 6. Demonstration of the ability of low energy electrons to selectivity cleave molecular bonds. At low energies ($\approx$ 0eV) electrons cleave the C-Cl bond in $CFCl_3$ whilst at circa 3ev they cleave the C-Cl bond [25].

Simultaneously the observation that the Earth is warming, and at a faster rate than at any time humanity has been present, has led to major scientific and political debate [19]. Global warming arises from certain molecules in the atmosphere 'storing' energy released from the surface and re-radiating it back to the Earth's surface. The main molecules responsible for global warming are CO_2, methane and N_2O but, once again, many industrially produced compounds were found to have a high 'Global Warming Potential' (GWP) many of which also contained halogens so had a large 'ozone depletion potential' (ODP). Largely these halogenated compounds with large GWPs were used in plasma processing to etch surfaces (e.g. C_2F_6, C_3F_8, CF_4, SiF_4, C_5F_{12}, CHF_3, CCl_2F_2, SF_6, HF) [20]. One of the most important 'plasma etching' processes was to build the 'chips' used in computers, the power and speed of which is inversely proportional to the size of the etched phenomena, the smaller the size of the 'integrated circuit' the faster and more powerful the computer with technology development .

In the 1990's industry developing next generation integrated circuits was informed that worldwide governments were preparing to phase out CFCs in order to prevent the ozone layer being irreversibly damaged and to slow global warming. The plasma industry would have to find new 'feed gases' with which to etch silicon and hence fabricate 'computer chips'! However in most cases the actual physico-chemical processes in the plasma were unknown and in replacing existing feed gases with ones with lower GWPs/ODPs in existing plasma reactors the effects on the efficiency (and hence commercial viability) of the plasma reactor were unknown. This highlighted the need to model the plasma reactor and build 'virtual reactors' within which different feed gases could be trialed and etch rates and microstructures studied. Such models required basic electron collision data for many molecular species for which little or no data existed. Both US and Japanese industry sought to engage the international atomic and molecular community to provide such data [21] leading to a revival in both experimental and theoretical work, the latter being particularly relevant since many of the most important species in these plasmas are radicals (CF_x where x =1 to 3) which could not be studied experimentally.

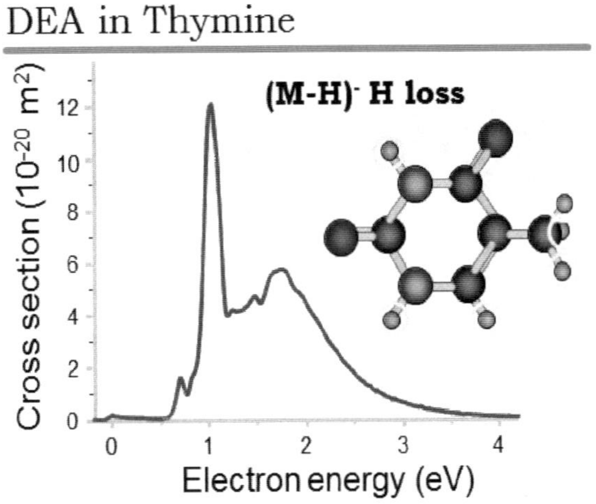

Figure 7. DEA to thymine illustrating selective bond cleavage using low energy electrons, H atoms are liberated from each site at distinct energies leading to fine structure in the DEA cross section [26].

The second field in which the study of electron collisions were found to be critical was radiation chemistry and the development of next generation 'radiotherapy'. In 2000 Sanche and co-workers published an astonishing result, they studied electron induced damage on stands of DNA [22]. The DNA helix can be ruptured in two ways one leading to a single strand break (ssbs), the other producing double strand breaks (dsbs). Sanche et al showed that, contrary to expectation, low energy (<10eV) electrons could produce both ssbs and dsbs and within these damage curves structures were revealed that are reminiscent of 'resonances' seen in electron-molecule scattering, figure 5. These results challenged many assumptions of the radiation chemistry community and our understanding of how DNA, the most important molecule in the cell nucleus, is damaged when irradiated by ionizing radiation. Instead of the primary radiation (X-ray, gamma ray, UV light) the secondary electrons produced by such radiation may induce 'direct' damage while also producing the free radicals that produce 'indirect damage'. Thus radiation chemistry and hence the models that underpin the radiotherapy protocols would need to be based on atomic and molecular interactions, operating on the nanoscale rather than the traditional macroscale picture. This led to an explosion of experimental studies that explored electron interactions with 'biomolecules' such as the nucleobases on which DNA is based [23] and provided theory with enormous challenges to model electron interactions with such complex molecular targets [24].

Electron induced chemistry.

Both plasma and radiation chemistry studies revealed the important role of resonance phenomena in electron scattering and highlighted the role of DEA. In plasmas operating with halogen containing species DEA produced anions that meant the plasma was 'electronegative'. while the low lying resonances observed in electron scattering from biomolecules demonstrated that DEA could fragment the molecule and hence lead to DNA rupture. Furthermore DEA was found to be a 'bond specific process' and by controlling the energy of the incident electron it was possible to control the molecular fragmentation pathways - the 'holy grail' of chemistry for more than a century. Figures 6 and 7 illustrate this electron induced chemical control. In figure 6 very low energy electrons ($\approx$ 0eV) incident upon $CFCl_3$ are able to cleave the C-Cl bond exclusively (producing a Cl^- anion and $CFCl_2$ neutral), whilst electrons at circa 3.2eV cleave the C-F bond with the electron being shared by the F and CCl_3 fragments [25].

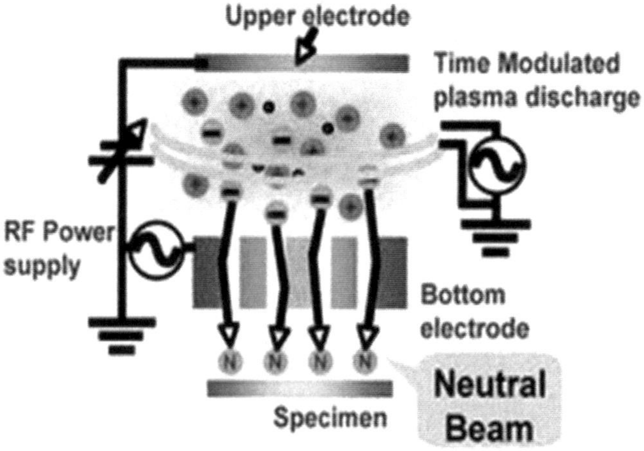

Figure 8. Schematic of a modulated plasma reactor that exploits electron induced chemistry to produce anions which are withdrawn from the plasma, charge exchanged with a carbon surface and then used to make a neutral beam for surface etching.

In biomolecules similar selective bond cleavage is observed, for example in thymine each of the bonds around the ring may be cleaved by electrons of a specific energy (figure 7) [26]. Accordingly is may be possible to design/select molecules that can be selectively fragmented to produce chemically reactive species that in turn produce new compounds or structures. Hence electrons may be the 'tool' for chemical control that femtosecond lasers and coherent control aimed to fulfil in last two decades of the 20th century and electron sources are much cheaper and more rugged that the complex laser systems used in coherent control.

The plasma industry was the first to exploit electron induced chemistry since the electron energy (temperature) in a plasma can be controlled by the voltages applied to the plasma generating electrodes. Samukawa and co-workers [27] demonstrated that, by pulsing the electrode voltages generating the plasma in an industrial etching reactor the electron temperature may be lowered and thus DEA channels opened, with the production of anions, Figure 8. These anions may be drawn out of the plasma and, by passing through a 'carbon filter', the anions converted to neutrals of the anionic species. Thus anions may be converted to neutral radicals which may then be used to etch the surface without the disadvantages of charging of the surface or UV damage from the plasma. Such neutral beam etching has been shown to produce well defined 20-30nm structures.

In radiation chemistry the recognition that DEA may play an important role in damage induced by ionizing radiation led to the proposal that molecules that have high DEA cross sections may be used as 'radiosensitizers' to increase the death of cancer cells [28]. Indeed many of the radiosensitizers' currently used in the clinic (such as bromouracil and cisplatin ($Cl_2H_6N_2Pt$)) contain halogenated species and thus have high DEA cross sections [29,30] – is this cause or effect ? New radiosensitizers have been proposed based on their DEA properties and ability to bind to DNA with higher probability of inducing ssb or dsb rupture when interacting with secondary electrons e.g 5-bromouridine and 5-nitrouracil [31,32]. More recently the introduction of metallic nanoparticles (normally gold or platinum) into cancer cells (ideally into the cell nucleus) has been suggested and tested with the expectation that such particles will be a ready source of secondary electrons which may then rupture DNA by DEA led processes [33,34].

This research is to be developed under a new pan-EU training network ARGENT [35]. Indeed electron induced processes are now being included in radiation damage modelling codes (e.g. LEPTS, GEANT4-DNA) which is a testimony to how far electron induced processes have been recognized within the radiation chemistry community in the 14 years since the pioneering work of Sanche and co-workers.

Future electron studies

The ubiquitous nature of electrons in nature and in the technology we use means that the need to study and exploit electron induced collisions will continue unabated in the 21st century. The recognition that we no longer need to be passive spectators of the electron collision but may manipulate and exploit such processes is opening up new opportunities and will provide the motivation for much of the future research. For example in astronomy it is now clear that we live in a 'molecular universe' with rich (astro)chemistry present in regions of star and planet formation [36]. Indeed anions have been observed in such regions (C_xH^- and C_xN^-) [37] and it is possible that many of the more complex molecules found in these regions are produced by electron irradiation of the ice layers formed on dust grains, molecules that themselves may be essential for development of life (e.g. the assembly of DNA) [38]. The recent Cassini-Huygens mission to Saturn and its moons has revealed that Titan (considered by many to be a representation of early earth before life evolved) has a rich ionospheric chemistry with large concentrations of anions formed by DEA of the nitriles present in the upper atmosphere [39]. Similarly electron induced processes are important in upper atmosphere of Jupiter and Saturn where aurora have been observed.

Figure 9. Schematic of the FEBID process in which a high energy beam of electrons (keV) irradiates a surface during deposition of a feed gas. Nanostructures are formed on the surface through electron induced dissociation and chemistry to the feed gas [42].

The next generation of electronic devices require even smaller (1nm or less) structures (transistors) to be created on surfaces. Such structures are apparently unable to be manufactured using current plasma etching techniques and require new methodologies. One such method is FEBID (Focused Electron Beam Induced Deposition) [40] in which a highly collimated beam of very high energy electrons is incident upon a substrate upon which a feed gas is directed, figure 9. These feed gases are typically large molecular compounds containing a metallic core. The electrons (or rather secondary electrons they liberate) release the metal by fragmenting the molecule and thus nm scale metallic structures can be fabricated. Central to this technology is the knowledge of electron interactions with the feed gas and how the electron induced chemistry both in the gas phase above the substrate and within the absorbate may be controlled to fashion the nanostructures, chemistry that is intriguingly different in the gas and solid (surface absorbate) phases [41]. Many of the feed gases proposed for

FEBID have never been studied before and are perhaps too large for current theoretical methods so have engendered a new set of experimental studies coordinated through the CELINA COST action [42].

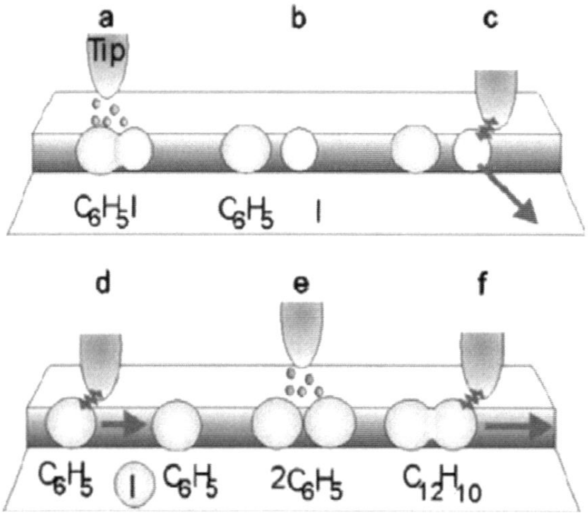

Figure 10. A STM is a bright electron source that may be used to induce chemistry on surfaces with the advantage that it may target one molecule as a specific spatial site. In this example it is used to fashion $C_{12}H_{10}$ from dissociation of C_6H_5I [44].

One of the major instrumental breakthroughs in the late 20th Century has been the development of the Scanning Tunnel Microscope (STM) allowing us to image atoms and molecules directly. A STM is *an electron source* and thus allows us to explore electron collisions in a new way. The 'electron beam' in a STM is highly focused and in terms of flux density very intense, indeed so intense that excitation of a molecule by two electrons is possible (analogous to two photon excitation in a strong laser field) [43]. However STM technology provides the opportunity to not only image the molecules but to move the molecules across the surface (e.g. to a reaction site) and may dissociate the molecule. The classic (and one of the earliest) example was the study of C_5H_5I (Figure 10) [44]. The STM electron 'beam' is used to cleave the molecule producing C_5H_5 + I the C_5H_5 fragment may then be moved across the surface to meet a second C_5H_5 fragment and under the STM tip the two fragments combine to form $C_{10}H_{10}$. It is now known that the iodine fragment is most likely the I^- anion formed by DEA. Thus STM can induce electron chemistry with the added advantage that it can be induced in a specific site and even with a specific molecule. Thus, if electron chemistry can be combined with STM technology, the 'holy grail' of chemistry achieving site and spatial specificity may be achieved using electrons. Therefore in future it is to be expected (required) that studies of electron interactions will draw on members of the atomic and molecular physics, surface science and STM/nanotechnology communities to explore electron collision dynamics and, perhaps, lead to the construction of new structure/materials – though it is to be noted that STMs are 'low throughput' machines for any industrial fabrication.

Looking ahead other areas of science and technology that will engage with electron collisions include the space community. Electron collisions play an important role in space craft re-entry phenomena and were a major topic in the EU Phys4Entry project [45], with the development of commercialization of space (e.g. in asteroid mining and space tourism) such modeling will become increasingly relevant. It is also becoming recognized that electrons are present in large numbers in flames and thus that electron induced processes are important in combustion studies [46]. Indeed the correlation between compounds used as fire retardants and their large DEA cross sections has been noted for some time [47]. Therefore one of the future areas ripe for development is exploration of electron chemistry in combustion.

Conclusions

In this review a brief history of the development of electron scattering from the pioneering experiment by Franck and Hertz a century ago to the resurgence of the field in the first decade of the 21st century has been presented. The discovery of the ability of electrons to cleave specific bonds in the target molecule has created a new opportunity to use electrons as a tool for controlling chemistry in different environments including plasma processing and radiotherapy. Electron induced chemistry is now believed to play an important role in chemical synthesis in space and in flames. Thus the study of electron collisions with an ever larger number of atomic and molecular targets will require the attention of many in the scientific community in the 21st century such that, on the occasion of the bicentenary of the Franck-Hertz experiment in 2114, it will still be seen as one of the great pioneering experiments in the history of Science founding the still active and exciting field of electron collisions.

Acknowledgements

This work is a review of the research of many people from many different fields. The author wishes to thank his many colleagues around the world for their comments and support. He recognizes the support of many funding agencies supporting his own work in the field including the EU Framework VII programme (particularly Marie Curie programme); the UK EPSRC and STFC and his host institution, The Open University.

References

[1] Franck J and Hertz G 1914 *Verhandlungen der Deutschen Physikalischen Gesellschaft* (in German) **16** 457; 1914 *Verhandlungen der Deutschen Physikalischen Gesellschaft* (in German) **16** 512
[2] Gingell J M, Mason N J, Zhao H, Walker I C and Siggel M R F 1999 *J Phys B* **32** 2729
[3] Ibach H and Mills D L 1982 *Electron energy loss spectroscopy and surface vibrations* (Academic Press)
[4] Hufner S 2003 *Photoelectron Spectroscopy: Principles and Applications* (Springer-Verlag)
[5] Ramsauer C 1921 *Annalen der Physik* **64** 513
[6] Mott N F 1930 *Proc. Math. Phys. Eng. Sci.* **127**(806) 658; Massey H S W 1930 *Proc. R. Soc. Lond. A.* **127** 666
[7] Mott N F and Massey H S W 1933, *The Theory of Atomic Collisions* (Clarendon Press, Oxford)
[8] Schulz G J 1963 *Phys. Rev. Lett.* **10**,104
[9] Schulz G 1973 *Reviews of Modern Physics* **45** 37873)
[10] 1984 *Electron-Molecule Interactions and Their Applications* vol 1, ed L Christophorou (Academic Press), chapter 5
[11] Kleinpoppen H, Lohmann B and Grum-Grzhimailo A 2013, *Perfect/Complete Scattering Experiments: Probing Quantum Mechanics on Atomic and Molecular Collisions* (Springer)
[12] 2010 *Electron Ionisation*, ed Maerk T and Dunn G H (Innsbruck University Press)
[13] Gay T J 2009 Physics and Technology of Polarized Electron Scattering from Atoms and Molecules *Advances In Atomic, Molecular, and Optical Physics* vol 57, chapter 4, p157
[14] Mason N J 1993 Laser Assisted Electron Atom Collisions *Reports on Progress in Physics* **56** 1275
[15] Burke P G 2011 *R-Matrix Theory of Atomic Collisions: Application to Atomic, Molecular and Optical Processes* (Springer-Verlag)
[16] Curik R and Carsky P 2012 *Low-Energy Electron Scattering from Molecules, Biomolecules and Surfaces* (CRC Press)
[17] Farman J C, Gardiner B G and Shanklin J D 1985 *Nature* **315** 207
[18] Holloway A M and Wayne R P 2010 *Atmospheric Chemistry* (RSc Publishing)

[19] Global Warming; The Hard ScienceHarvey L D D 2000 *Global Warming; The Hard Science* (Prentice Hall)
[20] Makabe T and Petrovic Z 2014 *Plasma Electronics: Applications in Microelectronic Device Fabrication* (Taylor and Francis)
[21] Christophorou L G and Olthoff J K 2004 *Fundamental Electron Interactions with Plasma Processing Gases* (Kluer Academic)
[22] Boudaiffa B, Cloutier P, Hunting D, Huels M A and Sanche L 2000 *Science* **287** 1658
[23] Baccarelli I, Bald I, Gianturco F A, Illenberger E and Kopyra J 2011 *Physics Reports* **508** 1
[24] 2006 *Advances in Quantum Chemistry: Theory of the Interaction of Radiation with Biomolecules* ed E Brandas and J R Sabin (Academic Press)
[25] Illenberger E, Scheunemann H U and Baumgärtel H 1999 *Chemical Physics* **37** 21
[26] Ptasińska S, Denifl S, Mróz B, Probst M, Grill V, Illenberger E, Scheier P and Märk T D 2005 *J. Chem. Phys.* **123** 124302
[27] Samukawa S 2006 *Japanese Journal of Applied Physics* **45** 2395
[28] Kopyra J, Keller A and Bald I 2014 *RSC Adv.* **4** 6825
[29] Abdoul-Carime H, Huels M A, Illenberger E and Sanche L 2001 *J. Am. Chem. Soc.* **123** 5354
[30] Hsing-Yin Chen, Hui-Fen Chen, Chai-Lin Kao, Po-Yu Yanga and Sodio C N Hsua 2014 *Phys. Chem. Chem. Phys* **16** 19290
[31] Abdoul-Carime H, Limao-Vieira P, Gohlke S, Petrushko I, Mason N J and Illenberger E 2004 *Chemical Physics Letters* **393** 442
[32] Ptasińska S, Alizadeh E, Sulzer P, Abouaf R, Mason N J, Märk T D and Scheier P 2008 *Int. Journal of Mass Spectroscopy* **277** 291
[33] Coulter J A, Hyland W B, Nicol J and Currell F J 2013 *Clinical Oncology* **25** 593
[34] Porcel E, Liehn S, Remita H, Usami N, Kobayashi K, Furusawa Y, Sech C and Lacombe S 2010 *Nanotechnology* **21** article number 085103
[35] http://www.mbnresearch.com/project-argent
[36] Tielens A G G M 2013 *Rev. Mod. Phys.* **85** 1021
[37] McCarthy M C, Gottlieb C A, Gupta H and Thaddeus P 2006 *Astrophys. J.* **652** L141
[38] Mason N J, Nair B, Jheeta S and Szymańska E 2014 *Faraday Discussions* **168** 235
[39] Vuitton V, Lavvas P, Yelle R V, Galand M, Wellbrock A, Lewis G R, Coates A J and Wahlund J E 2009 *Planetary and Space Science* **57**, 1558
[40] 2012 *Nanofabrication Using Focused Ion and Electron Beams: Principles and applications* ed Ivo Utke, Stanislav Moshkalev and Phillip Russell (Oxford University Press)
[41] Engmann S, Stano M, Matejčík S and Ingólfsson O 2012 *Phys Chem Chem Phys.* **14** 14611
[42] http://celina.uni-bremen.de/celina
[43] Sloan P A and Palmer R E 2005 *Nature* **434** 367
[44] Saw-Wai Hla, Ludwig Bartels, Gerhard Meyer and Karl-Heinz Rieder 2000 *Phys. Rev. Lett.* **85** 2777
[45] http://users.ba.cnr.it/imip/cscpal38/phys4entry
[46] Bisettia F and Mbark El Morslib 2014 *Combustion Theory and Modelling* **18** 148
[47] Mills R M 1968 *Combustion and Flame* **12** 513

Journal of Physics: Conference Series **565** (2014) 012002 doi:10.1088/1742-6596/565/1/012002

Optimized spectral analysis in magnetic resonance spectroscopy for early tumor diagnostics

Karen Belkić[1-3] and Dževad Belkić[1]
[1] Department of Oncology-Pathology, Karolinska Institute, Stockholm, Sweden
[2] School of Community and Global Health, Claremont Graduate University, USA
[3] Institute for Prevention Research, University of Southern California, USA

Email: Karen.Belkic@ki.se

Abstract. Molecular imaging through magnetic resonance spectroscopy (MRS) can provide information about key metabolites. Conventional applications of MRS are hampered by data analysis via the fast Fourier transform (FFT). Most MRS studies for cancer detection have relied upon estimations of a mere handful or even a single composite metabolite, e.g. total choline. These have yielded incremental improvements in diagnostic accuracy. In vitro studies reveal richer metabolic information for identifying cancer, particularly in closely-overlapping resonances. Among these are phosphocholine, a marker of malignant transformation. The FFT cannot assess these congested spectral components. This can be done by the fast Padé transform (FPT), an advanced, high-resolution, quantification-equipped method, applied to MRS time signals as encoded from patients with breast cancers and other cancers, with benign pathology and with normal tissue, as illustrated herein for the latter. With realistic noise levels, the FPT accurately computes the metabolite concentrations, including phosphocholine, which completely underlies phosphoethanolamine. In sharp contrast, the FFT produces a rough envelope spectrum with only a few shortened, broadened peaks, and key metabolites altogether absent. The FPT clearly separates true metabolites from spurious resonances. The efficiency and high resolution of the FPT translates into shortened examination time of the patient. These capabilities strongly suggest that by applying the FPT to time signals encoded in vivo from breast cancer and other malignancies, MRS will fulfill its potential to become a clinically-reliable, cost-effective method for early cancer detection.

1. Introduction

Magnetic resonance-based modalities are promising for early cancer detection, particularly since they do not entail exposure to ionizing radiation. Magnetic resonance imaging (MRI) is usually very sensitive, but is costly and not sufficiently specific. Molecular imaging through magnetic resonance spectroscopy (MRS) can potentially improve the specificity of MRI, by providing information about key metabolites. In this paper, we will focus upon MRS-based diagnostics and how these can be improved through optimization of signal processing methods. The actual problem area addressed will be related to screening and early detection of breast cancer. This problem is chosen in light of the fact that worldwide, breast cancer is the most commonly diagnosed malignancy among women and a major cause of death [1]. Survival is profoundly impacted by the stage at diagnosis [2]. Screening significantly improves survival, since with early detection, adequate and timely care can be provided [3]. However, consensus has not been reached as to the best strategies for early breast cancer detection

[4]. There is widespread recognition that screening programs need to be improved and made more cost effective [1]. We examine these topics from the standpoint of mathematical optimization via advanced signal processing, as this aspect of the problem pertains to MRS for breast cancer diagnostics.

1.1. Screening for breast cancer: The current status of diagnostic imaging

Currently, anatomical imaging through mammography together with ultrasound and magnetic resonance imaging (MRI) is the mainstay for breast cancer screening, with MRI generally showing the best sensitivity [5]. However, MRI has poor specificity and, as such, is associated with high biopsy rates, very often for benign lesions [3]. Expense is another drawback of MRI, such that it is currently considered cost-effective only for women at very high breast cancer risk [6].

Since these anatomic imaging modalities are insufficiently specific, functional and molecular imaging through magnetic resonance (MR) are garnering attention vis-à-vis breast cancer diagnostics [7]. Diffusion-weighted MRI (DWI) helps distinguish benign lesions and normal breast from cancer. Since malignancy is often associated with increased cellularity or intracellular density, the random motion of free water may be restricted, as reflected in a decreased apparent diffusion coefficient (ADC). However, a low ADC can occur with fibrotic tissue or benign proliferative changes, while necrotic tissue in a cancer is associated with increased ADC. Problems remain in selecting optimal cutpoints and recording parameters for DWI. Thus, breast cancer is not detected via DWI with full certainty [8].

Improved identification of breast cancer can potentially be achieved via molecular imaging through MRS or MR spectroscopic imaging (MRSI), whereby the metabolic characteristics of the examined tissue are assessed. When a single MRS voxel is inadequately representative, MRSI provides volumetric coverage. A recent meta-analysis of in vivo MRS and MRSI studies using 1.5T or 3T scanners reveals a pooled sensitivity of 73% and specificity of 88% for detecting over 700 breast cancers versus over 400 scans of non-cancerous breast [9]. These analyses were based on the total choline (tCho) peak, whose resonant frequency is ~3.2 parts per million (ppm). As an indicator of cell membrane turnover, tCho was used as a sign of cancer. Since signal-to-noise ratio (SNR) increases linearly with magnetic field strength, higher field scanners would be expected to improve SNR, to enhance MRS-based breast diagnostics. However, the diagnostic accuracy of MRS was not better with 3T compared to 1.5T scanners [9]. Using higher field (4T or 7T) MR scanners [10-12], tCho was detected in normal breast and benign lesions and was undetectable in several cancers. Overlap was seen among the ranges of tCho for these three categories of breast tissue. The high costs would also preclude high-field scanners from most breast cancer screening protocols. All told, tCho as a single composite spectral entity has not yielded adequate diagnostic accuracy for breast cancer screening. Further difficulties arise due to the various cutpoints of tCho used to define breast cancer.

1.2. Inadequacies due to reliance upon conventional data processing in MRS and MRSI

Whereas isolated resonances are better resolved with higher field scanners, widths are broadened when MRS time signals are encoded with stronger magnets. This adversely impacts on overlapping resonances that are abundant in MR spectra of the breast. Choline detection can be hampered with higher field magnets due to lipid-induced sidebands [13]. These lipids can be suppressed by prolonging the echo time (TE). However, potentially informative metabolites with short T_2 relaxation time will also have decayed with long TE. Furthermore, the lipid itself may be part of the actual disease process, such that diagnostic information could be lost with this encoding strategy [14].

Nuclear magnetic resonance performed in vitro on extracted breast specimens reveals more abundant metabolic information [15]. Components of tCho: phosphocholine (PC) resonating at ~ 3.22 ppm, glycerophosphocholine (GPC) at 3.23 ppm and free choline (Cho) at 3.21 ppm can better characterize breast tissue [16]. With malignant transformation in the breast, PC is increased in association with a "glycerophosphocholine to phosphocholine switch" due to the over-expressed enzyme choline kinase responsible for PC synthesis [16-18]. Besides the resonant frequencies of PC,

GPC and free choline being extremely close to one another, the far more abundant phosphoethanolamine (PE) also resonating at ~ 3.22 ppm, and which is not a cancer biomarker, obscures the PC peak on total shape spectra.

Encoded MRS time signals require mathematical transformation into their spectral representation in the frequency domain. Currently, this is done by the fast Fourier transform (FFT), built into all clinical MR scanners. The FFT is a low-resolution, linear, non-parametric processor. As a single polynomial, the FFT passes all the noise from the time domain into the frequency domain, with no way to suppress this noise corruption. The FFT can only produce a total shape spectrum; the underlying components cannot be generated via Fourier analysis. Thus, the overlapping resonances contained within the tCho peak cannot be disentangled by the FFT. No trustworthy information can be gleaned regarding the abundance of phosphocholine, PC, nor about the concentration of PE which, as mentioned, is completely superimposed on PC. Via the FFT, not even the number of overlapping resonances contained within a given peak can be determined. Post-processing via fitting is employed to surmise how many components underlie a given peak. Consequently, it is likely that false information will be generated by overfitting (overmodelling) and/or that physical resonances are missed because of underfitting (undermodelling). Fitting procedures cannot generate reliable quantitative information even about predetermined resonances, due to the likelihood of bias. Only by serendipity will a correct guess be made [19, 20]. It has therefore been concluded that the inability to autonomously reconstruct the spectral parameters, i.e. the fundamental complex frequencies and amplitudes, completely eliminates the FFT as a signal processor for solving the quantification problem, the "raison d'être" of MRS [19, 20].

1.3. The fast Padé transform: An advanced signal processor well-suited to MRS and MRSI

The fast Padé transform (FPT) is a high-resolution, non-linear signal processor with parametric and non-parametric forms. It is well-suited to MRS and MRSI [19-22], and can solve the above-described difficulties in breast cancer diagnostics. The number of genuine resonances contained in an encoded MRS time signal is handled as a parameter by the FPT, and is ascertained exactly. The complex-valued fundamental frequencies and amplitudes are precisely reconstructed for each resonance. With these parametric data, the key clinical information, the metabolite concentrations are accurately computed, even when resonances closely overlap [19, 21-22]. The spectrum generated by the FPT is the quotient of two polynomials (*P/Q*). There are two equivalent variants: $\mathrm{FPT}^{(+)}$ and $\mathrm{FPT}^{(-)}$ inside and outside the unit circle for the causal and anti-causal representation, respectively [19-22]. These two variants provide a fully self-contained checking procedure. The FPT algorithmically separates the genuine and spurious resonances by identifying pole-zero confluences. This is vital when there are many spurious resonances as in most MRS quantification problems, since over-determined systems of linear equations are encountered. The FPT-generated spectrum as the quotient of two polynomials (*P/Q*) produces a unique set of genuine spectral poles and zeros. The denominator polynomial Q is associated with the system poles and represents the positions (chemical shifts) and widths of peaks in a spectrum. The numerator polynomial P is associated with the valleys between adjacent peaks, and the system zeros are described thereby. Coincident poles and zeros, termed Froissart doublets [23], are unstable, roaming stochastically in the complex frequency plane with the slightest perturbation, and they never converge. Consequently, these spurious resonances are akin to noise which must be identified and discarded from the final results of the analysis. The two polynomials P and Q provide two degrees of freedom which facilitate noise suppression, as follows. In the Padé-reconstructed spectrum $P_{K'}/Q_{K'}$, the total number of resonances K' is larger than its genuine counterpart K in the unknown ratio P_K/Q_K. The overestimation $\Delta K = K' - K$ is equi-partitioned between $P_{K'}$ and $Q_{K'}$, such that $P_{K'}$ and $Q_{K'}$ will have an equal number of the same spurious resonances in $P_{K'}/Q_{K'}$. This excess, spurious content cancels out in $P_{K'}/Q_{K'}$. After this cancellation, only the true information P_K/Q_K remains in the reconstructed data. Further details are given Refs. [19, 23-28].

1.4. Applications of the FPT to MRS time signals from breast and rationale for the present study
We have applied the FPT to noise-free MRS time signals as encoded from extracted breast specimens from normal breast, fibroadenoma and breast cancer [15]. In all three cases, the FPT resolved and precisely quantified the nine resonances, including PC and PE that were almost completely overlapping [4, 19, 26]. Many spurious resonances were also produced; at convergence 750 resonances were retrieved, i.e. 741 spurious plus the 9 physical resonances. Although fewer than 2% of the resonances were genuine, the FPT delineated these two categories with certainty in all cases. The FPT was then applied to MRS time signals associated with breast cancer and benign breast pathology (fibroadenoma) with added noise (σ = 0.0289 RMS, where RMS is the root-mean-square of the noise-free time signal, RMS = $(N^{-1}\sum_{n=0}^{N-1}|c_n|^2)^{1/2}$ and σ is the standard deviation). Convergence was achieved in both cases at N_P = 1700, with accurate reconstruction of the spectral parameters for all 9 genuine metabolites. Also, 841 spurious resonances were generated and were recognized as Froissart doublets with pole-zero confluences and zero-valued amplitudes, together with marked instability in the face of changes in noise level or signal length [27, 28]. By this three-fold "signature" the non-physical content is distinguished from physical resonances [7, 23, 27, 28].

Based upon these results, we herein apply the FPT to noise-corrupted time signals from breast without pathology to provide normative data. Of particular interest is the capability of the FPT to resolve the resonances between 3.21 ppm and 3.23 ppm, where, as noted, elevated phosphocholine is a marker of malignant transformation [16, 18]. We assess the performance of the FPT in handling noise-corrupted MRS time signals in this controlled setting, as a key step towards clinical application of Padé-optimized MRS for breast cancer diagnostics.

2. Methods

2.1. The input data: an MRS time signal as encoded from normal, non-infiltrated breast
The time signal of the typical quantum-mechanical form was generated according to:

$$c_n = \sum_{k=1}^{K} d_k e^{in\omega_k\tau} \quad , \qquad \mathrm{Im}\,(\omega_k) > 0 \qquad (0 \le n \le N-1), \tag{1}$$

where N is the total signal length. These input data are based upon encoded time signals from tissue samples of normal, non-infiltrated breast from twelve patients, as per Ref. [15]. Each c_n is a sum of K = 9 damped complex exponentials $\exp\,(in\tau\omega_k)$ $(1 \le k \le 9)$ with complex amplitudes d_k. Herein, ω_k and d_k are the fundamental angular frequencies and amplitudes; $\omega_k = 2\pi\,\nu_k$, where ν_k is the linear frequency. The time signals of the form (1) were quantified via the FPT, as per Refs. [19-22].

Since ω_k is complex-valued, such that for Im(ω_k) > 0, as in (1), quantity c_n decreases exponentially over time $n\tau\,(n = 0, 1, 2,\ldots, N-1)$. Thus, $\exp\,(in\omega_k\tau) = \exp\,(-n\tau\,\lambda_k + i\,n\tau\,\mu_k)$, where $\lambda_k = \mathrm{Im}\,(\omega_k)$ and $\mu_k = \mathrm{Re}\,(\omega_k)$. The rhs of Eq. (1) is then a linear combination of K exponentially attenuated frequency-dependent cosinusoids and sinusoids each of which is premultiplied by the constant, real intensity $|d_k|$:

$$d_k e^{in\omega_k\tau} = \left(|d_k|e^{i\varphi_k}\right)e^{in\omega_k\tau} = |d_k|\{e^{-n\lambda_k\tau}[\cos(n\mu_k\tau + \varphi_k) + i\sin(n\mu_k\tau + \varphi_k)]\}, \tag{2}$$

where φ_k is the phase of the complex amplitude d_k.

Herein, for conciseness, we present explicitly only the reconstructions of the FPT$^{(-)}$. The converged findings are identical to those obtained from the FPT$^{(+)}$ [19-22]. The diagonal ($M = K$) and non-diagonal ($M \neq K$) complex-valued spectra in the FPT$^{(-)}$ are:

$$G^-_{M,K}(z^{-1}) = \frac{P^-_M(z^{-1})}{Q^-_K(z^{-1})} \quad ; \quad P^-_M(z^{-1}) = \sum_{r=0}^{M} p_r\, z^{-r}, \quad Q^-_K(z^{-1}) = \sum_{s=0}^{K} q_s z^{-s}. \tag{3}$$

Therein, z^{-1} is the harmonic variable $z^{-1} = \mathrm{e}^{-i\omega\tau}$ and $\{p_r^-, q_s^-\}$ are the expansion coefficients. The unique Padé spectrum $G_{M,K}^-(z^{-1})$ from Eq. (3) exists for the given (input) exact finite-rank Green's function $G^-(z^{-1})$ via:

$$G(z^{-1}) = G_{M,K}^-(z^{-1}) + \mathcal{O}(z^{-M-K-1}); \quad G(z^{-1}) = \sum_{n=0}^{N-1} c_n\, z^{-n}. \tag{4}$$

Here, the symbol $\mathcal{O}(z^{-M-K-1})$ signifies the error which is a series in powers $z^{-m}(M + K + 1 \le m \le \infty)$. Thus, the approximation $G(z^{-1}) \approx G_{M,K}^-(z^{-1})$ in the general FPT$^{(-)}$ includes exactly $M + K$ signal points $\{c_n\}$ from the input data $G(z^{-1}) = \sum_{n=0}^{N-1} c_n\, z^{-n}$ with $M + K \le N - 1$.

The MRS time signals from Ref. [15] of length N = 65536 were recorded at a Larmor frequency (ν_L) of 600 MHz and with a static magnetic field strength $B_0 \approx 14.1\mathrm{T}$. A bandwidth (BW) of 6 MHz was used, where the inverse of this BW is the sampling time τ. The nine resonances identified in Ref. [15] were grouped into two frequency bands. The first was from 1.3 ppm to 1.5 ppm and the second from 3.2 ppm to 3.3 ppm. Within the second band were seven resonances, including two which very closely overlapped at ~ 3.22 ppm: phosphocholine, PC resonance #4 and phosphoethanolamine, PE resonance #5 separated by 0.001 ppm. In the present study, we used a 32-fold shorter time signal of total length N = 2048. The amplitudes d_k were computed based upon the reported concentrations of metabolites $\mathrm{C_{met}}$ for normal non-infiltrated breast tissue from twelve patients [15], with $|d_k| = 2\mathrm{C_{met}}/\mathrm{C_{ref}}$, where $\mathrm{C_{ref}}$ = 0.05 mM/g. The internal reference used in Ref. [15] was TSP (3-(trimethylsilyl-) 3,3,2,2-tetradeutero-propionic acid), such that $|d_k| = \mathrm{C_{met}}/(25\mu\mathrm{M/g})$ of wet weight (ww) of the tissue. In Ref. [15], the T_2^* relaxation times were not reported. We set the line widths (full-widths at half-maximum, FWHMax) to be from 0.0008 to 0.0009 ppm. The peaks are represented by Lorentzians, consistent with the time signal from Eq. (1). In the absorption mode, we have used mainly the diagonal, $\mathrm{Re}(P_K^{\pm}/Q_K^{\pm})$ and paradiagonal, $\mathrm{Re}(P_{K-1}^{\pm}/Q_K^{\pm})$ forms, since these two variants for l = 0 and l = 1 are empirically observed to be more stable compared to those using other values of l (l = 2, 3, ...) in $P_{K-l}^{\pm}/Q_K^{\pm}$. The line widths are proportional to Im (ν_k). The smallest chemical shift difference of 0.001 ppm is only 1.11 to 1.25 times larger than the line widths of 8×10^{-4} to 9×10^{-4} ppm. The phases $\varphi_k (1 \le k \le 9)$ from complex-valued d_k were all set to zero. Thus, all the d_k's are real, $d_k = |d_k|$. The input data are displayed in the top panel (1) of table 1.

Through the parametric version of the diagonal FPT$^{(-)}$ to analyze the MRS time signal data from normal, non-infiltrated breast tissue, the coefficients $\{p_r^-, q_s^-\}$ of the polynomials $P_K^-(z^{-1})$ and $Q_K^-(z^{-1})$ from Eq. (3) were computed from the defining product $Q_K^-(z^{-1})G(z^{-1}) \approx P_K^-(z^{-1})$. When herein the coefficients of the same powers of z^{-1} are equated, two systems of linear equations are generated, one for $\{q_s^-\}$ and the other for $\{p_r^-\}$:

$$\sum_{j=0}^{K} c_{s-j}\, q_j^- = 0 \quad (0 \le s \le K, q_0^- = 1), \quad p_r^- = \sum_{j=0}^{r} c_{r-j}\, q_j^- \quad (0 \le r \le K). \tag{5}$$

After the set $\{q_s^-\}$ is extracted from the input data $\{c_n\}$ ($0 \le n \le N$–1), the other set $\{p_r^-\}$ becomes immediately available via the analytical expression $p_r^- = \sum_{j=0}^{r} c_{r-j}\, q_j^- \; (0 \le r \le K)$ from (5). Thus, the set $\{p_r^-, q_s^-\}$ is obtained by solving only one system of linear equations (that for $\{q_s^-\}$). Pole-zero coincidences come from correlation between the numerator (P_K) and the denominator (Q_K) polynomials in the Padé quotient P_K / Q_K. This correlation is evidenced by Eq. (5) where the expansion coefficients $\{p_r^-\}$ of P_K are derived from the expansion coefficients $\{q_s^-\}$ of Q_K. To extract the peak parameters, the characteristic equation $Q_K^-(z^{-1}) = 0$ is solved, leading to K unique roots $z_{k,Q}^{-1} = e^{-i\tau\omega_{k,Q}^-} (1 \le k \le K)$.

Thus, the sought eigen-frequency $\omega_{k,Q}^{-}$ is deduced via $\omega_{k,Q}^{-} = (i/\tau)\ln(z_{k,Q}^{-1})$. Hereafter, $z_{k,P}^{-1}$ and $z_{k,Q}^{-1}$ denote the zeros of $P_K^{-}(z^{-1})$ and $Q_K^{-}(z^{-1})$, respectively. The zeros $z_{k,Q}^{-1}$ are required for quantification. In addition, for signal-noise separation via Froissart doublets the zeros $z_{k,P}^{-1}$ are also needed. Through the FPT$^{(-)}$, the parameters $\{\omega_{k,Q}^{-}, d_k^{-}\}$ $(1 \leq k \leq K)$ of each physical resonance are obtained directly from the examined MRS time signal. The kth metabolite concentration is computed from the reconstructed amplitudes d_k^{-} via $C_{\text{met}}^{-} = (25\mu\text{M/g})\,|d_k^{-}|$, as mentioned.

2.2. *Addition of noise corruption to the input data*

We performed the reconstructions using the input data corrupted by noise. The latter were generated by adding complex-valued random zero-mean Gauss-distributed white noise of preassigned levels to the noiseless MRS time signal. The noise levels were σ = 0.00289 RMS and σ = 0.0289 RMS with, as noted, RMS being the root-mean-square of the noise-free time signal.

2.3. *Comparisons of the FPT and the FFT in non-parametric signal processing*

It is only for total shape spectra that the Padé and Fourier spectra can directly be compared, because as a non-parametric estimator, that is all the FFT can compute. The Fourier spectrum is given by:

$$F_k = \frac{1}{N}\sum_{n=0}^{N-1} c_n \exp(-2i\pi kn/N); \quad 1 \leq k \leq N\text{-}1\,, \tag{6}$$

where $2\pi k/T$ are fixed Fourier grid frequencies that are unrelated to ω_k from c_n. Here, T is the total duration or acquisition time of the signal, $T = N\tau$. The variable $\exp(-2i\pi kn/N)$ in (6) is an undamped harmonic. The non-parametric data are provided by the FPT when the Padé polynomials P_K^{-} and Q_K^{-} are extracted from the time signal $\{c_n\}$. The ratio P_K^{-}/Q_K^{-} is the complex-valued total shape spectrum whose real part is the absorption spectrum. The FPT generates the total shape spectrum P_K^{-}/Q_K^{-} at any given set of sweep frequencies that need not correspond to the preassigned Fourier grid [19, 20]. Thus, the FPT has interpolation capabilities relating directly to the features of the analyzed time signal.

2.4. *Quantification of MRS time signals from normal, non-infiltrated breast by the parametric FPT*

The MRS data are quantified by the FPT via polynomial rooting, whereby the roots of the characteristic equations of the numerator (P_K) and the denominator (Q_K) polynomials generate the zeros and poles of the Padé spectrum P_K^{-}/Q_K^{-}, respectively. Since the rational function P_K^{-}/Q_K^{-} is meromorphic, the zeros of Q_K^{-} are the poles of P_K^{-}/Q_K^{-}. Meromorphic functions have poles as the only singularities. Roots $z_{k,Q}^{-1}$ of equation $Q_K^{-}(z^{-1}) = 0$ reconstruct the fundamental or eigen-frequencies $\omega_{k,Q}^{-}$. The amplitude d_k^{-} is given via the analytical expression for the Cauchy residue of $P_K^{-}(z^{-1})/Q_K^{-}(z^{-1})$ taken at the kth pole $z^{-1} = z_{k,Q}^{-1}$. Although the quantification problem entails finding the eigen-set $\{\omega_{k,Q}^{-}, d_k^{-}\}$ $(1 \leq k \leq K)$, it is not the concluding part of spectral analysis, since this does not generate the eigenset $\{\omega_{k,Q}^{-}, d_k^{-}\}$ $(1 \leq k \leq K)$ with K being the true number of resonances. Rather, because an over-determined system of linear equations is set up, a larger collection is produced $\{\omega_{k,Q}^{-}, d_k^{-}\}$ $(1 \leq k \leq K')$, where K' is the sum of K genuine and K'' spurious resonances $(K' = K + K'')$. Thus, we obtain $P_{K'}^{-}/Q_{K'}^{-}$ instead of P_K^{-}/Q_K^{-}. To omit the K'' false resonances that are not the constituents of the K input eigen-frequencies and the K associated eigen-amplitudes, we root the numerator polynomial via $P_{K'}^{-}(z^{-1}) = 0$ to obtain the zeros $\{z_{k,P}^{-1}\}$ $(1 \leq k \leq K')$ of the spectrum $P_{K'}^{-}(z^{-1})/Q_{K'}^{-}(z^{-1})$. A subsequent binning is carried out according to stable versus unstable resonances as a function of the varying degree K' of polynomials $P_{K'}^{-}$ and $Q_{K'}^{-}$. Finally, the exact number K of genuine resonances is reconstructed from $K = K' - K''$ after discarding the K'' unstable resonances as spurious.

3. Results

We systematically augmented the signal length for the same bandwidth (i.e. 3 acquisition times) to check the constancy of the spectral parameters. The reconstructed spectral parameters and metabolite concentrations are shown in panels (2) to (4) of table 1 at total orders K = 350, 850 and 1020. The partial signal length N_P is given by $N_P = 2K$ for total signal length $N = 2048$. In panel (2) of table 1, at $N_P = 700$, only 8 of the 9 genuine resonances were resolved. Between 3.220 ppm and 3.221 ppm only 1 resonance was generated at 3.221 ppm, whose computed concentration is the sum of PC + PE ("4 + 5"). At N_P =1700 convergence was attained for all the resonances, such that both PC and PE are fully resolved with their correct amplitudes and computed concentrations (panel (3)). At all higher N_P and at full signal length N convergence remained stable, as illustrated in the bottom panel (4) of table 1 at N_P = 2040.

Figure 1 summarizes key clinical MRS data for normal, non-infiltrated breast tissue. In panel (1) the input data, the real part (Re) of the complex time signal is shown, as per the encoded data from normal non-infiltrated breast reported in Ref. [15]. To the noiseless time signal, we added Gaussian-distributed noise with σ = 0.0289 RMS. The imaginary part (Im) of the time signal is similar (omitted). The output data are shown either for the interval between 3.2 ppm and 3.3 ppm or [3.16, 3.34] ppm. These are presented as spectra in panels (2), (3), (4) and as metabolite concentrations in panel (5). The absorption total shape spectrum reconstructed via the FFT in panel (2) at full signal length (N = 2048) is totally inaccurate and non-convergent, with only a few rough peaks, while the FPT has converged at N_P = 1700 on panel (3) (total shape spectrum) and panel (4) (component shape spectra) where PC is seen to completely underlie PE. The converged Padé-generated concentration map at N_P = 1700 is presented on panel (5), wherein the concentrations of all the metabolites, including PC are accurately computed. Signal-noise separation is presented in panels (6) and (7) where FWHMax (full width at half maximum: spectral poles) equals FWHMin (full width at half minimum: spectral zeroes) for non-physical data symbolized as ⊙ (Froissart doublets). These show marked instability at the two noise levels: panel (6) (σ = 0.00289 RMS) and panel (7) (σ = 0.0289 RMS). The zero peak heights in (8) (the abscissa crosses the centers of the empty circles) are the 3rd "signature" of spurious resonances. In contradistinction, for physical resonances FWHMax $\neq$ FWHMin. For such results, the reconstructed and input poles (O and ×) coincide, as indicated by ⊗. These are stable against the two noise levels differing by a factor of 10 in panels (6) and (7). Also, the related peak heights are non-zero (panel (8)). Although such true structures may show pole-zero near-coincidences and some of the peak heights (GPC, PC and CHO) are close to zero, they show no change for different noise levels, as seen in the dash-lined box in panels (6) and (7).

4. Discussion and Conclusions

The algorithmic success of the FPT is a direct result of its quantum mechanical underpinnings, whereby the frequency spectrum is represented as the unique ratio of two polynomials as in the finite-rank Green's function [19, 20]. Its superior resolution compared to the FFT demonstrated herein and elsewhere [19, 20, 22, 27, 28] has several reasons. One is the interpolation capability of the FPT. This occurs because a spectrum in the FPT$^{(-)}$, given by a polynomial quotient $P_K^-(z^{-1})/Q_K^-(z^{-1})$ where $z^{-1} = e^{-i\omega\tau}$, can be taken at any sweep frequency ω. Consequently, frequency ω can be selected independently of the employed number of signal points $\{c_n\}$. Thus, the FPT achieves interpolation in spectra, based on the actual analyzed time signal, whereas the minimal separation $\omega_{\min}$ in an FFT spectrum is determined by the total acquisition time T. The FFT spectrum is defined only on the Fourier grid points $2\pi k/T$ (k = 0,1,2,3,..., N − 1), where N is the total signal length. Thus, the FFT cannot interpolate. The superfluous data (zeros) added in the encoding domain do not improve resolution in the Fourier spectrum. Further, the FPT can extrapolate beyond the N input signal points because the implicitly present inverse of Q_K^- is an infinite sum (Maclaurin series). The FPT has an extra degree of freedom to cancel out noise from the numerator P_K^- and the denominator Q_K^-, whereas

CONVERGENCE of QUANTIFICATION in $FPT^{(-)}$; PARTIAL SIGNAL LENGTHS: N_P = 700, 1700, 2040

(1) Input Data: Spectral Parameters, Concentrations and Metabolite Assignments

N_k or # k	$Re(\nu_k)$ (ppm)	$Im(\nu_k)$ (ppm)	$\|d_k\|$ (au)	C_k(μM/g)	M_k
1	1.332	0.0008	0.020	0.500	LAC
2	1.471	0.0008	0.004	0.100	ALA
3	3.212	0.0008	0.001	0.025	CHO
4	3.220	0.0008	0.001	0.025	PC
5	3.221	0.0009	0.007	0.175	PE
6	3.232	0.0008	0.001	0.025	GPC
7	3.251	0.0008	0.018	0.450	β-GLC
8	3.273	0.0008	0.006	0.150	TAU
9	3.281	0.0008	0.011	0.275	M-INS

(2) Pade-Reconstructed Data: N_P = 700, N = 2048 (Partial Convergence)

N_k^- or # k	$Re(\nu_k^-)$ (ppm)	$Im(\nu_k^-)$ (ppm)	$\|d_k^-\|$ (au)	C_k^-(μM/g)	M_k^-
1	1.332	0.0008	0.020	0.500	LAC
2	1.471	0.0008	0.004	0.100	ALA
3	3.212	0.0008	0.001	0.025	CHO
4+5	3.221	0.0009	0.008	0.200	PC+PE
6	3.232	0.0008	0.001	0.025	GPC
7	3.251	0.0008	0.018	0.450	β-GLC
8	3.273	0.0008	0.006	0.150	TAU
9	3.281	0.0008	0.011	0.275	M-INS

(3) Pade-Reconstructed Data: N_P = 1700, N = 2048 (Full Convergence)

N_k^- or # k	$Re(\nu_k^-)$ (ppm)	$Im(\nu_k^-)$ (ppm)	$\|d_k^-\|$ (au)	C_k^-(μM/g)	M_k^-
1	1.332	0.0008	0.020	0.500	LAC
2	1.471	0.0008	0.004	0.100	ALA
3	3.212	0.0008	0.001	0.025	CHO
4	3.220	0.0008	0.001	0.025	PC
5	3.221	0.0009	0.007	0.175	PE
6	3.232	0.0008	1.001	0.025	GPC
7	3.251	0.0008	0.018	0.450	β-GLC
8	3.273	0.0008	0.006	0.150	TAU
9	3.281	0.0008	0.011	0.275	M-INS

(4) Pade-Reconstructed Data: N_P = 2040, N = 2048 (Full Convergence)

N_k^- or # k	$Re(\nu_k^-)$ (ppm)	$Im(\nu_k^-)$ (ppm)	$\|d_k^-\|$ (au)	C_k^-(μM/g)	M_k^-
1	1.332	0.0008	0.020	0.500	LAC
2	1.471	0.0008	0.004	0.100	ALA
3	3.212	0.0008	0.001	0.025	CHO
4	3.220	0.0008	0.001	0.025	PC
5	3.221	0.0009	0.007	0.175	PE
6	3.232	0.0008	0.001	0.025	GPC
7	3.251	0.0008	0.018	0.450	β-GLC
8	3.273	0.0008	0.006	0.150	TAU
9	3.281	0.0008	0.011	0.275	M-INS

Table 1: Quantification of MRS time signals typical of normal breast [15]. Shown are spectral parameters, concentrations and assignments of metabolites. The real and imaginary frequencies are in the 2nd and 3rd column. The amplitudes, concentrations and metabolite assignments are in the 4th, 5th and 6th column. The input data are in panel (1). The reconstructions by the fast Padé transform, $FPT^{(-)}$, are in panels (2), (3) and (4) at partial signal lengths $N_P = 700$, 1700 and 2040. The total signal length is $N = 2048$. Full convergence by the $FPT^{(-)}$ is seen in panels (3) and (4) at $N_P = 1700$ and 2040. For more details, see the main text.

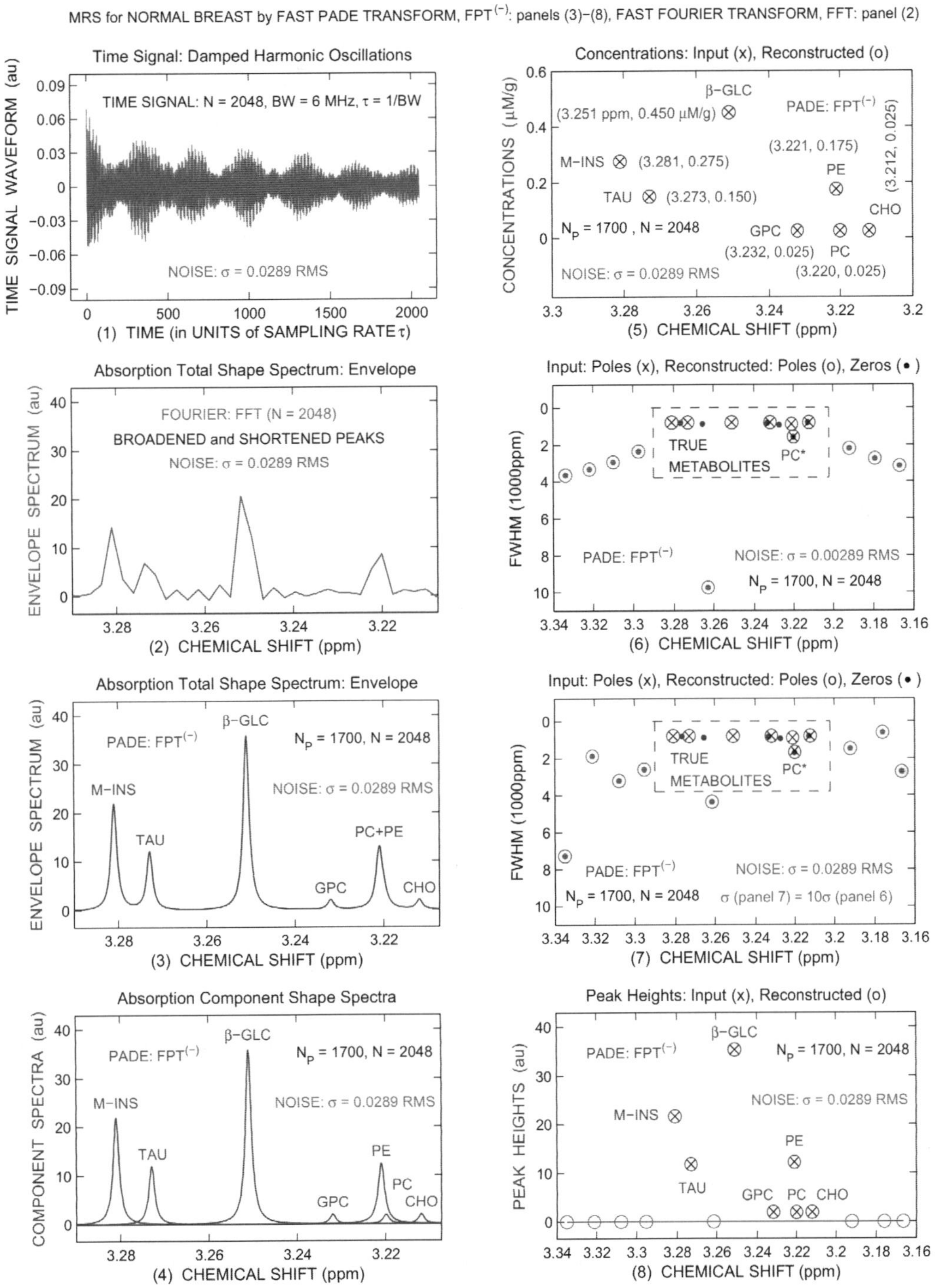

Figure 1: (Color online) The fast Padé transform, $FPT^{(-)}$, applied to MRS time signals typical of normal breast [15]. Input data: time signal in panel (1). Output data: absorption spectra in panels (2), (3) and (4); concentrations of metablites in panel (5); signal-noise separation via Froissart doublets in panels (6), (7) and (8). Symbols: input data (+) and output data (O , •). To avoid clutter in panels (6) and (7), the full width at the half maximum, FWHM, is doubled at frequency 3.220 ppm (as PC*). For more details, see the main text.

in the FFT, as a single polynomial and being linear, noise is imported intact from the measured time domain to the computed frequency spectrum [19, 20]. Due to the superior resolution performance of the FPT, fewer and shorter MRS time signals could be encoded, thereby diminishing examination time. This translates into greater comfort for the patient and lower cost.

For normal, non-infiltrated breast, the concentration of PC was ~15% that of the overlying PE, whereas for the other two cases (fibroadenoma and breast cancer), the relative concentrations were ~30% [15]. Moreover, the PC concentration was less than 20% of that for fibroadenoma [19, 26]. Nevertheless, the PC concentration, albeit very low, was computed precisely by the FPT in the presence of moderate levels of noise. This genuine metabolite was clearly distinguished from the numerous spurious resonances by its stability in the face of different noise levels, as well as at different partial signal lengths after convergence. All the other genuine metabolites were also confidently identified, with their concentrations precisely computed by the FPT, even though several (notably the other components of total choline) were likewise only sparsely present. Precisely the capabilities of the FPT demonstrated here help pave the way for developing urgently-needed normative data bases, to which comparisons could be confidently made with various benign breast pathologies, as well as with breast cancer of various types.

The results presented herein provide further evidence that mathematical optimization through the FPT will be of vital importance for realizing the potential of magnetic resonance spectroscopy, MRS, and spectroscopic imaging, MRSI, for early cancer diagnostics, in particular for breast cancer. As long as MRS and MRSI rely upon the FFT, plus post-processing via fitting, breast cancer cannot be distinguished from normal, non-infiltrated breast nor from benign breast lesions with sufficient confidence. Moreover, the low resolution of the FFT renders MRS and MRSI of the breast extremely time-consuming and therefore uncomfortable for the patient and costly. Encoding for MRS entails poor SNR as a major problem. Exclusive reliance upon the FFT requires encoding many transients (~ 200) to be averaged for improving SNR. This keeps the patient longer in the scanner, resulting in increased inhomogeneity of the static magnetic field. During this extended encoding time, the patient moves, so that more and more transients are likely not to be associated with the same intended voxel. This leads to incoherent averaging, when adding together these different time signals. Hence, this illustrates the need to reformulate the entire concept of encoding in MRS by relying upon another theory which requires fewer encoded transients (~ 20) and yet improves SNR and resolution. The FPT is optimally poised to serve this purpose critical to the use of MRS in daily clinical practice.

The closely-overlapping resonances, notably the components of total choline, within the MR spectra of the breast appear to be of crucial diagnostic importance for early cancer diagnostics. The FFT plus non-unique fitting procedures cannot provide any certainty about these overlapping resonances. Thus, even at high magnetic field strengths, proton MRS has thus far only generated estimates of total choline. With these, only incremental improvement are seen compared to anatomical imaging with MRI, in the accuracy with which breast cancer is distinguished from non-malignant breast tissue.

The high-resolution, quantification-equipped FPT represents a paradigm shift for signal processing. This has been confirmed in interdisciplinary research fields and applications, where generic time signals comprised of complex damped harmonics are routinely encountered. These capacities of the FPT are also shown herein. Namely, in the presence of realistic noise levels, the FPT clearly distinguishes spurious resonances from genuine metabolites and exactly computes the concentrations of the latter, including phosphocholine, PC, albeit only minimally present for normal, non-infiltrated breast. The high resolution of the FPT together with its enhancement of SNR yields shorter examination time, thus increasing patient comfort plus lowering costs. Particularly with closely-overlapping resonances in the MR spectra, convergence is attained by a greatly over-determined system of linear equations. The signal-noise separation procedure uniquely linked to the FPT is the means by which genuine i.e. physical resonances are identified and the metabolite concentrations precisely computed. Simultaneously, the same procedure identifies the much more numerous spurious resonances via coincident poles and zeros (Froissart doublets), accompanied by zero or near zero amplitudes and marked instability against various levels of truncation and/or noise. The results presented in this paper in which normative data for the breast are generated by the FPT in the presence of noise, bring us one step closer to applying the fast Padé transform to time signals encoded in vivo from normal, benign and cancerous breast in order for MRS and MRSI to realize their potential to, at last, become a reliable, cost-effective method for breast cancer diagnostics.

Acknowledgements
This work was supported by Radiumhemmets Research Fund, Cancerfonden and FoUU.

References

[1] Eccles S, Aboagye E, Ali S, Anderson A, Armes J and Berditchevski F 2013 *Breast Cancer Res.* **15** R92
[2] Sant M *et al* 2003 *Int. J. Cancer* **106** 416
[3] Saslow D *et al* 2007 *CA. Cancer J. Clin.* **57** 75
[4] Belkić K and Belkić Dž 2011 *Isr. Med. Assoc. J.* **13** 236
[5] Sardanelli F *et al* 2011 *Invest. Radiol.* **46**, 94
[6] Feig S 2010 *Radiol. Clin. N. Am.* **48** 879
[7] Belkić Dž and Belkić K 2013 *Technol. Cancer Res. Treat.* DOI: 10.7785/tcrt.2012.500401
[8] Malayeri A *et al* 2011 *RadioGraphics* **31** 1773
[9] Balzter P and Dietzel M 2013 *Radiology* **267** 735
[10] Haddadin I *et al* 2009 *NMR Biomed.* **22** 65
[11] Bolan P *et al* 2003 *Magn. Reson. Med.* **50** 1134
[12] Klomp D *et al* 2011 *NMR Biomed.* **24** 1337
[13] Bolan P *et al* 2002 *Magn. Reson. Med.* **48** 215
[14] Stanwell P and Mountford C 2007 *Radiographics* **27** S253
[15] Gribbestad I, Sitter B, Lundgren S, Krane J and Axelson D 1999 *Anticancer Res.* **19** 1737
[16] Katz-Brull R, Seger D, Rivenson-Segal D, Rushkin E and Degani H 2002 *Cancer Res.* **62** 1966
[17] Eliyahu G, Kreizman T and Degani H 2007 *Int. J. Cancer* **120**, 1721
[18] Glunde K, Jiang J, Moestue S and Gribbestad I 2011 *NMR Biomed.* **24**, 673
[19] Belkić Dž and Belkić K 2010 *Signal Processing in Magnetic Resonance Spectroscopy with Biomedical Applications* (London, Taylor & Francis)
[20] Belkić Dž 2005 *Quantum Mechanical Signal Processing and Spectral Analysis* (Bristol, Institute of Physics Publishing)
[21] Belkić Dž 2006 *Phys. Med. Biol.* **51** 2633
[22] Belkić Dž 2006 *Phys. Med. Biol.* **51** 6483
[23] Belkić Dž 2009 *Adv. Quantum Chem.* **56** 95
[24] Ojo A 2010 The Analysis and Automatic Classification of Nuclear Magnetic Resonance Signals. PhD Thesis, The University of Edinburgh Research Archive (http://hdl.handle.net/1842/4109)
[25] Zhang J 2013 Human Brain Glutamate, Glutamine, γ-Aminobutyric Acid: Proton Magnetic Resonance Spectral Quantification with The Fast Padé Transform. PhD Thesis University of California Los Angeles (escholarship.org/uc/item/1np12339)
[26] Belkić Dž and Belkić K 2009 *J. Math. Chem.* **45** 790
[27] Belkić Dž and Belkić K 2014 *J. Math. Chem.* **52** 2680
[28] Belkić Dž and Belkić K 2014 *Technol. Cancer Res. Treat.* DOI: 10.1177/1533034614547446

27th Summer School and International Symposium on the Physics of Ionized Gases (SPIG 2014) IOP Publishing
Journal of Physics: Conference Series **565** (2014) 012003 doi:10.1088/1742-6596/565/1/012003

The Role of High-Energy Ion-Atom/Molecule Collisions in Radiotherapy

Dževad Belkić

Department of Oncology and Pathology, Karolinska Institute, P.O. Box 260, SE-171 76, Stockholm, Sweden

E-mail: Dzevad.Belkic@ki.se

Abstract. The need for ions in radiotherapy stems from the most favorable localization of the largest energy deposition, precisely at the tumor site with small energy losses away from the target. Such a dose conformity to the target is due to heavy masses of ions that scatter predominantly in the forward direction and lose maximal energy mainly near the end of their path in the vicinity of the Bragg peak. The heavy masses of nuclei preclude noticeable multiple scattering of the primary ion beam. This occurrence is responsible for only about 30% of ion efficiency in killing tumor cells. However, ionization of targets by fast ions yields electrons that might be of sufficient energy to produce further radiation damage. These $\delta-$electrons, alongside radicals produced by ion-water collisions, can accomplish the remaining 70% of tumor cell eradication. Electrons achieve this chiefly through multiple scattering due to their small mass. Therefore, energy depositions by both heavy (nuclei) and light (electrons) particles as well as highly reactive radicals need to be simultaneously transported in Monte Carlo simulations. This threefold transport of particles is yet to be developed for the existing Monte Carlo codes. Critical to accomplishing this key goal is the availability of accurate cross section databases. To this end, the leading continuum distorted wave methodologies are poised to play a pivotal role in predicting energy losses of ions in tissue as discussed in this work.

1. Introduction

There could be no adequate particle transport theory without relying upon the most accurate collisional cross sections. As to electromagnetic interactions, the existing *ab initio* quantum-mechanical theories of fast ion-atom collisions [1–5] can abundantly provide cross section databases that are indispensable for proper Monte Carlo simulations of energy losses in matter. The use of these theories derived from first principles of physics would obviate the need for customary resorting to empirical formulae for cross sections. One such fitting formula is due to Rudd [6] who suggested a phenomenological expression by adjusting the involved free parameters to approximately reproduce the available experimental data for ionization in proton-water collisions. For other ions of nuclear charge $Z_{\rm P} > 1$, this formula relies upon the $Z_{\rm P}^2-$scaling rule borrowed from the first Born approximation. Such a scaling is not justified because it neglects the effect of electron capture to continuum, which scales like $Z_{\rm P}^3$ [7, 8]. Another *ad hoc* formula proposed by Hansen, Kocbach and Stolterfoht (HKS) [6] empirically modifies the semi-classical version of the binary encounter approximation. The reason for which none of the existing empirical formulae is warranted is that they all lack the predictive power without which theoretical modeling is void of substance.

Energy losses of ions are needed for dose-planning systems in hadron therapy of deep-seated tumors. Ions possess certain special features of a particular relevance to hadron therapy because of their Bragg-tailored dose distributions [9–11] showing that most energy is deposited at the treated target due to a negligible role of multiple scattering effects. Hence conformity of ion dose distributions to the tumor topology. In their passage through the tissue from the entrance and within the plateau region, impact energies of ions are high and ionization dominates. However, at their range, i.e. in the close vicinity of the Bragg peak, the initially fast ions abruptly slow down such that electron capture and electron loss processes become the dominant channel for energy losses. Therefore, especially near the Bragg peak, reliable particle transport theories must account for electron transfer and electron loss from dressed ions. In the latter process, above 150 keV, the largest contribution comes from double ionization via simultaneous emission of one electron from the target and the other from the dressed projectile, as seen in Fig. 1. Within the last few millimeters around the Bragg peak, thousands of electron capture and electron loss processes take place. The Bragg peak on the plot of stopping power versus the traversed path is reminiscent of the Massey peak of the cross section as a function of the incident energy. Thus, collisions near the Bragg peak in ion transport phenomena in tissue are associated with the Massey maximum as a resonant effect, which is manifested whenever the incident velocity matches the classical orbiting velocity of the active electron from the target. It is these and several other related circumstances that promote the role of quantum-mechanical theories from atomic and molecular collisions in providing the necessary non-phenomenological databases of cross sections for track structure computations and detailed Monte-Carlo simulations of ion transport in tissue.

2. Energy deposition by fast ions

Over the years, the subject of hadron therapy has been inspired by the seminal work of Wilson [16] who in the 1940s was the first to suggest that protons, alpha particles or carbon nuclei and some other light ions, with their dose distributions ending with the Bragg peak, should provide optimal target conformity and tumor control. The significance of Wilson's proposal was in having been conveyed in such a convincing manner that physicians were quick to accept it. Moreover, physicians wasted no time in beginning patient treatments by light ions already in the 1950s in Berkeley [17] and Uppsala [18].

High-energy light ion beams from protons to carbon nuclei, as a powerful part of hadron therapy of deep-seated tumors, are increasingly in demand worldwide, and this motivates construction of hospital-based accelerators. Positron-emitting secondary ions generated from collisions of primary projectiles with the treated tissue provide the possibility for special tomographic verifications of dose depositions at the targeted sites. Such information is of utmost importance, since any inaccuracy in dose delivery to the tumor relative to the prescribed dose planning system can be used to update the input data to the algorithmic codes for energy losses in order to provide the corrected doses in the subsequent fractionated treatments of the patient. This adaptive dose delivery and a proper inclusion of the relative biological effectiveness (RBE) are among the major constituents of biologically-optimized and, indeed, personalized radiotherapy which is suited to the specific needs of each patient.

The challenges for the theory are multifaceted, especially at intermediate impact energies where most of the otherwise powerful perturbation methods are of limited applicability. On the other hand, non-perturbative methods are expected to be more adequate for this particular energy region. The other level of challenges for theoreticians encompasses collisions with participation of two and more active electrons as these processes may become increasingly important near the Bragg peak. Such collisions are characterized by electron-electron correlation effects that are known to play an important role at low and high energies, but have been reported in the past literature to be of minor significance at intermediate energies. Yet another challenging

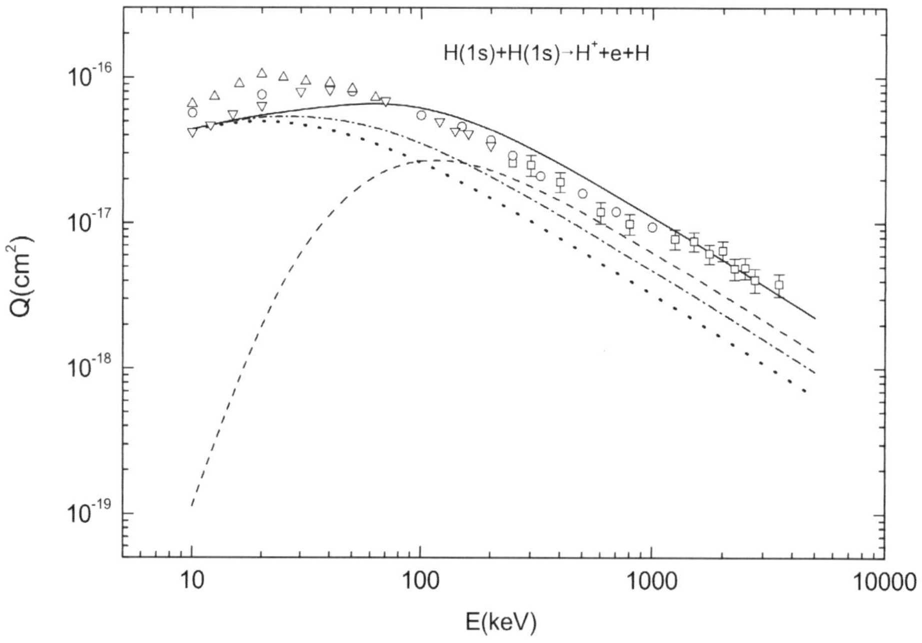

Figure 1. Total cross sections Q as a function of the impact energy E for electron loss process $H(1s) + H(1s) \longrightarrow H^+ + e_1 + H(f)$. Theory (CB1-4B: the four-body boundary-corrected first Born approximation [3]): (i) Dotted curve: cross section Q_1 for target $H(1s)$ unaffected by collision with $f = 1s$ i.e. $H(1s) + H(1s) \longrightarrow H^+ + e_1 + H(1s)$. (ii) Chained curve: cross sections $Q_{\sum_{n=1}^{6}}$ for all the target bound states symbolized as $H(\sum_{n=1}^{6})$ with $f = \sum_{n=1}^{6}$ where the contributions from the principal quantum numbers $1 \leq n \leq 6$ are summed up i.e. $H(1s) + H(1s) \longrightarrow H^+ + e_1 + H(\sum_{n=1}^{6})$ (also included here are the sums over all the pertinent degenerate levels of the angular and magnetic quantum numbers). (iii) Dashed curve: cross sections Q_c for H(c) with $f = c$ where c stands for continuum of the target to denote ionization of the target electron via $H(1s)+H(1s) \longrightarrow H^+ + e_1 + H^+ + e_2$ (the projectile and target electrons are denoted by e_1 and e_2, respectively). (iv) Full curve: cross sections $Q_{c+\sum_{n=1}^{6}}$ for all the target states $H(c + \sum_{n=1}^{6})$. No formation of the negative hydrogen ion H^- by electron capture from the target $H(1s)$ by the projectile $H(1s)$ is taken into account in the computations. Experimental data (all the target states): $\triangle$ [12], $\square$ [13], $\bigtriangledown$ [14] (without electron capture) and $\circ$ [15].

aspect of the theories is the area of molecular targets, among which water molecule plays a pivotal role in medical physics. In radiotherapy, water is considered as a tissue-like matter because of its abundance in the human body. For this reason, most benchmark computations on ion energy losses in tissue are carried out with water as a prototype tissue-like target. Here, the challenges are enhanced by a large number of open channels, on top of the difficulties in handling the accompanying particle transport phenomena, such as energy as well as range fluctuations called straggling. One of the end points here is to evaluate precisely the dose to be delivered by energetic ions at a given position located deeply in the treated body where the tumor resides. This is a very complex problem, which demands many-layered expertise from nuclear and atomic collision physics.

3. Plethora of cross sections for ion-atom/molecule collisions for hadron therapy

The atomic physics research field of high-energy collisions of ions with matter provides its versatile databases (cross sections, rate coefficients, ...) to other branches of physics (astrophysics, plasma physics, particle transport physics, ...) and applied disciplines (fusion research, radiation protection in manned space missions, hadron therapy). Such data describe various inelastic scattering events, including electron capture (or equivalently, electron transfer or charge exchange), excitation, electron loss (projectile ionization), ionization (target ionization) and their combinations (transfer excitation, transfer ionization, loss-excitation, loss-ionization) [1–3]. Probabilities for multiple electron transitions are enhanced for the increased nuclear charge of projectiles. Both sets of the mentioned data, those for single- and multiple-electron transitions, need to be very accurate in their role as the input data for subsequent stochastic simulations that employ deterministic cross sections for fast ion-atom collisions in various applications. These data from atomic collisions of ions with matter can be obtained from theories and measurements. A variety of such experimentally determined and theoretically predicted cross sections has extensively been accumulated in the literature particularly over the last three decades [1, 2].

In exhaustive comparisons with measurements, past experience conclusively established that the most adequate quantum-mechanical theory for high-energy ion-atom collisions is the continuum distorted wave (CDW) method [19–21] which is valid in the region slightly above the Massey resonance peak [22]. The name of this method comes from correlating the projectile with the target through the allowance of electronic continuum intermediate states that act as a distortion in the total scattering wave function of the whole system. For example, in the entrance channel, the active electron is simultaneously bound to the target nucleus and free in the field of the projectile nucleus. Therefore, in the initial state of the system, the distortion associated with continuum intermediate states of the active electron stems from the Coulomb wave function centered on the impinging nuclear charge. Distorted waves are of especially critical importance to the exit channel for ionizing collisions, as they describe the two-center effects [20] that are completely missing from the first Born (B1) [23–27] and the Bethe-Bloch [28,29] approximations. The two-center effects include intermediate stages of collision and describe active electrons as simultaneously propagating in the two Coulomb fields, one of the projectile and the other of the target nuclei. In ionization of a target, the B1 approximation employs only one full Coulomb wave function centered on the target nucleus in the exit channel for the final scattering state Ψ_f^-, whereas the electron motion in the projectile field is inadequately described by a plane wave [23–27]. By contrast, the CDW method describes the whole ionization collision in a more physically adequate manner by using three Coulomb wave functions. Two of these latter wave functions are for the electronic motions and they are centered on the projectile and target nuclei for Ψ_f^-. The third Coulomb wave describes the relative motion of the two heavy nuclei. The physical significance of the additional electronic Coulomb wave function in the CDW method for ionization is in leading to a new source for production of emitted electrons due to the already mentioned effect of electron capture to continuum (ECC) of the projectile. Here, the electron is considered as being "captured" by the projectile albeit not in a bound, but rather in a continuum state. Through this mechanism, which has been experimentally detected in single ionization [30–33], the projectile and electron travel together in the same direction with nearly equal velocities vectors $\vec{v}$ and $\vec{\kappa}$, respectively ($\vec{\kappa} \approx \vec{v}$). This is recognized as a resonance effect (hence the resulting enhanced cross sections), which is manifested in an asymmetric cusp-shaped angular distribution for the forward emitted electrons. The description of the exit channel for ionization in the B1 approximation could be roughly satisfactory only for very slow ejected electrons, that are near the target nucleus and simultaneously far away from the nucleus of the projectile. The cusp mechanism has also been observed experimentally in solid state targets and in this area the name "convoy electrons" is interchangeably used for ECC electrons [34]. This latter name stems from the circumstance that ECC electrons are aligned with and accompanied by scattered projectiles as reminiscent of a convoy-like moving away from the target rest. Emitted slow electrons provide the major contribution to the total cross sections. However, for thorough studies of collisional dynamics, it is imperative to look beyond total (integrated) cross sections to describe properly the differential cross sections that give the energy and angular distributions of electrons ionized from the target [20, 35, 36]. Both these distributions manifestly exhibit the two-center effect. The other two mechanisms for production of electrons are the emissions leading to the forward and binary peaks (both described by the CDW and B1 methods). The forward emission occurs at a zero-valued angle of the ejected electron, $\theta_e \approx 0$. The binary encounter (BE) peak corresponds to $\kappa \approx 2v\cos\theta_e$.

The most remarkable feature of an interplay between the BE and ECC mechanisms of ionization is manifested when passing from single to multiple ionization in the case of non-hydrogenlike targets. This has been evidenced in experimental measurements of doubly differential cross sections for single and multiple ionization of Ar by H^+ [37,38] as well as of He, Ne and Ar by $F^{8+,9+}$ and $I^{23+,26+}$ [39–41]. The ECC cusp effect involving two-electron continua was studied via simultaneous electron loss (projectile ionization) and target ionization in He – He collisions [32,33]. The striking observation reported in Refs. [37–41] is that the ECC cusp can even dominate the BE peak for ionization. Specifically, the coincidence experiment of Afrosimov *et al* [38] (simultaneous measurements of double differential cross sections for near-forward-cone ejected electrons and charge state of the target rest) demonstrated that the BE electrons were produced almost entirely by single ionization of Ar by protons. Moreover, determination of the position of the BE peak in coincidence with detection of single-charged recoil ions showed that the emitted BE electrons stem from the valence M-shell of Ar. Among all the electron orbital velocities of Ar, the one which is associated with the M-shell of this atom matches most closely the condition $v_e \approx 2v$ for the forward emitted BE electrons ($\cos\theta_e \approx 1$). Furthermore, in the same experiments [37, 38], the group of ECC electrons were found to be significant only when detected in coincidence with multiple charged Argon ions Ar^{k+} ($k = 2, 3, 4, ...$), whereas totally negligible counts were recorded for Ar^+. In other words, the ECC electrons become a dominant mechanism for multiple ionization. To see from which target shell these electrons were emitted, it is sufficient to compare the binding electronic energies of various shells in Ar and find the level which matches best the ejected electron energy from the ECC peak at which the orbital electron velocity equalizes the projectile speed ($v_e \approx v$, $\cos\theta_e \approx 0$). This was found to be the L-shell energy of Ar [37, 38]. Here, it is pertinent to emphasize certain remarkable similarities between the experimental data for ionization [38] and charge exchange [42]. Thus e.g. the charge-state distributions near the ECC peak was estimated to be 13%, 42%, 37% and 8% for the charge states 1+, 2+, 3+ and 4+, respectively [38]. This is very close to the corresponding charge-state distributions 13%, 50%, 30% and 7% for the same charge states 1+, 2+, 3+ and 4+ in the case of charge exchange [42]. The direct implication of the findings from the experiments reported in Refs. [37–41] is that theories for multiple ionization must include the two-center effect in order to account properly for emission of fast electrons that are predominantly from the ECC source. On the other hand, regarding various applications of such theories aimed at determination of energy losses of charged particles during their passage through matter, the main issue is the relative importance of slow versus fast electrons in single versus multiple ionization. Overall, these experiments with non-hydrogenlike atomic targets prove that the BE and ECC electrons play a remarkably complementary role as they appear to be generated by two completely different mechanisms (single and multiple ionization) and originate from the two well-separated target energy levels (outer and inner shells). Thus, while for single ionization the ECC mechanism is conspicuous, it became inconspicuous for multiple ionization. Single ionization is associated with slow electrons from the one-center BE mechanism which is a direct collision of the projectile and the target active electron, such that the target nucleus and the remaining electrons are considered as being passive as if they were mere spectators. On the other hand, multiple ionization is mediated by fast electrons due to the ECC mechanism which necessitates two centers thus activating both the projectile and target nucleus. As opposed to a single collision (projectile nucleus – target electron) in the BE effect, a double collision of the active electron on the projectile and target nucleus is required to produce the ECC cusp as reminiscent of the Thomas billiard-type two successive elastic encounters occurring in capture of an electron from the target by an energetic projectile. This type of correlation between electron capture and ionization permits obtaining the same angular distributions in a given theory for the Thomas peak in charge exchange by using the corresponding ionization transition amplitude in the vicinity of the ECC cusp [43].

In the CDW method [20], the distortions due to electronic continuum intermediate states are properly included in the entrance and exit channels. Nevertheless, the corresponding full Coulomb wave function within the initial scattering state was found in applications to lead to overestimation of experimentally measured total cross sections near and below the Massey peak. Typically, all total cross sections computed by the CDW method keep on rising as the impact energy E decreases, whereas the corresponding experimental data generally decline in the same region exhibiting the Massey peak. This bending of the curve for the total cross sections can also be obtained within the CDW methodologies if the full Coulomb wave function for the distortion of the initial state in the entrance channel is approximated by its long range asymptotic eikonal form. The resulting simplification of the CDW method is acronymed as MCB (modified Coulomb Born) [44, 45], or alternatively, CDW-EIS [46], where EIS stands for the eikonal initial state. In this way, the Massey peak is systematically reproduced by the MCB or CDW-EIS method leading to quantitative agreement with measurements, as illustrated in Fig. 2 and reviewed in Refs. [4, 47, 48]. It is not necessary to start from the CDW method to introduce its simplified variant in the form of the CDW-EIS approximation as originally done by Crothers and McCann [46]. Rather, an alternative derivation [44, 45] yielding the same result is also possible without any reliance upon the CDW method. Irrespective of the way of deriving the transition amplitudes, the CDW and CDW-EIS methods share precisely the same description for the final total scattering state. They differ in the initial scattering state, where the CDW-EIS method uses the ansatz in the form of the logarithmic Coulomb phase factor for the electronic continuum intermediate states. With such an ansatz, the perturbation potential in the entrance channel is unequivocally defined by the application of the operator $H - E$ onto the initial state in the matrix element for the prior form of the transition amplitude T_{if}^{-}. Here, H and E are the full Hamiltonian and the total energy of the whole system (projectile plus target). The CDW and CDW-EIS methods are computationally attractive since the fully analytical expressions are available for the transition amplitudes for inelastic collisions involving one active electron. The detailed illustrations of these methods for collisions of positively charged projectiles and either neutral targets or positively charged ions are abundantly available in the literature [1–5, 20, 43–52].

4. Adaptation of theories for ion-atom to ion-molecule collisions

Regarding electromagnetic interactions, the available state-of-the-art quantum-mechanical scattering theories from ion-atom collisions [1,2] can be extended to a wider class of biomolecular targets of direct relevance to cancer therapy by energetic light ions. This can be accomplished by using molecular wave functions in terms of linear combinations of atomic orbitals (LCAO) in the form of the Slater-type orbitals (STOs). With such a convenient ansatz, especially with single center orbitals located on the heavy atomic constituent of a given molecule (e.g.

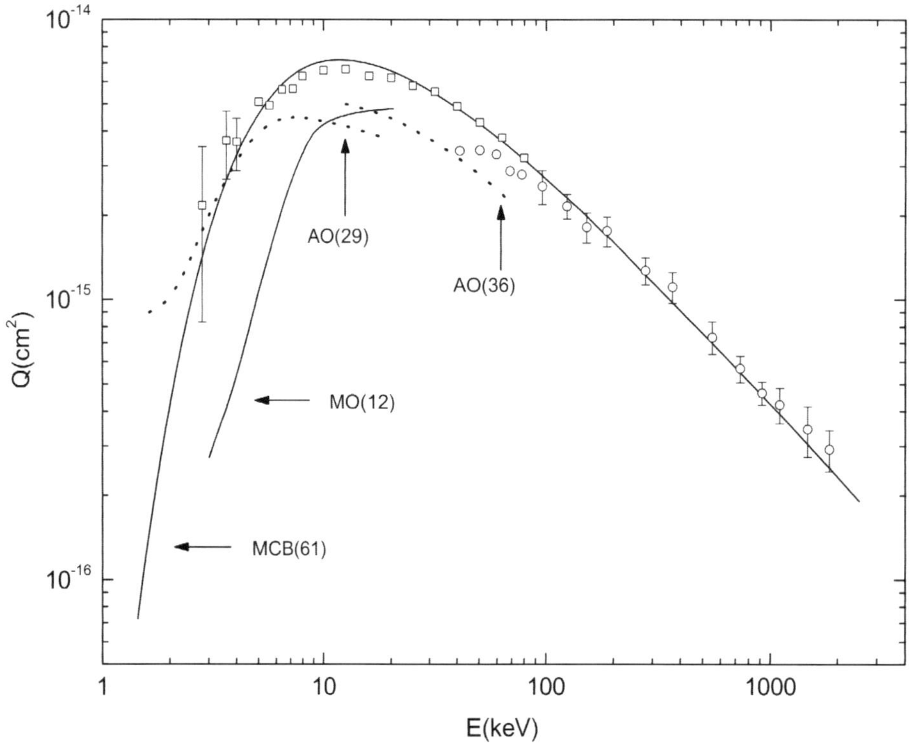

Figure 2. Total cross sections Q as a function of the incident energy E for electron detachment process $H^+ + H^-(1s^2) \longrightarrow H^+ + H + e$. Theory (only the final ground state $H(1s)$ of the atomic hydrogen): (i) Dashed curves (Atomic Orbitals – AO method [53] with N=29, 36). (ii) Lower full curve $E \in [3, 20]$ keV (Molecular Orbitals – MO method [54] with N=12). (iii) Upper full curve $E \in [1.4, 2498]$ keV (MCB, or equivalently, CDW-EIS method [44, 45] with N=61). Integer N associated with the acronyms is the number of variational parameters in the target Configuration Interaction (CI) wave function from Joachain and Terao [55] (N=61) as well as from atomic [53] (N=29, 36) and molecular [54] (N=12) orbitals. Experiment: ○ [56] and □ [57].

Oxygen in the case of H_2O), these theories for ion-atom collisions can be directly adapted to ion-molecule scattering. This is possible because the matrix elements as the form factors for electron capture and ionization involving the STOs for bound states and hydrogenlike continuum with an effective charge in the Sommerfeld parameter for the Coulomb wave function [21] can be analytically calculated precisely in the same way for both atomic and molecular targets. The only difference is that in the case of a molecular target an additional triple integration is required over the three Euler angles for the orientation of the molecular axis. This is due to the fact that the variational parameters of the molecular orbitals are obtained for a particular orientation of the internuclear axis giving the minimum of the expectation value of the exact Hamiltonian of the given molecule. In general, experiments are performed with no reference to

Journal of Physics: Conference Series **565** (2014) 012003 doi:10.1088/1742-6596/565/1/012003

any special orientation of the internuclear axis within the studied molecular target. Therefore, in order to compare such experimental data with theories, the latter must perform averaging over all the orientations of the internuclear axis of a molecular target. For the above-specified molecular orbitals used within e.g. the CDW and CDW-EIS approximations this averaging, which amounts to the said three integrals over Euler angles, can be calculated analytically by using the well-known properties of the Wigner rotation matrix. Thus, in this setting the usual codes for the CB1, CDW and CDW-EIS approximation from ion-atom collisions can readily be adapted for direct applications to ion-molecule scattering [58–66]

5. Major ion-tissue interactions in high-energy collisions

When ions enter the tissue, they are said to be transported through it up to a certain distance (called range R) which is determined primarily by the initial energy E of the projectile beam, its energy loss, range straggling and multiple scattering [6, 67]. Ions lose their energy because they collide with electrons of the constituents of the traversed tissue. Due to their heavy mass, which is about $2000N$ times larger than the electron mass ($N \geq 1$ being the number of protons and neutrons in a nucleus), ion trajectories deviate only slightly from their incident directions. In other words, they scatter mainly in the direction of a narrow forward cone ($\theta_i \approx 0$). Many small angle scatterings occur while the ions pass by the nuclei of atoms/molecules of the encountered tissue. Such a cumulative effect is called multiple scattering and this phenomenon is more enhanced in thicker than in thinner targets. A beam contains a large number of ions and this leads to certain statistical effects. One of them is range straggling, which represents fluctuations in the range R of individual ions. All the ions of the same energy do not attain the same range, since their collisions (by which they produce many secondary particles along their paths) are of a probabilistic i.e. statistical rather than deterministic nature. Thus, such ions are viewed as straggling to reach their range. Hence the term "range straggling".

In particle transport physics, one of the most important observables that characterizes the penetration capability of particles is the stopping power $S(z)$, which represents the loss of ion energy per unit of the traversed path length z along the particle track, as denoted by $S(z) \equiv -\mathrm{d}E/\mathrm{d}z$. At high incident energies $E \geq 350$ MeV/amu of interest for hadron therapy on deep seated tumors of the corresponding range of about ~ 25 cm, the main scattering events that lead to energy loss of ions are nuclear collisions (mainly with neutron losses from projectile nuclei) and target ionizations. As mentioned, ions scatter mainly forward, meaning that their paths are practically straight lines along which they transfer their energy to the tissue which becomes ionized or excited (at high energies, the probability for ionization is about three times larger than that of excitation). Ions deposit several times more energy at their range R (i.e. in the vicinity of the Bragg peak) than elsewhere such that this inverse dose-depth distribution is in sharp contrast to electrons and photons [9–11]. The measurement of Bragg and Kleeman [11] was the first to report that $\alpha-$particles deposit nearly all their energy when they are about to stop i.e. very close to their range. Dose-depth distributions or profiles are the curves that display the stopping power $S(z)$ as a function of depth z. A dose-depth profile of an ion beam is mostly a plateau-like curve, which ends abruptly with a prominent maximum as the Bragg peak situated at the target position. By contrast, electrons and photons deposit maximum energy near the beginning of their track, so that afterwards their dose curves decrease with distance all the way up to the target which is tumor in radiotherapy [68]. The tumorous tissue needs to be destroyed to stop the uncontrollable production of clonogenic cells and ions can partially achieve this through nuclear and ionizing collisions with the tissue. However, as stated, only some 30% of ionization of the tissue's main target, the deoxy-ribonucleic acid (DNA) molecule, is accomplished by direct hits of ions counting all the primaries, secondaries and ions of higher-order generations. The remaining 70% of ionization rate of DNA is achieved by radicals and the secondary $\delta-$electrons that are produced by the impact of ions on tissue. If the electrons emitted from a target are sufficiently energetic to be able on their own to lead to further ionzations they are called $\delta-$electrons.

6. Discussion and conclusion

The Bragg curve $S(z)$ represents the stopping power as a function of the traversed length z. Given a beam which penetrates a fixed medium, the stopping power is the energy loss per unit of the traversed path length i.e. $-\mathrm{d}E/\mathrm{d}z$ so that $S(z) = -\mathrm{d}E/\mathrm{d}z$. Due to their heavy mass, energetic nuclei lose very small amount of their initial kinetic energy per each collision with electrons from the traversed medium. Therefore, the track of the primary particles is practically a straight line in the incident beam direction. Consequently, the stopping power varies slowly and gradually with z until the range R is reached at which point a sudden release of energy of the beam occurs resulting in the emergence of the Bragg peak. As such, qualitatively, the Bragg curve behaves like a hyperbola for all but the last couple of centimeters of z prior to reaching the range R at which point it sharply drops following a linear trend with the increased z. Such a behavior is reminiscent of a resonant cross section with a sharp maximum superimposed on a long background-type tail due to 'potential scattering' as encountered in resonant scattering theory. More concretely, the 'inverse Bragg function' (obtained via the replacement of the length z on the abscissa by the corresponding energy E extracted from the usual energy-range relationship) appears precisely as a curve typical for ionization or excitation total cross section as a function of impact energy. In this latter curve, the 'inverse Bragg peak' is recognized as the usual Massey peak from the field of ion-atom collision theory. On the other hand, the mechanism for the Massey peak is the well-known adiabatic hypothesis, which explains the resonance by a near matching of the incident velocity v and the classical velocity v_0 of the electron in the orbit of the target atom from which ionization or excitation takes place. Hence, reference to the 'inverse Bragg function' is useful, since it determines that the underlying mechanism for the usual Bragg peak is resonance via the velocity-matching condition ($v \approx v_0$).

Specifically, in hadron therapy, clinical dose planning systems need very accurate predictions of the amount of the delivered radiation to tumorous tissues at a fixed depth. This would increase the tumor control by the clinician. The main task of radiotherapy is to maximally destroy the targeted tumor with clonogenic cells, and simultaneously spare the healthy surrounding tissue, which is under risk for collateral damage by the same beam of particles. Energetic heavy nuclei (protons, α-particles and similar hadrons) are particularly well-suited for this task, since, by reference to the concept of the Bragg peak, they can, in principle, faithfully deliver the prescribed dose at any given location in the body of the treated patient. This assumes that the dose is computed by reliance upon accurate theoretical data for all the relevant energy losses suffered by the incident beam. Such losses are due to nucleus-nucleus and nucleus-electron collisions. At present, one of the most advanced Monte Carlo codes SHIELD-HIT for modeling of the passage of ions through tissue is excellent in nuclear stopping powers due to the availability of the best state-of-the-art data bases for nuclear reactions including transmutations[1]. However, the same code uses the electronic stopping powers of limited accuracy because of the exclusive reliance upon the high-energy Bethe-Bloch formula $\propto [(1/E)\ln(E) + \cdots]$, which is by way of two fitting parameters empirically sewn to the Lindhart-Scharf low-energy asymptote $\propto E^{1/2}$ to crudely cover the region down to $\sim 10 - 20$ keV/amu instead of having 1 MeV/amu as the usual cut-off energy of the projectile.

The Bethe-Bloch formula, as the high-energy asymptotic total cross section to the undistorted first Born approximation, includes energy losses of the primary particles using only single

[1] The original Monte Carlo code SHIELD [69, 70] was designed for nuclear physics as well as for radiation protection purposes in manned space research programs with the main goal to assess the ways to improve shielding against intense radiations from galactic nuclei. More recently [71], this algorithm was adapted to applications within hadron therapy where it is called SHIELD-HIT with the acronym HIT standing for heavy ion transport.

ionization and excitation of targets. However, electron capture and electron loss channels are open with their enhanced probabilities when the penetrating beam of heavy particles is sufficiently slowed down to reach its range near the Bragg peak. For example, by directing a beam comprised of only $\alpha-$particles to traverse a given matter towards a target, charge-changing phenomena take place several thousand times, so that alongside the primaries (He^{2+}), the beam would also contains the secondaries (He^{+} and He). Importantly, these changes of the initial charge of the primary particles occur almost entirely within the last few millimeters of the range within the Bragg peak which is clinically most relevant.

The field of the well-established theories on ion-atom collisions, such as the CDW and CDW-EIS methods, adapted to ion-molecule collisions by way of molecular orbitals in terms of linear combination of atomic orbitals, as discussed in this work, can come to the rescue for overcoming the incompleteness of the Bethe-Bloch descriptions of electromagnetic interactions. This can significantly improve the needed cross section databases for Monte Carlo simulations of energy losses of ions in molecular targets of the traversed tissue. Such *ab initio* cross sections can readily be precomputed and stored as tables from which direct sampling can be made in stochastic simulations. In this way the predictions of Monte Carlo simulations would become much more accurate than any of the customary estimates using the Bethe-Bloch expression or employing some empirical formulae with fitting parameters. A direct clinical usefulness of this envisaged upgrade of particle transport theories and truck structure predictions would be far more reliable dosimetric estimates of energy depositions in tissue traversed by ions. This, in turn, would improve dose planning system in conjunction with hadron therapy.

Acknowledgments

This study was supported by the Swedish Cancer Society Research Fund (Cancerfonden), King Gustav the Fifth's Jubilee Foundation at Radiumhemmet of the Karolinska University Hospital and the Stockholm City Council through FoUU.

References

[1] Belkić Dž 2008 *Quantum Theory of High-Energy Ion-Atom Collisions* (London, Taylor and Francis)
[2] Belkić Dž, Mančev I and Hanssen J 2008 *Rev. Mod. Phys.* **80** 249
[3] Belkić Dž 2009 *Adv. Quantum Chem.* **56** 251
[4] Belkić Dž 2010 *J. Math. Chem.* **47** 1366
[5] Belkić Dž 2010 *J. Math. Chem.* **47** 1420
[6] Stopping powers and ranges for protons and alpha particles, 1993 *ICRU Report 49: International Commission of Radiation Units and Measurements* (Bethesda MD, Academic Press)
[7] Belkić Dž 1980 *J. Phys. B* **13** L589
[8] Bernardi G C, Fainstein P D, Garibotti C R and Suárez S 1990 *J. Phys. B* **23** L139
[9] Bragg W H 1904 *Phil. Mag.* **8** 719
[10] Bragg W H and Kleeman R D 1904 *Phil. Mag.* **8** 726
[11] Bragg W H and Kleeman R D 1905 *Phil. Mag.* **10** 318
[12] Stier P M and Barnett C F 1956 *Phys. Rev.* **103** 896
[13] Barnett C F and Reynolds H K 1958 *Phys. Rev.* **109** 355
[14] McClure G W 1968 *Phys. Rev.* **166** 22
[15] Hvelplund P and Andersen A 1982 *Phys. Scr.* **26** 370
[16] Wilson R R 1946 *Radiology* **47** 487
[17] Malis L I, Loevinger R, Kruger L and Rose J L 1957 *Science* **126** 302
[18] Larsson B, Leksell L, Rexed B, Sourander P, Mair W and Andersson B 1958 *Nature* **182** 1222
[19] Cheshire I M 1964 *Proc. Phys. Soc.* **84** 89
[20] Belkić Dž 1978 *J. Phys. B* **11** 3529
[21] Belkić Dž, Gayet R and Salin A 1979 *Phys. Rep.* **56** 279
[22] Massey H S W 1949 *Rep. Prog. Phys.* **12** 248
[23] Bates D R and Griffing G W 1953 *Proc. Phys. Soc. A* **66** 961
[24] Boudrioua O, Champion C, Dal Cappello C and Popov Y V 2007 *Phys. Rev. A* **75** 022720

[25] Champion C, Dal Cappello C, Boudrioua O, Lekadir H, Sato Y and Ohsawa D 2007 *Phys. Rev. A* **75** 032724
[26] Champion C, Boudrioua O and Dal Cappello C 2008 *J. Phys. Conf. Ser.* **101** 012010
[27] Dal Cappello C, Champion C, Boudrioua O, Lekadir H, Sato Y and Ohsawa D 2009, *Nucl. Instr. Meth. Phys. Res. B* **267** 781
[28] Bethe H 1930 *Ann. Phys. Lpz.* **5** 325
[29] Bloch F 1933 *Ann. Phys. Lpz.* **16** 285
[30] Crooks G B and Rudd M E 1970 *Phys. Rev. Lett.* **25** 1599
[31] Lucas M W and Harrison K G 1972 *J. Phys. B* **5** L20
[32] Sarkadi L and Orbán A 2008 *Phys. Rev. Lett.* **100** 133201
[33] Sarkadi L and Orbán A 2009 *Nucl. Instr. Meth. Phys. Res. B* **267** 270
[34] Koschar P 1984 in: *Forward Electron Ejection in Ion Collisions*, Book Series "Lecture Notes in Physics" (Berlin, Springer) **213** 129
[35] Rudd M E, Kim Y-K, Madisson D H and Gray T J 1992 *Rev. Mod. Phys.* **64** 441
[36] Stolterfoht N, DuBois R D and Rivarola R D 1997 *Electron Emission in Heavy Ion-Atom Collisions* (Berlin, Springer)
[37] Fastrup B, Horsdal-Pedersen E, Afrosimov V V, Basalaev A A and Panov M N 1999 *21st Int. Conf. Phys. Elec. Atom. Coll.* AIP Conference Proceedings, 500 (Eds. Itikawa Y, Okuno K, Tanaka H, Yagishita A and Matsuzawa M) **2** 462
[38] Afrosimov V V, Basalaev A A, Fastrup B, Horsdal-Pedersen E, Kashnikov K V and Panov M N 2000 *J. Phys. B* **33** 4237
[39] Hagmannn S and Ali I 1999 *Phys. Scr.* **T80** 329
[40] Hagmannn S, Ali I, Lüdde H-J and Wagner E 1999 http://www/jrm.phys.ksu.edu/Resource/Pubs/Progress/a-1-5.pdf (posted: August 31, 1999)
[41] Hagmannn S and Ali I 2000 http://www/jrm.phys.ksu.edu/Resource/Pubs/Progress/a-1-4.pdf (posted: January 18, 2000)
[42] Pedersen E H and Larsen L 1999 *J. Phys. B* **12** 4085
[43] Gulyás L, Fainstein P D and Shirai T 2001 *J. Phys. B* **34** 1473
[44] Belkić Dž 1997 *Nucl. Instr. Meth. Phys. Res. B* **124** 365
[45] Belkić Dž 1997 *J. Phys. B* **30** 1731
[46] Crothers D S F and McCann J F 1983 *J. Phys. B* **16** 3229
[47] Fainstein P D, Ponce V H and Rivarola R D 1991 *J. Phys. B* **24** 3091
[48] O'Rourke S F C, McSherry D M and Crothers D S F 2002 *Adv. Chem. Phys.* **121** 311
[49] Gulyás L and Fainstein P D 1998 *J. Phys. B* **31** 3297
[50] Fainstein P D and Gulyás L 2005 *J. Phys. B* **38** 317
[51] Igarashi A, Gulyás L and Fainstein P D 2007 *Eur. Phys. J. D* **42** 73
[52] Belkić Dž, Mančev I and Milojević N 2013 in: *Fast Ion-Atom and Ion-Molecule Collisions* (Ed. Belkić Dž) 189 (Singapore, World Scientific Publishing)
[53] Ermolaev A M 1988 *J. Phys. B* **21** 81
[54] Sidis V, Kubach C and Fussen D 1983 *Phys. Rev. A* **27** 2431
[55] Joachain C J and Terao M 1991 *Private communication*
[56] Peart B, Walton D S and Dolder K T 1970 *J. Phys. B* **3** 1346
[57] Melchert F, Krüdener S, Huber K and Salzborn E 1999 *J. Phys. B* **32** L139
[58] Olivera G H, Martínez A E, Rivarola R D and Fainstein P D 1995 *Radiat. Res.* **144** 241
[59] Olivera G H, Martínez A E, Rivarola R D and Fainstein P D 1996 *Nucl. Instr. Meth. Phys. Res. B* **111** 7
[60] Houamer S, Popov Yu V, Dal Cappello C, Champion C 2009, *Nucl. Instr. Meth. Phys. Res. B* **267** 802
[61] Galassi M E, Rivarola R D, Beuve M, Olivera G H, Fainstein P D 2010 *Phys. Rev. A* **62** 022701
[62] Champion C, Galassi M E, Weck P F, Fojón O A, Hanssen J and Rivarola R D 2012 *J. Phys. Conf. Ser.* **388** 102003
[63] Rivarola R D, Galassi M E, Fainstein P D and Champion C 2013 *Adv. Quantum Chem.* **65** 231
[64] Champion C, Hanssen J and Rivarola R D 2013 *Adv. Quantum Chem.* **65** 269
[65] Tachino C A, Monti J M, Fojón O A, Champion C and Rivarola R D 2014 *J. Phys. Conf. Ser.* **488** 102029
[66] Monanari C C and Miraglia J E 2014 *J. Phys. B* **47** 015201
[67] Stopping powers and ions heavier than helium, 2005 *ICRU Report 73: International Commission of Radiation Units and Measurements* (Oxford, Oxford University Press)
[68] Schardt D, Elsässer T and Schulz-Ertner D 2010 *Rev. Mod. Phys.* **82** 383
[69] Sobolevsky N M 1970 Report JINR (Joint Institute for Nuclear Research) B1-2-5458 Dubna 1 (in Russian)
[70] Sobolevsky N M 2001 Proceeding of the 3rd Yugoslav Nuclear Society International Conference *YUNSC* (Belgrade, The VINČA Institute) 539
[71] Gudowska I, Sobolevsky N, Andreo P, Belkić Dž and Brahme A 2004 *Phys. Med. Biol.* **49** 1933

Single electron capture in fast ion-atom collisions

Nenad Milojević
Department of Physics, Faculty of Sciences and Mathematics, University of Niš, P.O.Box 224, 18000 Niš, Serbia
E-mail: nenad81@pmf.ni.ac.rs

Abstract. Single-electron capture cross sections in collisions between fast bare projectiles and heliumlike atomic systems are investigated by means of the four-body boundary-corrected first Born (CB1-4B) approximation. The prior and post transition amplitudes for single charge exchange encompassing symmetric and asymmetric collisions are derived in terms of two-dimensional real integrals in the case of the prior form and five-dimensional quadratures for the post form. The dielectronic interaction $V_{12} = 1/r_{12} \equiv 1/|\mathbf{r}_1 - \mathbf{r}_2|$ explicitly appears in the complete perturbation potential V_f of the post transition probability amplitude T_{if}^{+}. An illustrative computation is performed involving state-selective and total single capture cross sections for the p − He (prior and post form) and $He^{2+}, Li^{3+}, Be^{4+}, B^{5+}, C^{6+}$ − He (prior form) collisions at intermediate and high impact energies. We have also studied differential cross sections in prior and post form for single electron transfer from helium by protons. The role of dynamic correlations is examined as a function of increased projectile energy. Detailed comparisons with the measurements are carried out and the obtained theoretical cross sections are in reasonable agreement with the available experimental data.

1. Introduction

The investigation and differentiation electron capture from atomic and molecular targets by the impact of fast ions has been a topic of considerable interest in atomic collision physics for many years. This occurs because of the critical importance of charge transfer cross sections in a number of applications in astrophysics, plasma physics, thermonuclear fusion research and medical accelerators for hadron radiotherapy. In the present work we report on our theoretical investigation of single electron capture from heliumlike atomic systems interacting with bare projectiles (p, $He^{2+}, Li^{3+}, Be^{4+}, B^{5+}$, and C^{6+}), with all four particles actively participating in these processes. Hence, this is a pure four-body problem that involves laborious analytical and numerical theoretical computations within any type of first order approximations. Total and differential cross sections in prior and post form are calculated by using four-body boundary-corrected first Born (CB1-4B) approximation. The first Born approximation with the correct boundary conditions has been introduced within the three-body formalism [1]. On the other hand, the CB1-4B method was adapted and applied to single-electron capture [2]-[4].

The CB1-4B method is a fully quantum mechanical four-body formalism, and it explicitly considers each individual particle and all the interactions among them in the collision under investigation. It strictly preserves the correct boundary conditions in both collisional channels, and it is well known that the boundary conditions (the asymptotic convergence problem [5]-[9]) are of essential importance for atomic collisions whenever the aggregates are charged in the asymptotic channels.

Atomic units will be used throughout unless otherwise stated.

2. Theory

We consider single electron capture in collisions of a completely stripped projectile with a helium-like target:

$$X^{Z_P+} + (Z_T; e_1, e_2)_{1s^2} \longrightarrow (X^{Z_P+}, e_1)_{nlm} + (Z_T, e_2)_{1s}\,, \tag{1}$$

where Z_K is the charge of the K th nucleus (K = P, T) and nlm is the usual set of three quantum numbers of hydrogenlike atomic systems. Here, the parentheses symbolize the bound states. Let $\vec{s}_1$ and $\vec{s}_2$ ($\vec{x}_1$ and $\vec{x}_2$) be the position vectors of the first and second electron (e_1 and e_2) relative to the nuclear charge of the projectile Z_P (target Z_T), respectively. Further, let $\vec{R}$ be the position vector of Z_T with respect to Z_P.

The prior and post form of the transition amplitude for process (1) in the CB1-4B approximation without the term $(\rho v)^{2iZ_p(Z_T-2)/v}$ read as [2]:

$$T^{\pm}_{if}(\vec{\eta}) = \int\int\int \mathrm{d}\vec{x}_1 \mathrm{d}\vec{x}_2 \mathrm{d}\vec{R} \varphi^*_{nlm}(\vec{s}_1)\varphi^*_{100}(\vec{x}_2) V^{\pm} \varphi_i(\vec{x}_1, \vec{x}_2)\, \mathrm{e}^{-i\vec{\alpha}\cdot\vec{R} - i\vec{v}\cdot\vec{x}_1} (vR + \vec{v}\cdot\vec{R})^{i\xi}\,, \tag{2}$$

$$V^+ \equiv V_f = \Delta V_{P2} + \Delta V_{T1} + \Delta V_{12}\,, \quad \Delta V_{12} = V_{12} - V^{\infty}_{12}\,, \quad V_{12} = \frac{1}{r_{12}}\,, \quad V^{\infty}_{12} = \frac{1}{x_1}\,, \tag{3}$$

$$V^- \equiv V_i = \Delta V_{P1} + \Delta V_{P2}\,, \tag{4}$$

where

$$\Delta V_{Pi} = Z_P\left(\frac{1}{R} - \frac{1}{s_i}\right), \quad (i = 1, 2)\,, \quad \Delta V_{T1} = (Z_T - 1)\left(\frac{1}{R} - \frac{1}{x_1}\right). \tag{5}$$

Here, $\xi = (Z_P - Z_T + 1)/v$ where v is the velocity of the projectile and $\vec{\rho} = \vec{R} - \vec{Z}$, $\vec{\rho}\cdot\vec{Z} = 0$. The vector of the distance between the two electrons e_1 and e_2 is denoted by $\vec{r}_{12} = \vec{x}_1 - \vec{x}_2 = \vec{s}_1 - \vec{s}_2$, and we have $r_{12} = |\vec{r}_{12}|$. The momentum transfer $\vec{\alpha}$ and transverse momentum transfer $\vec{\eta}$ are defined by:

$$\vec{\alpha} = \vec{\eta} - (v/2 - \Delta E/v)\hat{\vec{v}}\,, \quad \vec{\eta} = (\eta\cos\phi_\eta, \eta\sin\phi_\eta, 0)\,, \quad \vec{\eta}\cdot\vec{v} = 0\,, \tag{6}$$

where $\Delta E = E_i - E_f$ with E_i being the binding energy of the two-electron target and $E_f = -Z_P^2/[2n^2] - Z_T^2/2$. The function $\varphi_i(\vec{x}_1, \vec{x}_2)$ denotes the two-electron ground state wave function of the atomic system $(Z_T; e_1, e_2)_{1s^2}$. The functions $\varphi_{nlm}(\vec{s}_1)$ and $\varphi_{100}(\vec{x}_2)$ in Eq. (2) represent the bound state wave functions of the hydrogen-like atomic systems $(X^{Z_P+}, e_1)_{nlm}$ and $(Z_T, e_2)_{1s}$, respectively.

The original nine-dimensional integral for transition amplitude can be reduced to a two-dimensional integral and five-dimensional integral over real variables in prior [3] and post [4] form, respectively.

The differential (DCS) and total (TCS) cross sections (prior and post form) in the CB1-4B method are given by:

$$\frac{dQ^{\pm}_{if}}{d\Omega}\left(\frac{a_0^2}{sr}\right) = \frac{\mu^2}{4\pi^2}|T^{\pm}_{if}(\vec{\eta})|^2, \quad Q^{\pm}_{if}(a_0^2) = \int_{4\pi} \frac{dQ^{\pm}_{if}}{d\Omega}\mathrm{d}\Omega = \frac{1}{2\pi v^2}\int_0^{\infty} \mathrm{d}\eta\eta |T^{\pm}_{if}(\vec{\eta})|^2\,, \tag{7}$$

where $\mu_i \simeq \mu_f \simeq \mu = M_P M_T/(M_P + M_T)$ is the reduced mass of the incident and target nuclei with the mass M_P and M_T respectively. All numerical integrations are performed by means of the Gauss-Legendre and Gauss-Mehler quadrature. The numbers of integration points are varied until convergence to two decimal places has been attained for state-selective and total cross sections.

Total cross sections $Q^{\pm}(\Sigma)$ for electron capture into all the final states for reaction (1) are obtained by applying the Oppenheimer (n^{-3}) scaling law [10, 11] via:

$$Q^{\pm}_{\rm tot} \equiv \sum_{n=1}^{\infty} Q^{\pm}_n \simeq Q^{\pm}(\Sigma_N) = \sum_{n=1}^{N-1} Q^{\pm}_n + \gamma(3, N-1) Q^{\pm}_N, \tag{8}$$

where

$$\gamma(3, N) = 1 + (N+1)^3 \zeta(3) - \sum_{n=1}^{N+1} \left(\frac{N+1}{n}\right)^3, \tag{9}$$

with $\zeta(3)$ being the Riemann zeta function $\zeta(3) = \sum_{n=1}^{\infty} n^{-3} \approx 1.202$, and partial cross sections

$$Q^{\pm}_n = \sum_{l=0}^{n-1} Q^{\pm}_{nl}, \quad Q^{\pm}_{nl} = \sum_{m=-l}^{+l} Q^{\pm}_{nlm}. \tag{10}$$

The computations are stopped when the total cross sections $Q^{\pm}_{\rm tot}$ become insensitive to inclusion of more partial (state-selective) cross sections $Q^{\pm}_n$.

3. Results and discussions

First we shall examine electron capture from He(1s^2) by proton (asymmetric reaction):

$$\mathrm{p} + \mathrm{He}(1\mathrm{s}^2) \longrightarrow \mathrm{H}(nlm) + \mathrm{He}^+(1\mathrm{s}) \tag{11}$$

in the energy range from 20 to 10000 keV. The total CB1-4B cross sections in the prior form obtained with the complete perturbation V_i by using Eq. (8) (where is $N = 4$):

$$Q^{-}_{\rm tot} \simeq Q^{-}(\Sigma_4) = Q^{-}_1 + Q^{-}_2 + Q^{-}_3 + 2.561 Q^{-}_4, \tag{12}$$

are plotted in Fig.1. The full line shows the results obtained by means of the two-parameter wave function of Silverman *et al.* [12] for the ground state of the helium atom:

$$\varphi_i(\vec{x}_1, \vec{x}_2) = (N/\pi)(\mathrm{e}^{-\alpha_1 x_1 - \alpha_2 x_2} + \mathrm{e}^{-\alpha_2 x_1 - \alpha_1 x_2}), \quad N^{-2} = 2[(\alpha_1\alpha_2)^{-3} + (\alpha_1/2 + \alpha_2/2)^{-6}], \tag{13}$$

whereas the dashed line represents the cross sections obtained by using uncorrelated one-parameter Hylleraas wave function [13] for the ground state of the helium atom:

$$\varphi_i(\vec{x}_1, \vec{x}_2) = (\alpha^3/\pi)\,\mathrm{e}^{-\alpha(x_1+x_2)}, \quad \alpha = Z_T - 5/16. \tag{14}$$

The total cross sections obtained with these wave functions are close to each other, and the dashed curve slightly overestimates the full curve at impact energies in the range from 100 to 1000 keV, but outside this range the behaviour is the opposite. The relative difference $\chi = |Q^{-}_S - Q^{-}_H|/Q^{-}_S$ (where Q^{-}_S and Q^{-}_H are the present cross sections obtained by using wave functions of Silverman *et al.* [12] and Hylleraas [13], respectively) of the prior cross sections between these wave functions does not exceed 22%. These results are seen to be in very good agreement with all the available measurements in an energy range which remarkably covers nearly three orders of magnitude.

The total cross sections for reaction (11) computed in the prior and post version of the CB1-4B method by using the four-parameter wave function of Löwdin [22] for ground state are displayed in Fig. 2. This function is given by $\varphi_i(\vec{x}_1, \vec{x}_2) = (N/\pi)(\alpha_1 e^{-\beta_1 x_1} + \alpha_2 e^{-\beta_2 x_1})(\alpha_1 e^{-\beta_1 x_2} + \alpha_2 e^{-\beta_2 x_2})$, where the normalization constants $N^{-1} = [\alpha_1^2/\beta_1^3 + \alpha_2(\alpha_2/\beta_2^3 + 16\alpha_1/(\beta_1+\beta_2)^3)]$. The total cross sections have been obtained by summing over all contributions from the individual shells and sub-shells up to $n = 3$: $Q^{\pm}_{\rm tot} \simeq Q^{\pm}(\Sigma_3) = Q^{\pm}_1 + Q^{\pm}_2 + 2.081 Q^{\pm}_3$.

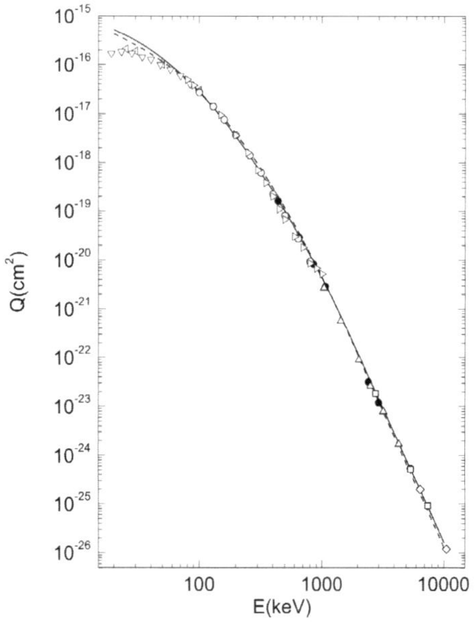

Figure 1. Total cross sections $Q^-(\Sigma_4)$ (in cm^2) as a function of the laboratory incident energy for reaction (11). The full and the dashed curves are the present cross sections obtained by using the helium ground state wave functions of Silverman *et al.* [12] and Hylleraas [13], respectively. Experimental data: ▽ Shah *et al* [14], △ Schryber [15], ∘ Shah and Gilbody [16], □ Horsdal-Pedersen *et al.* [17], ◇ Berkner *et al.* [18], ▷ Williams [19], ◁ Martin *et al.* [20] and • Welsh *et al.* [21].

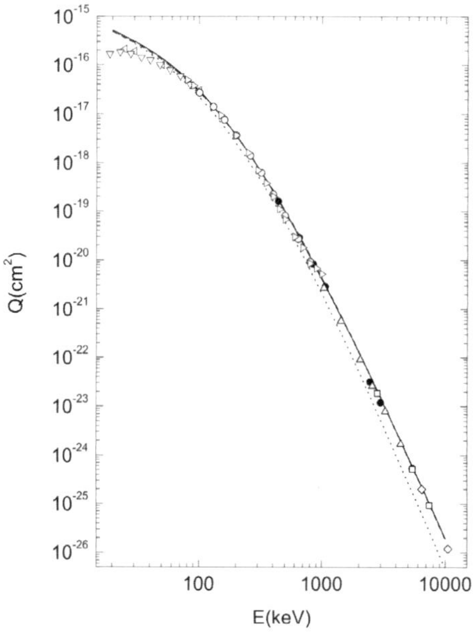

Figure 2. Total cross sections $Q^{\pm}(\Sigma_3)$ (in cm^2) as a function of the laboratory incident energy for reaction (11). The full and the dotted curves are the present cross sections $Q^+(\Sigma_3)$ in the post form with and without ΔV_{12} in V_f, respectively. The dashed curve is the present cross sections $Q^-(\Sigma_3)$ with the complete potential V_i. For all the three curves, the initial ground state of the helium atom is described by means of the Löwdin [22] orbital. Experimental data: the same as in Fig. 1.

Potential ΔV_{12} in the complete perturbation V_f from (3) represents a source of the dynamic electron correlation effect, which is included throughout $1/r_{12}$. From Fig. 2 we can see that above 100 keV, where the CB1-4B model is expected to be valid, the discrepancy between the results with the complete perturbation V_f (full line) and the corresponding cross sections without ΔV_{12} in V_f (dotted line) is increasing. These discrepancies increase even further with augmentation of the incident energy. Such a pattern indicates that the role of the screened dynamic correlations ΔV_{12} becomes more significant at higher incident energies.

Also we examine the so-called post-prior discrepancy, which arises from the unequal perturbation potentials ($V_i \neq V_f$) in the transition amplitudes, as well as from the unavailability of the exact bound-state wave function of heliumlike atomic systems. The difference between the results for the prior and post cross sections is very small (the relative difference $\chi_p = |Q^+ - Q^{+(0)}|/Q^+$, where Q^+ and $Q^{+(0)}$ are the present cross sections for capture with and without ΔV_{12} respectively, is up to 6.6%), as can be seen in Fig. 2. This is an excellent property of the CB1-4B approximation, since the same physical assumptions are involved in the prior (dashed line) and post (full line) forms of this theory. Namely, electron-electron interaction is indirectly included through $-Z_P/s_2$ ($-Z_P/s_2 = -Z_P(1/r_{12} - \vec{r}_{12}\cdot\vec{s}_1/r_{12}^3 + \cdots)$ - the Taylor series around $\vec{s}_1$) term in the complete potential V_i.

Differential cross sections (DCS) for single electron capture in process (11) are studied at 100 keV projectile energy.

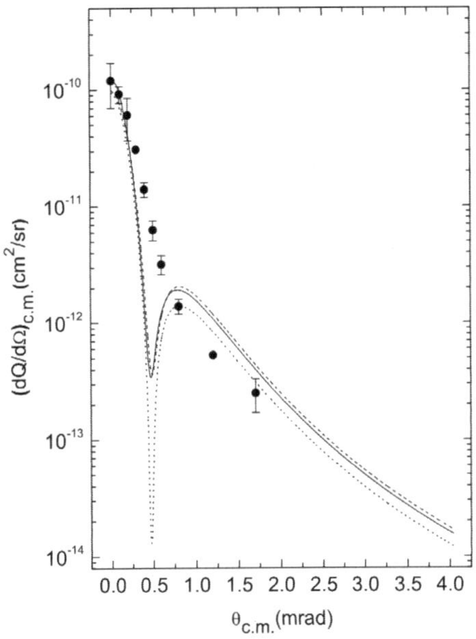

Figure 3. DCS for single electron capture for reaction (11) at 100 keV projectile energy. The full and dashed lines are the present $dQ^-_{tot}/d\Omega$ obtained by using the wave functions of Silverman *et al.* [12] and Hylleraas [13], respectively. The dotted line presents $dQ^-_{1s}/d\Omega$ with wave functions of Silverman *et al.* [12]. Exp. data: • Martin *et al.* [20].

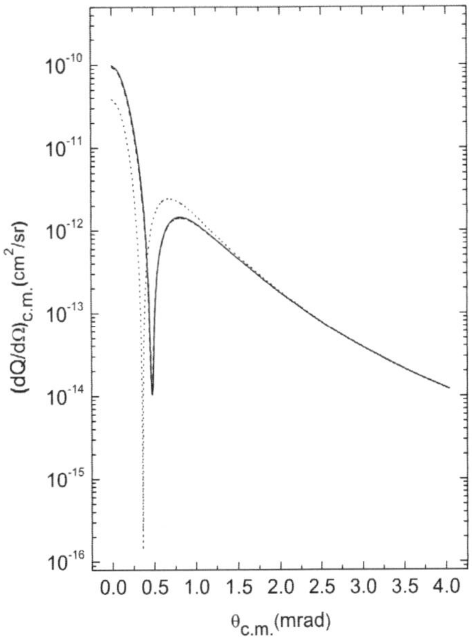

Figure 4. DCS for single electron capture for p + He(1s^2) $\longrightarrow$ H(1s) + He$^+$(1s) collisions at 100 keV projectile impact energy. The full, dashed and dotted lines are the present $dQ^+_{1s}/d\Omega$, $dQ^-_{1s}/d\Omega$ and $dQ^{+(0)}_{1s}/d\Omega$, respectively. All calculations are performed by using the helium ground-state wave functions of Silverman *et al.* [12].

Theoretical results with the available experimental data are shown in Fig. 3. Differential cross sections summed over the final states of atomic hydrogen according relation: $dQ^-_{tot}/d\Omega \simeq dQ^-_1/d\Omega + dQ^-_2/d\Omega + dQ^-_3/d\Omega + 2.561dQ^-_4/d\Omega$ are in excellent agreement with the experimental

Journal of Physics: Conference Series **565** (2014) 012004 doi:10.1088/1742-6596/565/1/012004

results at scattering angles from 0 to 0.12 mrad, but at angles above 0.79 mrad theoretical cross sections overestimate the measurements. The results obtained with Hylleraas [13] wave function overestimate the results which have been calculated with Silverman *et al.* [12] function at higher scattering angles. In the same figure, the cross sections $dQ^-_{1s}/d\Omega$ for capture into the $1s$ state are depicted. We can see that the contribution from the excited states is more significant at higher than at lower scattering angles. Figure 4 depicts the prior and post DCS of the CB1-4B method for single-electron capture into the $1s$ state of atomic hydrogen using the helium wave function of Silverman *et al.* [12]. As expected, the difference between prior $dQ^-_{1s}/d\Omega$ (dashed line) and post (full line) $dQ^+_{1s}/d\Omega$ forms DCS with complete perturbation potentials V_i and V_f respectively is very small at all the considered angles. Comparing the results for the post form DCS $dQ^{+(0)}_{1s}/d\Omega$ without ΔV_{12} in V_f and results for $dQ^-_{1s}/d\Omega$ and $dQ^+_{1s}/d\Omega$, we can conclude that the screened dynamic correlations ΔV_{12} are very important at lower angels.

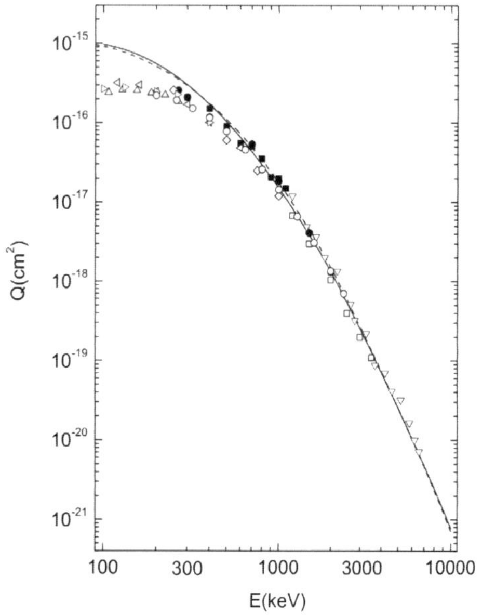

Figure 5. TCS $Q^-(\Sigma_4)$ (in cm^2) as a function of the laboratory incident energy for reaction $He^{2+} + He(1s^2) \longrightarrow He^+(\Sigma) + He^+(1s)$. The full and the dashed curves are the present TCS obtained by using the wave functions of Silverman *et al.* [12] and Hylleraas [13], respectively. Exp. data: △ Shah *et al.* [14]; ○ Shah and Gilbody [16]; ◇ Mergel *et al.* [23]; ▽ Hvelplund *et al.* [24]; ● DuBois [25]; □ de Castro Faria *et al.*[26]; ■ Pivovar *et al.* [27, 28]; ◁ Rudd *et al.* [29]; ▷ Alessi *et al.* [30].

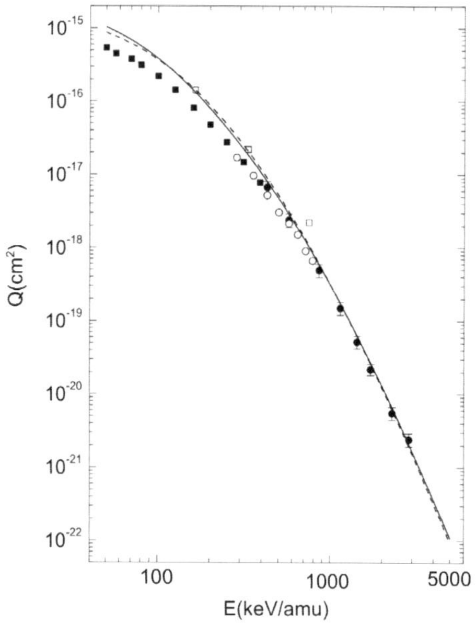

Figure 6. TCS $Q^-(\Sigma_4)$ (in cm^2) as a function of the laboratory incident energy for reaction $Li^{3+} + He(1s^2) \longrightarrow Li^{2+}(\Sigma) + He^+(1s)$. The full and the dashed curves are the present TCS obtained by using the helium ground state wave functions of Silverman *et al.* [12] and Hylleraas [13], respectively. Experimental data: ■ Shah and Gilbody [16]; ● Woitke *et al.* [31]; ○ Sant'Anna *et al.* [32]; □ Dmitriev *et al.* [33].

As seen from Figs. 3 and 4, the CB1-4B method exhibits an unphysical and experimentally unobserved dip at $\theta_{c.m.} \simeq 0.47$ mrad. This sharp dip is due to strong cancellation of the opposite contributions coming from the attractive and repulsive potentials in Eqs. (2)-(4).

Cross sections in prior form for electron capture for $He^{2+} + He(1s^2)$ collisions are calculated in the energy range from 100 to 10000 keV. These results are obtained by applying the Eq. (12) and they are depicted in Fig. 5. The dashed line (Hylleraas [13]) slightly overestimates the full line (Silverman *et al.* [12]) in the range 500 to 3000 keV. The relative difference between these curves does not exceed 16.4%, for all the considered energy. It can be seen from Fig. 5 that agreement between the present CB1-4B cross sections (full and dashed curves) and a number of measurements is excellent above 400 keV.

In Fig. 6, we make a comparison of measurements and the theoretical total cross sections for electron capture from $He(1s^2)$ by Li^{3+} obtained by using Eq. (12) in energy range from 50 to 5000 keV/amu. The present cross sections, the full line (Silverman *et al.* [12]) and the dashed line (Hylleraas [13]), are in excellent agreement with the measurements [16], [31]-[33]. These curves are also very close to each other and the relative difference is up to 16.3%.

The results ($Q^-_{tot} \simeq Q^-(\Sigma_5) = Q^-_1 + Q^-_2 + Q^-_3 + Q^-_4 + 3.049Q^-_5$) for charge exchange in $Be^{4+} + He(1s^2)$ collisions at energies from 100 to 3000 keV/amu are depicted in Fig. 7, by the full (Silverman *et al.* [12]) and dashed curves (Hylleraas [13]). The dashed line in this case also

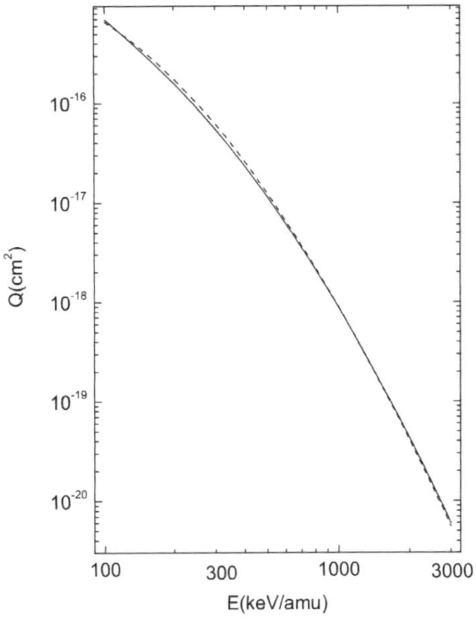

Figure 7. TCS $Q^-(\Sigma_5)$ (in cm^2) as a function of the laboratory incident energy for reaction $Be^{4+} + He(1s^2) \longrightarrow Be^{3+}(\Sigma) + He^+(1s)$. The full and the dashed curves are the present TCS obtained by using the helium ground state wave functions of Silverman *et al.* [12] and Hylleraas [13], respectively.

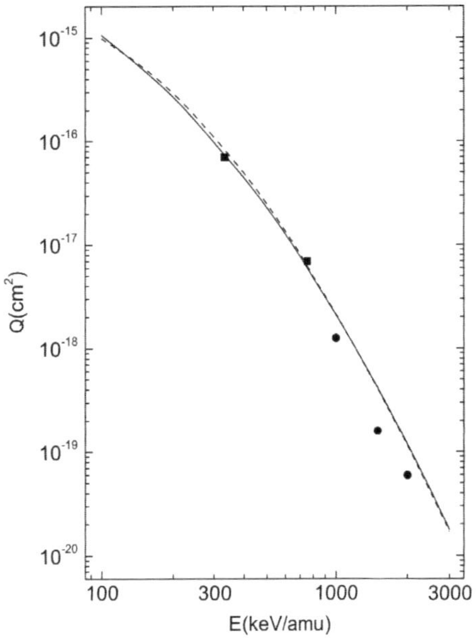

Figure 8. TCS $Q^-(\Sigma_6)$ (in cm^2) as a function of the lab. incident energy for reaction $B^{5+} + He(1s^2) \longrightarrow B^{4+}(\Sigma) + He^+(1s)$. The full and the dashed curves are the present TCS obtained by using the wave functions of Silverman *et al.* [12] and Hylleraas [13], respectively. Exp. data: ■ Dmitriev *et al.* [33]; • Hippler *et al.* [34].

slightly overestimates full line at energies from 120 to 850 keV/amu, where the relative difference between them does not exceed 13.1%. Unfortunately, no experimental data are available for comparison. The qualitative behaviours of the total cross sections are very similar to those

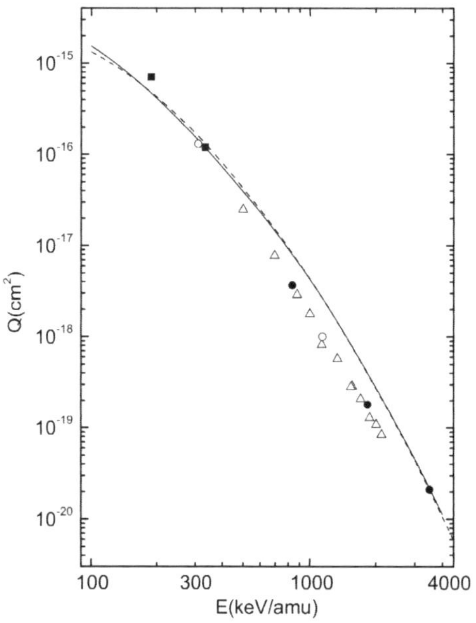

Figure 9. TCS $Q^-(\Sigma_7)$ (in cm^2) as a function of the laboratory incident energy for reaction C^{6+} + He(1s^2) $\longrightarrow$ $C^{5+}(\Sigma)$ + He$^+$(1s). The full and the dashed curves are the present TCS obtained by using the helium ground state wave functions of Silverman *et al.* [12] and Hylleraas [13], respectively. Experimental data: ■ Dmitriev *et al.* [33]; • Anholt *et al.* [35]; ○ Graham *et al.* [36]; △ Dillingham *et al.* [37].

corresponding to those for the other considered ions.

Total cross sections for single electron capture in the B^{5+} + He(1s^2) $\longrightarrow$ $B^{4+}(nlm)$ + He$^+$(1s) process obtained by using $Q^-_{\text{tot}} \simeq Q^-(\Sigma_6) = Q^-_1 + Q^-_2 + Q^-_3 + Q^-_4 + Q^-_5 + 3.541 Q^-_6$ are presented by the full (Silverman *et al.* [12]) and dashed curves (Hylleraas [13]) in Fig. 8. At all the considered energies from 100 to 3000 keV/amu their relative difference is below 13%. The present CB1-4B theory in the prior form describes the experimental data of Dmitriev *et al.* [33] quite successfully, whereas it slightly overestimates the measurements of Hippler *et al.* [34].

A similar trend in behaviour is noticed in the C^{6+}+He(1s^2) $\longrightarrow$ $C^{5+}(nlm)$+He$^+$(1s) collisions, where the total cross sections obtained by applying $Q^-_{\text{tot}} \simeq Q^-(\Sigma_7) = Q^-_1 + Q^-_2 + Q^-_3 + Q^-_4 + Q^-_5 + Q^-_6 + 4.035 Q^-_7$ at energies from 100 to 4000 keV/amu. These results are shown in Fig. 9 by full (Silverman *et al.* [12]) and dashed (Hylleraas [13]) curves. Now the relative difference is smaller than 11% at energy above 150 keV/amu. It follows from Fig 9. that CB1-4B approximation is in satisfactory agreement with available experimental data [33], [35]-[37]. At higher energies the results from CB1-4B method overestimate experimental data, except those at the largest energy 3500 keV/amu. The discrepancy between the present calculations and the experimental total cross sections for the last two processes at higher energies is not unexpected. Namely, the present CB1-4B method does not include the intermediate ionization channels (electronic continuum intermediate states), which dominate over charge exchange at high energies.

4. Conclusions

We have studied the single charge exchange in collisions between bare nuclei (p, He^{2+}, Li^{3+}, Be^{4+}, B^{5+} and C^{6+}) and He targets at intermediate and high impact energies, within the framework of CB1-4B approximation. The results for the prior and post cross sections (both differential and total) for p+He collisions are nearly identical, provided that the repulsive electron-electron interaction $1/r_{12}$ is taken into account in V_f. Moreover, we proved that the contribution from ΔV_{12} increases with augmenting impact energy and decreasing scattering angles. The prior form of the total cross sections has been also calculated by using Silverman *et al.* [12] and Hylleraas [13] wave functions for ground state of helium atom. The relative difference between these results is small, and decreases with increasing the charge of the projectile. On the other side discrepancy between DCS obtained with Silverman *et al.* [12] and Hylleraas [13] wave functions increases with augmenting scattering angle. The contribution from the excited states is more significant at higher than at lower scattering angles. The CB1-4B method is found to be systematically in very good agreement with the available experimental data for p, He^{2+}, Li^{3+} projectiles, but for multiply-charged projectiles B^{5+} and C^{6+} the CB1-4B approximation slightly overestimates measurements at higher impact energy. This behaviour is mainly attributed to the neglect of the intermediate ionization channels which are important in this region.

Acknowledgments

The author thanks the Ministry of Education, Science and Technological Development of the Republic of Serbia for support through Project No. 171020. Thanks are due to Professor Dževad Belkić for helpful discussions and a critical review of the manuscript.

References

[1] Belkić Dž, Gayet R and Salin A 1979 *Phys. Rep.* **56** 279
[2] Mančev I and Milojević N 2010 *Phys. Rev. A* **81** 022710
[3] Mančev I, Milojević N and Belkić Dž 2012 *Phys. Rev. A* **86** 022704
[4] Mančev I, Milojević N and Belkić Dž 2013 *Phys. Rev. A* **88** 052706
[5] Dollard D 1964 *J. Math. Phys.* **5** 729
[6] Cheshire I M 1964 *Proc. Phys. Soc. London* **84** 89
[7] Belkić Dž 2004 *Principles of Quantum Scattering Theory* (Bristol, Institute of Physics)
[8] Belkić Dž, Mančev I and Hanssen J 2008 *Rev. Mod. Phys.* **80** 249
[9] Belkić Dž 2009 *Quantum Theory of High-Energy Ion-Atom Collisions* (London, Taylor & Francis)
[10] Oppenheimer J R 1928 *Phys. Rev.* **31** 349
[11] Brinkman H C and Kramers H A 1930 *Proc. K. Ned. Akad. Wet.* **33** 973
[12] Silverman J N, Platas O, Matsen F A 1960 *J.Chem.Phys.* **32** 1402; Eckart C 1930 *Phys. Rev.* **36** 878
[13] Hylleraas E 1929 *Z. Phys.* **54** 347
[14] Shah M B, McCallion P and Gilbody H B 1989 *J. Phys. B* **22** 3037
[15] Schryber U 1967 *Helv. Phys. Acta* **40** 1023
[16] Shah M B and Gilbody H B 1985 *J. Phys. B* **18** 899
[17] Horsdal-Pedersen E, Cocke C and Stockli M 1983 *Phys. Rev. Lett.* **50** 1910
[18] Berkner K H, Kaplan S N, Paulikas G A and Pyle R V 1965 *Phys. Rev.* **140** A729
[19] Williams J F 1967 *Phys. Rev.* **157** 97
[20] Martin P J, Arnett K, Blankenship D M, Kvale T J, Peacher J L, Redd E, Sutcliffe V C, Park J T, Lin C D and McGuire J 1981 *Phys. Rev. A* **23** 2858
[21] Welsh L M, Berkner K H, Kaplan S N and Pyle R V 1967 *Phys. Rev.* **158** 85
[22] Löwdin P 1953 *Phys. Rev.* **90** 120
[23] Mergel V, Dörner R, Jagutzki O, Leneinas S, Nüttgens S, Spielberger L, Unverzagt M, Cocke C L, Olson R E, Schultz M, Buck U, Zanger E, Theisinger W, Isser M, Geiss S and Schmidt-Böcking H 1995 *Phys. Rev. Lett.* **74** 2200
[24] Hvelplund P, Heinemei J, Horsdal-Pedersen E and Simpson F R 1976 *J. Phys. B* **9** 491
[25] DuBois R D 1987 *Phys. Rev. A* **36** 2585
[26] de Castro Faria N V, Freire Jr F L and de Pinho A G 1988 *Phys. Rev. A* **37** 280
[27] Pivovar L I, Tabuev V M and Novikov M T 1962 *Sov. Phys. JETP* **14** 20

[28] Pivovar L I, Tabuev V M and Novikov M T 1961 *Zh. Eksp. Teor. Fiz.* **41** 26
[29] Rudd M E, Goffe T V and Itoh A 1985 *Phys. Rev. A* **32** 2128
[30] Alessi M, Otronto S and Focke P 2011 *Phys. Rev. A* **83** 014701
[31] Woitke O, Závodszky P A, Ferguson S M, Houck J H and Tanis J A 1998 *Phys. Rev. A* **57** 2692
[32] Sant'Anna M M, Santos A C, Coelho L F, Jalbert G, de Castro Faria N V, Zappa F, Focke P and Belkić Dž 2009 *Phys. Rev. A* **80** 042707
[33] Dmitriev I S, Teplova Ya A, Belkova Ya A, Novikov N V and Fainberg Yu A 2010 *Atomic Data and Nuclear Data Tables* **96** 85
[34] Hippler R, Datz S, Miller P D, Pepmiller P L and Dittner P F 1987 *Phys. Rev. A* **35** 585
[35] Anholt R, Xu X-Y, Stoller Ch, Molitoris J D and Meyerhof W E 1988 *Phys. Rev. A* **37** 1105
[36] Graham W G, Berkner K H, Pyle R V, Schlachter A S, Stearns J W and Tanis J A 1984 *Phys. Rev. A* **30** 722
[37] Dillingham T R, Macdonald J R and Richard P 1981 *Phys. Rev. A* **24** 1237

27th Summer School and International Symposium on the Physics of Ionized Gases (SPIG 2014) IOP Publishing
Journal of Physics: Conference Series **565** (2014) 012005 doi:10.1088/1742-6596/565/1/012005

Absolute differential cross sections for elastic electron scattering from small biomolecules

Jelena Maljković

Laboratory for Atomic Collision Processes, Institute of Physic Belgrade, University of Belgrade, Pregrevica 118, 11080, Belgrade Serbia

E-mail: jelenam@ipb.ac.rs

Abstract. The results of an experimental investigation of electrons colliding with a set of biomolecules that are assumed to be analogues of the building blocks of DNA (furan, 3-hydroxytetrahydrofuran and pyrimidine) and proteins (formamide, N-methylformamide) are presented. Absolute differential cross sections at medium incident electron energies 40 eV–300 eV are presented and compared for these different targets. The experimental results are also compared with available calculations, based on the corrected form of independent atom model and show good agreement over the energy range studied.

1. Introduction

The role of low energy secondary electrons in providing a route to damage of DNA and other biomolecules during irradiation of cellular systems has only recently been recognised [1] This has led to the need to understand how electrons interact with such large macromolecules however there are both experimental and theoretical limitations in undertaking such studies (for example DNA and most membrane compounds cannot be studied in the gas phase. Accordingly it is proposed to model DNA by considering it as an ensemble of smaller molecules which are themselves analogues of the constituent biomolecules (e.g. the nuclear bases) such as Tetrahydrofuran (THF) and tetrahydrofurfuryl alcohol (THFA). This approach has resulted in an extensive investigation of electron interaction with such 'analogue' molecules [2-21]. Although secondary electrons are rather low energy, Scifoni et al [22] has shown that in the tail of the secondary electron distribution many electrons have energies of circa 100 eV. Therefore in the present studies we have explored 50 to 300 eV electron interactions with several biomolecular analogues including; furan (C_4H_4O) [8], 3-hydroxytetrahydrofuran (3HTHF, $C_4H_8O_2$) [4], which are analogues to deoxyribose in DNA (Figure1.) pyrimidine (Py, $C_4H_4N_2$) [5] which is an analogue of pyrimidine bases (Figure 1.) and formamide and N-methylformamide [6,7] the smallest molecules that contain peptide bond (Figure 1.).

Several of these analogue molecules have been explored previously. VUV absorption spectra and electron energy loss spectra of Py have been reported by [9], as have electron induced vibrational and electron excitations of a Py condesate on thin film [10]. Tetrahydrofuran (THF) and tetrahydrofurfuryl alcohol have been investigated extensively [11, 12], since at standard temperature and pressure they are in a liquid state but with sufficient vapour pressure to make measuremnts in 'conventional' electron scattering apparatus. For furan very recently Khakoo and coauthors have published differential cross sections for elastic electron scattering on furan, from 1-50 eV [13]. While

Smytkowski and coauthors have published absolute total cross sections from 0.6 -400eV and ionizing and integral cross sections up to 4keV [14]. Integral and differential cross sections for electron scattering from furan have also been calculated by Bettega and Lima [15] and energy loss energy loss spectra have been reported by Guliani and coauthors with high energy resolution [16]. Electron interactions with the 3HTHF molecule has been studied by Viscaino et al [17] and differential cross sections for low energy electrons have been published.

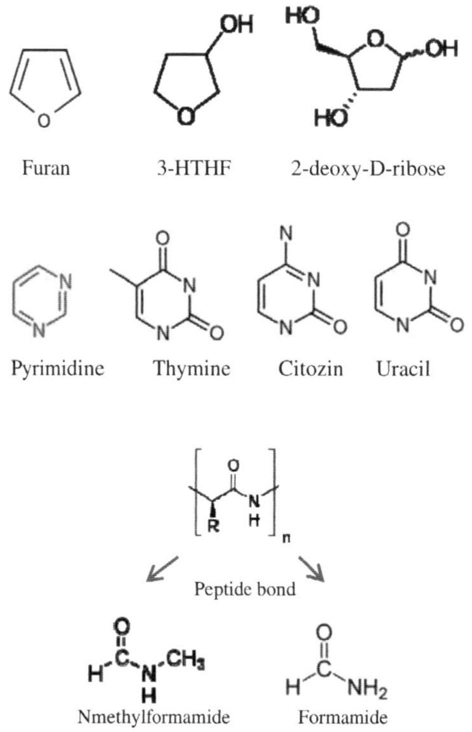

Figure 1. Schematic drawing of: furan, 3HTHF, deoxyribose, pyrimidine, pyrimidine bases, formamide, NMF and peptide bond.

Formamide (CH_3NO) and N-methylformamid (NMF, C_2H_5NO) (Figure 1) are interesting as they are the simplest models of the peptide bond NH-C=O. Bettega has published [18] theoretical cross sections for low energy electron scattering (1-12eV) from formamide. Recently photoionisation spectra of formamide and N-methylformamide have been studied because of their discovery in the interstellar medium [19] while the NMF molecule has received increased attention since it has been shown to have an important role in antitumor activity [20,21].

2. Experimental methodology

The present experimental results have been obtained using the cross beam electron spectrometer UGRA [23] within the Institute of Physics in Belgrade. The main components of the spectrometer are an electron gun, a double cylindrical mirror analyzer (DCMA) and a channeltron as detector. Primary electrons, produced by thermionic emission from a hairpin cathode, are extracted and focused by a system of cylindrical electrodes to cross, perpendicularly, a beam of target molecules exiting from a stainless steel needle. The scattered electrons are retarded and focused with a 4-element lens into the DCMA and then focused using a 3-element lens into the channeltron. The electron gun can be rotated around the target gas beam from -30° to 110°. The whole experiment is conducted within a vacuum chamber pumped by a turbomolecular pump. A base pressure was around 4×10^{-7} mbar and the working pressure was around 2×10^{-6} mbar. The electron energy resolution was 0.5 eV determined by the thermal spread of the primary electrons. The angular resolution was about ±2° and was obtained by comparing our measured experimental results with previously published ones for noble gases [24] at the incident electron energies where DCS minima are most pronounced, for example for Ar at 40 eV and 50 eV.

2.1. Absolute differential cross sections

The measured relative differential cross sections are placed on an absolute scale using the relative flow method [25]. In this method the intensity of elastically scattered electrons from the target molecules and from a reference gas are compared, at fixed scattering angle and incident energy, under the same experimental conditions. Therefore, in the experiment the same beam profiles for both targets need to be achieved. Nickel and coauthors [25] have showed that the same beam profiles will be obtained if mean free paths for both gases are the same. According to the gas kinetic theory, the connection between the mean free path, the pressure and the squared gas kinetic diameter is:

$$\lambda = \frac{RT}{\pi D^2 N_A P\sqrt{2}} \qquad (1)$$

Where λ is the mean free path, R the gas constant, T the gas temperature, D^2 the squared gas kinetic diametar, N_A is Avogadro's number and P the gas pressure. This means that in order to have the same mean free paths, one needs to adjust the pressures behind the needle according to the ratio of the squared gas kinetic diameters: $P_x : P_{ref}=D^2_{ref} : D^2_x$. If these conditions are achived absolute DCSs can be obtained by using the formula:

$$DCS_x(E,\theta) = DCS_{ref}(E,\theta)\frac{N_x F_{ref}}{N_{ref} F_x}\sqrt{\frac{M_{ref}}{M_x}} \qquad (2)$$

$DCS_x\,(E_0,\,\theta)$ and $DCS_{ref}\,(E_0,\,\theta)$ are the absolute differential cross sections for elastic electron scattering on target molecules and reference gas, respectively. N_x and N_{ref} are measured intesities of scattered electrons, F_x and F_{ref} are the gas flow rates, M_x and M_{ref} are the molecular mases of the target and the reference gas, respectively.

Beside the intesities of the elasticaly scattered electrons, N_x and N_{ref} , a quantity that must be measured in the experiment is the flow rate of the target molecule and the reference gas (F_x and F_{ref}). In the experiment the target molecule (anf the reference gas) is introduced into a constant volume and the presure rise is measured using a baratron. According to gas kinetic theory, if the volume (V_0) and the temperature (T_0) are constant, the pressure rise in time can be connected to the flow, in a region of the linear dependence:

$$\frac{dP}{dt} = \frac{kT_0}{V_0}\frac{dn}{dt} = cF \qquad (4)$$

Practically, a home-made LabView program has been used for measuring the pressure dependence as a function of time. It should be noted that biomolecules can absorb on the walls of the gas line and this can influence the measured flow rates and thus, according to (2), the absolute DCSs. These effects have been investigated by Homem et al [26] and they conclude that adsorption was stronger when the method of the pressure increase was used, which is far from equilibrium, therefore the absorption and desorption cannot compensate each other. In order to reduce this effect in the present work, all elements of the experimental setup were heated (sample, pipes, valves). The obtained experimental DCSs have been compared with the theoretical results obtained by the group in Madrid [27,28].

3. Results

Absolute differential cross sections for elastic electron scattering from formamide and N-methylformamide at 100 eV, 150 eV and 300 eV are compared in Figure 2. Absolute DCSs for formamide and NMF show very similar angular dependences – both DCS curves show broad minima at 90°. The cross sections for NMF are slightly higher than formamide , which is to be expected considering the relative size of the molecules. It is interesting to note that the measured difference between the absolute DCSs match very nicely with the absolute DCS for methane [7] suggesting that in this energy range the 'building blocks' apporoach can be used to estimate the elastic DCS of larger. Theoretical and experimental results for both molecules are in very good agreement both in shape and on the absolute scale. The theory is based on the Independent Atom Model (IAM) [27, 28], where the problem of collisions of electrons with molecules is reduced to the collision of electrons with individual atoms, assuming that scattering from each atom is independent. This technique gives good results for middle and high incident electron energies, but for low energy electrons the effects of screening need to evaluated. Screening coefficients are introduced in the present Screen Corrected Additivity Rule (SCAR) method, resulting in a corrected cross section, at a given incident energy, calculated from the atomic cross sections. Relatively small differences between experimental results and SCAR theory for formamide and NMF can be seen at small scattering angles, which may be consequence of higher background contributions and less accurate volume corrections in the experiment and high contributions from inteference terms at small scattering angles in the SCAR approximation [29, 30]. Therefore both, experiment and theory need improvement to adequately represent forward elastic scattering.

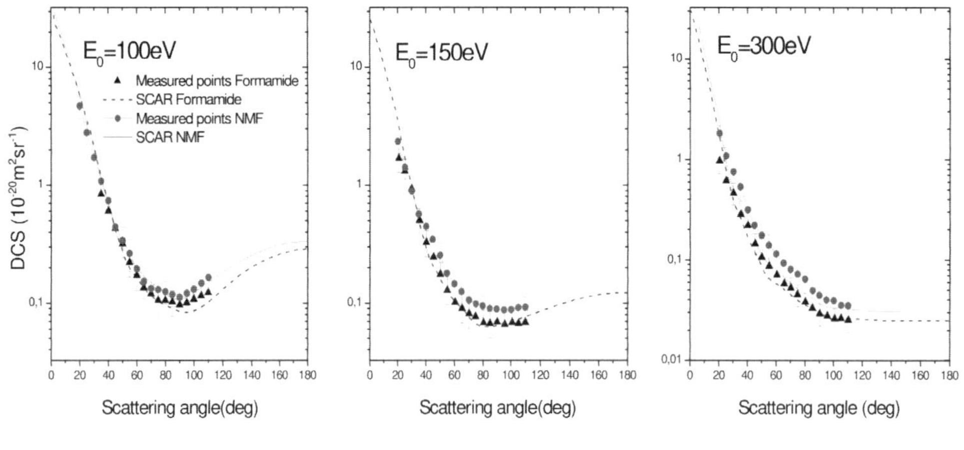

Figure 2. Comparison of absolute DCSs for formamide [6] and NMF [7]

Absolute experimental and theoretical DCSs for elastic electron scattering from Py, Furan and 3HTHF are presented in Figure 3. The experimental results are compared at 50 eV, 100 eV and 300 eV incident electron energies, for scattering angles between 8° and 110° for 3HTHF, between 20°and 110° for furan and from 30° to 110° for pyrimidine. The DCSs are very close on the absolute scale, showing similar redistribution of elastically scattered electrons from all targets. At 50 eV the absolute DCSs for 3HTHF are higher than for furan at small scattering angles, and this discrepancy decreases with increasing incident electron energy. Beside experimental imperfections at smaller scatering angles, this effect can be also explained [27, 28] by very different dipole moments between furan and 3HTHF, 0.71D and 1.67D, respectively. A higher dipole moment induces more intensive inelastic rotational excitations, which contribute to the measured "elastic" absolute DCSs, due to the limited energy resolution. Formamide and NMF DCS curves at lower incident energies (50 eV and 100 eV) show a broad minima at 90°, since both are analogues of DNA this may suggest a similar angular distribution in electron scattering from DNA. Once again there is good agreement between experiment and theory both in shape and in absolute cross section, suggesting that the SCAR calculations can be used as a reliable estimation of the DCSs for elastically scattered electrons from these molecules over the present angular and energy ranges.

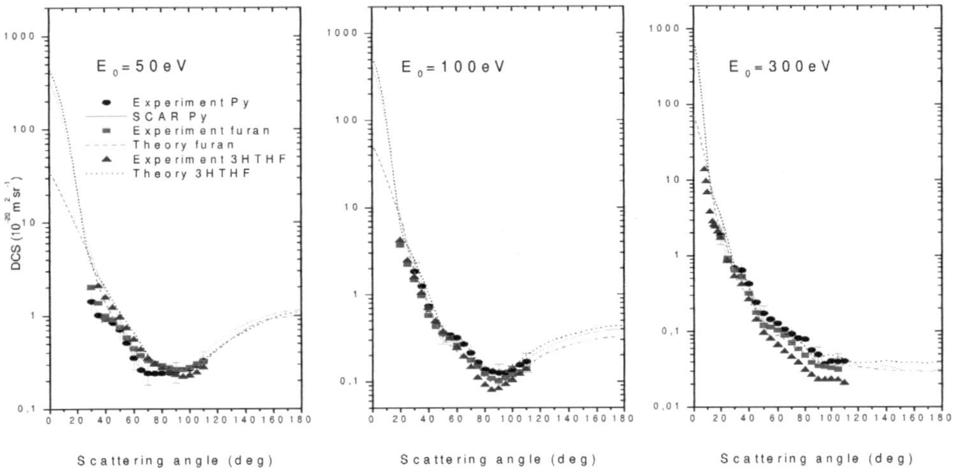

Figure 3. Comparison of absolute DCSs for pyrimidine [5], furan [8] and 3HTHF [4].

4. Conclusion

In summary, the expeirmental technique used to measure absolute DCSs for electron scattering from different molecules that are considered suitable analogues for some components of DNA or peptides has been reviewed. The recently published absolute DCSs for different molecules: formamide and NMF (analogues for a peptide bond) and furan, 3HTHF and Py (analogues of the building blocks of DNA) have been compared both on the absolute scale and in shape . Finally, the experimental results are compared with SCAR calculations and results are in very good agreement both in shape and magnitude.

Acknowledgments

The work was supported by the Ministry of Education and Science of Republic of Serbia (Project No. 171020) and motivated by the COST Action MP 1002 (Nano-IBCT). I would like to acknowledge Prof G. Garcia and P. Blanco for the theoretical results and their collaboration and A. R. Milosavljević and Prof N. Mason for their critical reading of the paper.

5. References

[1] Boudaiffa B, Cloutier P, Hunting D, Huels M A and Sanche L 2000, *Science* **287,** 1658

[2] Milosavljević A R, Giuliani A, Šević D, Hubin-Franskin M J and Marinković B P 2005, *Eur. Phys. J. D.* **35**, 411

[3] Milosavljević A R, Blanco F, Šević D, García G and Marinković B P 2006, *Eur. Phys. J. D.* **40**, 107

[4] Milosavljević A R, Blanco F, Maljković J B, Šević D, García G and Marinković B P 2008, *New J. Phys.* **10**, 103005
[5] Maljković J B, Milosavljević A R, Blanco F, Šević D, García G and Marinković B P 2009, *Phys. Rev. A* **79,** 052706
[6] Maljković J B, Blanco F, García G, Marinković B P and Milosavljević A R 2012, *Nucl. Instrum. Meth. B.* **279** 126
[7] Maljković J B, Blanco F, García G, Marinković B P and Milosavljević A R 2012, *Phys. Rev. A* **85,** 042723
[8] Maljković J B, Blanco F, Čurik R, García G, Marinković B P and Milosavljević A R 2012, *J. Phys. Chem.* **137** 064312
[9] Ferreira da Silva F, Almeida D , Martins G , Milosavljević A R , Marinković B P , Hoffmann S V, Mason N J, Nunes Y, Garcia G and Limão-Vieira P 2010, *Phys. Chem. Chem. Phys.*, **12**, 6717
[10] Levesque P L, Michaud M and Sanche L 2005, *J.Chem.Phys* **122**, 094701
[11] Allan M 2007, *J. Phys. B.* **40**, 3531
[12] Winstead C and McKoy V 2006, *J. Chem. Phys.* **125**, 074302
[13] Khakoo M A, Muse J, Ralphs K, Da Costa R F, Bettega M H F and Lima M A P 2010, *Phys. Rev. A.* **81**, 062716
[14] Smytkowski C, Możejko P, Ptasinska-Denga E and Sabisz A, 2010, *Phys. Rev. A.* **82**, 032701
[15] Bettega M H F and Lima M A P 2007, *J. Chem. Phys.* **126**, 194317
[16] Guiliani A and Hubin-Franskim M J 2001, *Int. J. Mass Spectrosc.* **205**, 163
[17] Viscaino V, Roberts J, Sullivan J P, Brunger M J, Buckman S J, Winstead C and McKoy V 2008, *New. J. Phys.* **10**, 053002
[18] Bettega M 2010, *Phys.Rev.A.* **81,** 062717
[19] Leach S, Jochims H W and Baumgärtel H 2010, *J. Phys. Chem. A*, **114**, 484
[20] Iwakawat M, Tofilon P J, Hunter N, Stephens L C and Milas L 1987, *Clinical and experimental metastasis,* **5,** 289
[21] Solov'yov A V, Surdutovich E, Scifoni E, Mishustin I and Greiner W 2009, *Phys. Rev. E* **79**, 011909
[22] Scifoni E, Surdutovich E and Solov'yov A V 2010, *Phys. Rev. E* **81,** 021903
[23] Milosavljević A R, Mandžukov S, Ševic D, Čadež I and Marinković B P 2006, *J. Phys. B.* **39**, 609
[24] Williams J F and Willis B A 1975, *J. Phys. B.* **8**, 1670
[25] Nickel J C, Zetner P V, Shen G, and Trajmar S 1989, *J. Phys. E.* 22, 730
[26] Homem M G A, Iga I, Sugohara R T, Sanches I P and Lee M T 2011, *Rev. Sci. Instrum.* 82, 013109
[27] Blanco F and García G 2003, *Phys. Lett. A* **317,** 458
[28] Blanco F and García G 2004, *Phys. Lett. A* **330,** 230

Journal of Physics: Conference Series **565** (2014) 012006 doi:10.1088/1742-6596/565/1/012006

Electron-indium atom scattering and analysis of electron and optical spectra

Maja S Rabasović
Institute of Physics, Pregrevica 118 Belgrade, University of Belgrade, Serbia

E-mail: majap@ipb.ac.rs

Abstract. Experimental study of indium atom using electron and optical spectroscopy is presented in this paper. Both experimental techniques including experimental setups are described. Differential and integrated cross sections on elastic and inelastic electron scattering by indium atom are measured using electron spectrometer. The measurements are performed at incident electron energies of E0 = 10, 20, 40, 60, 80 and 100 eV within the large scattering angles ranging from 10° to 150° in steps of 10°. The experimental results are presented and comparison with the values predicted by calculated optical potentials method is conducted, showing good agreement. The differential cross sections (DCSs) for electron-impact excitation of the resonant state 6s 2S1/2 of Indium atom are measured at small and large angles. The forward scattering function method has been used for normalizing the generalized oscillator strengths (GOS) to determine optical oscillator strength and obtaining the absolute DCS values. Optical spectrum of In I and In II lines has been acquired by a streak camera. The experimental results regarding indium lines obtained by time resolved laser induced breakdown spectroscopy (LIBS) could be useful for obtaining the important plasma parameters such as temperature, electron density as well as plasma-expansion velocity and plasma starting times.

1. Introduction

Indium (atomic number 49) is the metallic element of group IIIa of the Periodic Table. It has open shell structure with the 5s2 5p configuration in the ground state. Term of the ground state according to Russell-Saunders coupling is 5p 2P1/2. The metastable (5p 2P3/2) and resonant (6s 2S1/2) levels are positioned at 0.274 and at 3.022 eV from the ground level, respectively.

Indium has a low melting point, 429.75 K. Owing to that indium is a very attractive in the production of semiconductor and optoelectronic devices. Indium oxide is a transparent ceramic material, when doped with Tin (~ 10%) it becomes conducting but stays transparent. Indium tin oxide (ITO) can be applied by coating to a wide range of glass and plastic materials to make transparent conducting panels [1, 2]. ITO is a part of each liquid crystal displays (LCD). Uses of indium are rapidly increasing, but on the other hand it is relatively rare material. It ranks the 61st in crustal abundance. It is about three times as abundant as silver.

Another very important role of indium is the application of an indium single ion as a reference for an optical frequency standard of the excellent stability and accuracy. The narrow 5s2 1S0 – 5s5p 3P0 transition in In+ at a wavelength of 236.5 nm could be used as a reference clock transition [3]. Further, indium is a suitable as a tracer for two line atomic fluorescence (TLAF) thermometry measurements [4]. This method works by using diode lasers based laser induced fluorescence with wavelength of 410 and 451 nm, respectively, to excite the temperature sensitive 5p 2P1/2 – 6s 2S1/2 and 5p 2P3/2 – 6s 2S1/2 transitions of indium atoms seeded into a flame.

Advantage of the indium atoms is that the split-orbit splitting in the 5p ground state of indium leads to an energy spacing that is equal to kT in typical combustion environments (1000-3000 K) [4]. Later, Chrystie et al [5] demonstrated the capability of this technique for time-resolved measurements in an acoustically perturbed flame.

In this paper we present experimental results obtained using electron and optical spectroscopy. The results that concern the elastic differential cross sections and resonant excitation of indium are obtained using electron spectroscopy in our laboratory [6, 7]. The other techniques that are implemented in our laboratory are based on laser induced phenomena (laser induced fluorescence and laser induced breakdown spectroscopy (LIBS)). The later mentioned is a type of optical emission spectroscopy that uses a highly energetic laser pulse to create a plasma plume on the sample surface [8]. At the end of the laser pulse, the plasma cools and excited ions and atoms emit characteristic optical radiation which could be used to analyze the spectral lines of the component in material [9, 10]. LIBS can detect all elements, limited only by the energy of the laser pulse as well as the sensitivity and wavelength range of the spectrograph and detector.

2. Experimental techniques

2.1. Electron spectrometer (ESMA)

A detailed description of the apparatus ESMA and has been given elsewhere [6, 7]. A crossed beam technique was exploited in electron spectroscopy measurements. Electron spectrometer ESMA consists of the hemispherical selectors in monochromator and analyser, the cylindrical lenses, the rotating plate on which the analyser is mounted and that provides assessment of scattering angles from -30° to +150° in respect to incident electron beam. Perpendicularly to the scattering plane defined by the incident and scattered view cones of the monochromator and analyser, the resistively heated oven is placed on together with the several layers of tantalum shields and copper shield in outermost radius as well as the copper cold finger cup for collecting metal atom vapours (Figure. 1(a)).

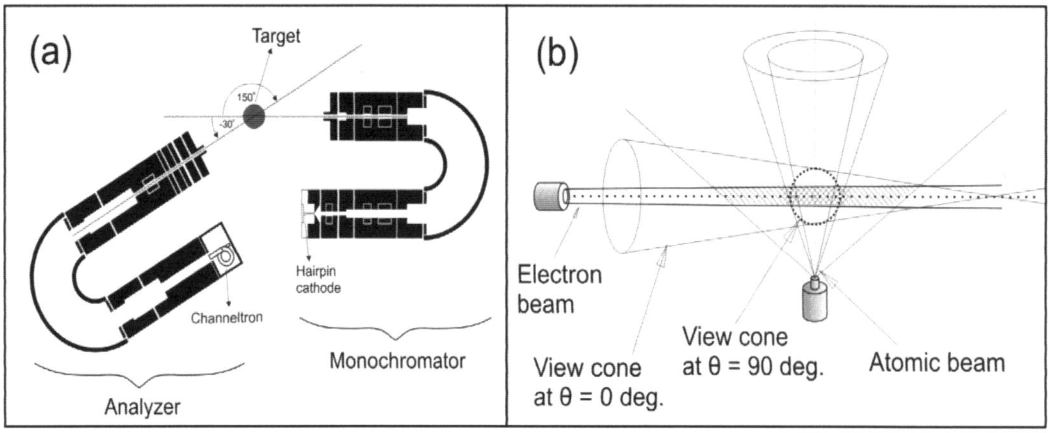

Figure 1. (a) Schematic overview of electro-optical system in spectrometer ESMA. **(b)** Scattering geometry in electron-atom beam experiments. The intersection of the indium atom beam, electron beam and view cone of the detector presents the interaction volume.

The indium vapour beam was produced by heating the oven crucible containing In metal. The oven has been modified in order to achieve higher temperatures. The oven consists of two separate heaters with the same diameters, one for the top of the stainless steel crucible and nozzle, and the other for the body of the crucible, and their temperatures were controlled by two thermocouples. The heating currents of those heaters were 5.3 A and 3.4 A, respectively. They provided a variable temperature difference between the top and the bottom. The nozzle was maintained at approximately 100° higher temperature. The atomic beam was effused through a L = 20 mm long channel in the cap of the oven crucible that has an inner diameter of 2a = 2.5 mm. The aspect ratio was large (γ = 0.125). The evaporation temperature of the indium was approximately 1300 K and the metal-vapor pressure was about 10 Pa (0.07 Torr).

2.2. Time resolved LIBS system

The system for LIBS measurements consists of the excitation and detection parts. A schematic diagram of the experimental system is shown in Figure 2. Nd:YAG laser and Optical Parametric Oscillator (OPO, Vibrant 266) are used as the excitation part. The OPO system, pumped by a pulsed Q switched Nd:YAG laser (Briliant B) includes the second and fourth harmonic generator (SHG and FHG). In this paper the fundamental output at 1064 nm and pulse energy of 17 mJ is used to create an optical breakdown in ambient air. Focusing of the laser beam is obtained using lens with the focal length of 40 mm. The OPO system is controlled by OPOTEK software installed on PC.

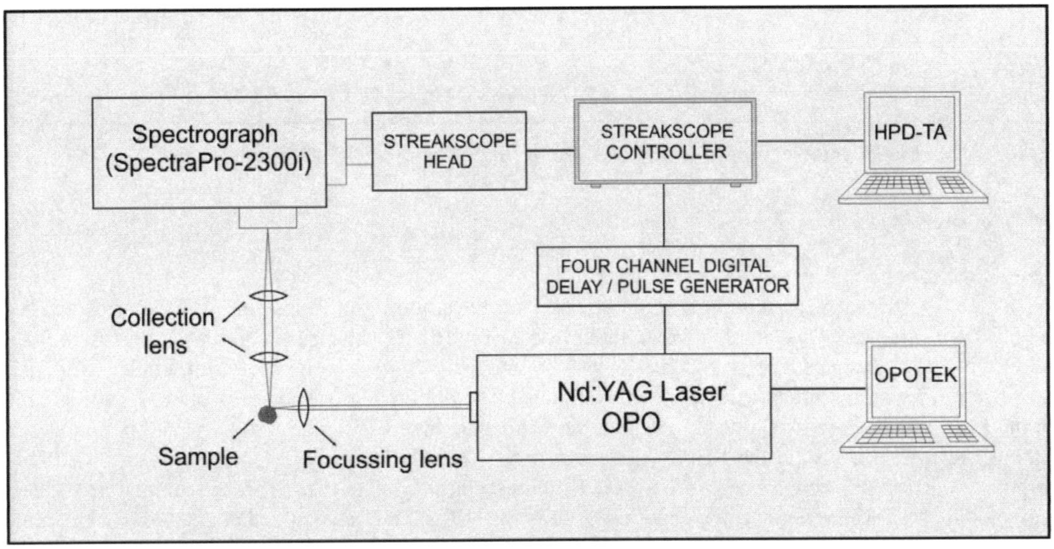

Figure 2. Time-resolved laser-induced breakdown: experimental setup.

The optical emission from the plasma plume is collected by using a spectrograph (SpectraPro 2300i) and recorded with a Hamamatsu streak camera (model C4334-01) with integrated video streak camera (Figure 3). The fundamental advantage of the streak camera is its two dimensional nature (wavelength and time), that is especially important in measuring time-resolved LIBS spectra. This enables monitoring of temporal evolution of the ionic and atomic emission lines or spatial development of the optical emission of the plasma plume. The data are acquired and analyzed using High Performance Digital Temporal Analyzer (HPD-TA) software, provided by Hamamatsu.

The spectrograph contains the triple grating turret. The diffraction gratings of 50, 150 and 300 g/mm were installed. For measurements presented in this paper the grating of 50 g/mm was used covering a 330 nm spectral range. The spectrograph was connected to the streak camera (Figure 3). In the streak camera, the light passing through the input slit is focused onto the photocathode of the streak tube by the input optics. The photocathode converts the light into photoelectrons that are accelerated by an electric field onto a phosphor screen. An electro-optical focusing system focuses the electrons on the screen. After passing the anode aperture the photo-electrons are deflected by a transversal electric field that increases linearly with time before reaching the phosphor screen. In this way, the input spectra from different points of time are spatially separated on the phosphor screen. The phosphor layer converts the electrons into light that can be observed by a coupled CCD camera. The CCD chip has resolution of 640 x 480 pixels, and camera has the spectral range from 200 nm to 850 nm. Because the deflection of streak sweep is not completely straight, there is a geometric distortion of the streak image in the sweep direction, visible here as a tilted spectrum [11]. This distortion is easily removed by a curvature correction tool of the HPD-TA.

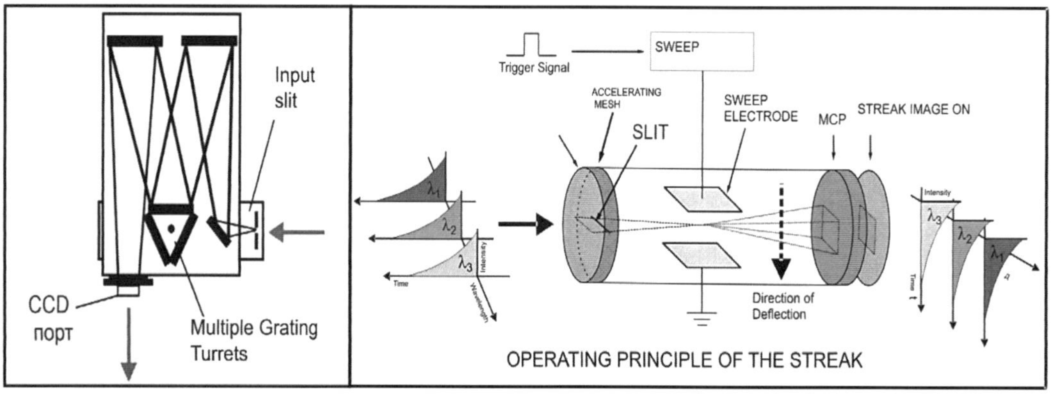

Figure 3. Schematic intersection of detection system (Spectrograph and streaktube).

3. Experimental results

3.1. Electron scattering by indium atom

We present experimental results regarding elastic electron indium atom scattering in the intermediate range. The results were presented and explained in detail in [6]. The measurements were performed at incident electron energies of E_0= 10, 20, 40, 60, 80 and 100 eV within the scattering angles ranging from 10° to 150°. Theoretical calculations have been carried out using the complex phenomenological optical potential (OP) with allowance of the spin-orbit interaction [6]. The real part of this potential without absorption is called the SEPSo approximation. The McCarthy type potential is used as the imaginary part of OP, which takes into account the absorption effects (SEPASo approximation). The relative values of experimental differential cross sections (DCSs) are normalized at 20° to the SEPASo theory. The three types of absorption potentials (VA_i , i=1, 2, 3) are used: two with summarized electron densities $\rho_1=\rho_{5p}+\rho_{5s}$ (VA_1) and $\rho_2=\rho_{5p}+\rho_{5s}+\rho_{4d}$ (VA_2) in subshells 5p, $5s^2$ and $4d^{10}$ and one with a partial influence of the electron density in $4d^{10}$ subshell (VA_3).

The comparison of the present experimental results with the present theoretical DCSs calculated in SEPSo and SEPASo approximations are presented in Figure 4. Note that at 10 eV and 20 eV, calculations with VA_1 and VA_3 coincide. As seen from Figure 4, the SEPSo calculation for all energies, except for 10 eV, reproduces the shape of the angular dependence of experimental cross sections, the number of minima and their angular positions. For 10 eV in theoretical calculations, as well as in the experiment, one minimum was obtained, but at 86°, that is about 20° shifted toward the lower angles than in the experimental DCS. For 40 eV and 60 eV DCS could be calculated using the absorption potential VA_1. At 80 eV and 100 eV the inclusion into absorption potential of the density ρ_{4d} (VA_2) gives a better agreement with the experimental data at intermediate and large scattering angles as compared to calculation using VA_1.

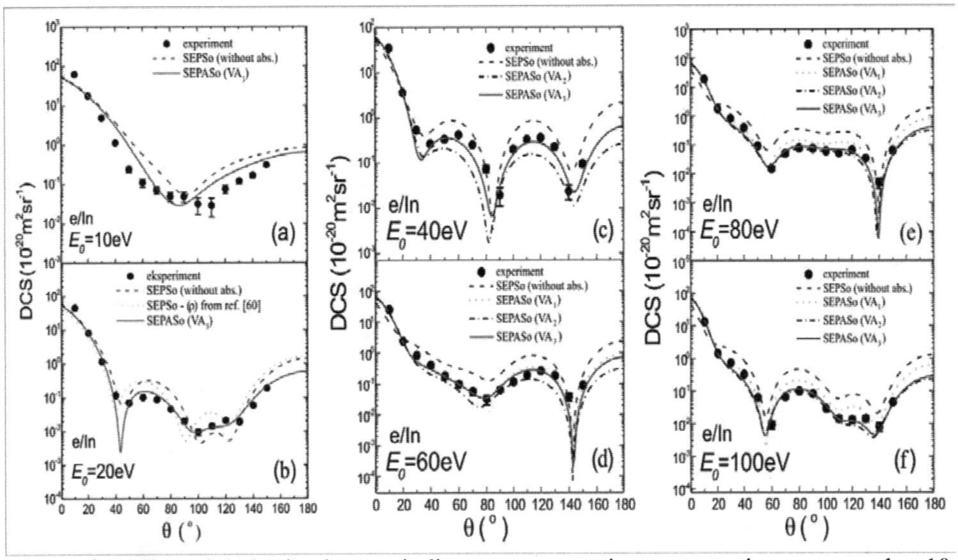

Figure 4. Differential elastic electron-indium atom scattering cross sections measured at 10, 20, 40, 60, 80 and 100 eV.

The experimental (circles) and theoretical (lines) results concerning the electron impact excitation of the resonant state 6s $^2S_{1/2}$ of indium atom are presented in Figure 5. We performed measurements for the electron-impact energies at 10, 20, 40, 60, 80 and 100 eV and at small scattering angles from 1° to 10° [7]. We normalized the relative DCSs by using the FSF method to absolute DCS values. Detailed explanation of this procedure could be seen in [7]. In Figure 5 we presented the absolute GOS values versus K^2 on a log-log graph together with the FSF for the resonant transition 5p $^2P_{1/2}$ – 6s $^2S_{1/2}$ in indium atom. The theoretical procedure was carried out within the framework of relativistic distorted-wave (RDW) calculations [12]. This RDW method is especially suited for the present study since it contains the effect of spin-orbit coupling for all the electrons in the system, both bound and free. In particular, the two fine structure levels of the ground state with separated wave functions and energy levels have been distinguished within the RDW method.

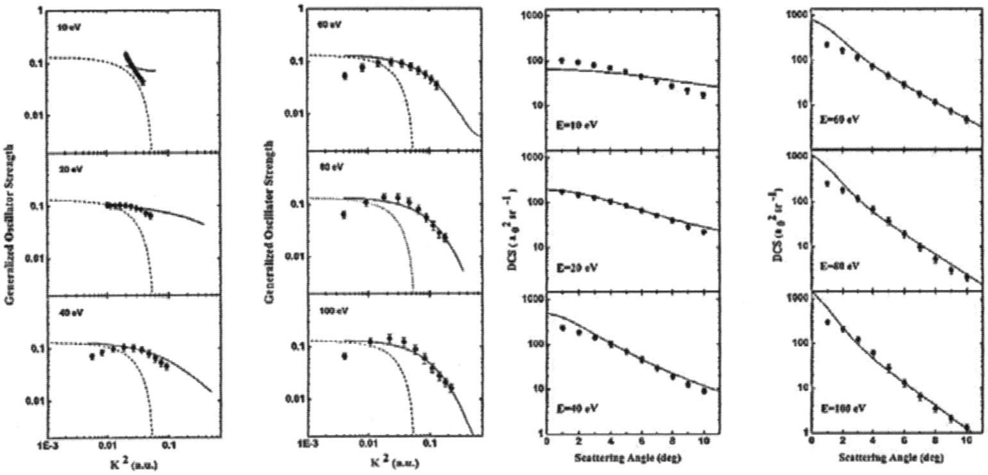

Figure 5. GOSs and DCSs for excitation of the resonant 6s 2S state of the indium atom at: 10 eV, 20 eV, 40eV, 60 eV, 80 eV and 100 eV electron-impact energies and at small scattering angles.

3.2. Optical emission spectra of indium using LIBS

A set of time resolved optical emission spectra for almost pure indium sample are presented in Figure 6. Time resolved measurements are performed in the time domain from 500 ns to 20 μs. In the early stage of laser induced plasma, the emission spectra are dominated by background continuum spectra due to bremsstrahlung radiation and recombination. The LIBS plasma obtained with nanosecond lasers exhibits a very intense continuum that decreases with time to allow detection of elemental emission. In order to reduce continuum it was utilized a delay unit for minimizing continuum to a reasonable level. Plasma shielding and subsequently the continuum emission could be reduced using lasers with femtoseconds or picoseconds pulse widths [13].

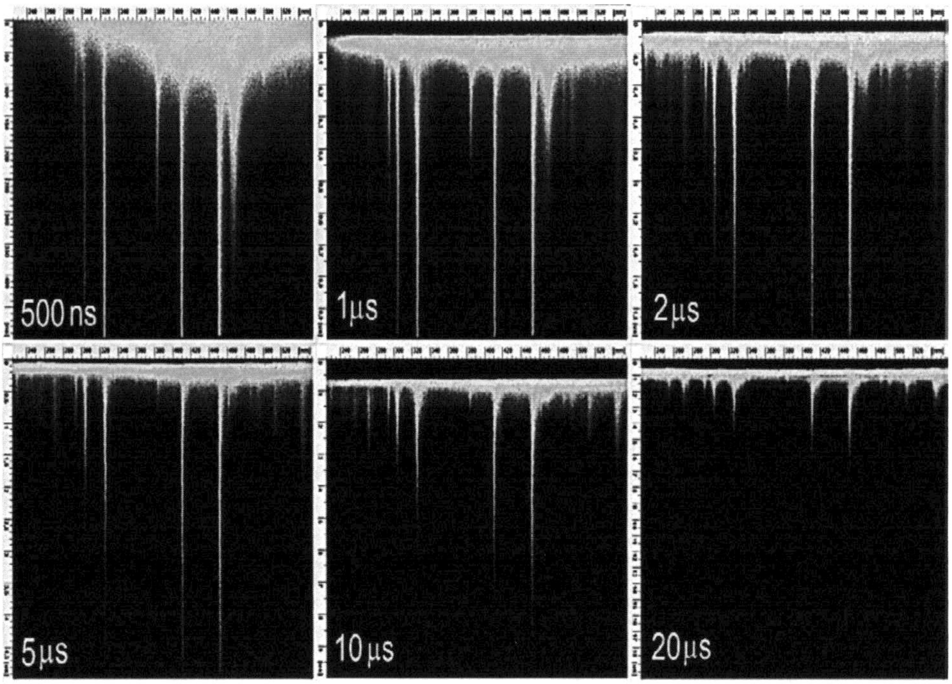

Figure 6. Time resolved LIBS streak images of pure indium sample with time range from 500 ns to 20 μs.

The temporal evolution of laser ablation plume from the image with time range of 1 μs (Figure 6) has been used for determination of the plasma parameters. We obtained the line profiles at 100 ns, 200 ns, 500 ns and 700 ns by the integration of the narrow range of time. From these profiles it can be seen how ionic lines (In II) dominate in the early stage of the plasma together with continuum radiation. Ionic lines are concentrated near the surface of the target. However, during cooling plasmas ionic lines are diminished while remaining the atomic (In I) lines as a result of recombination. The elemental analysis and plasma parameters (electron number density and electron temperature) could be determined from the emission spectra. The calculations show that both these parameters increase with the increase in the laser irradiance. The electron temperature (T_e) could be estimated using Boltzmann method or two lines method (271 and 451 nm) by assuming a local thermal equilibrium (LTE) and also that the plasma is optically thin (absorption and scattering can be neglected) [8, 10]. Electron temperatures obtained from line profiles in Figure 6 are typically ranged in the interval of 5000-20000K. The estimated electron density obtained using Saha Boltzmann equation amounts to ~ 10^{17} cm^{-3}. These values depend on the laser used and on the type of sample considered [10]. Variations of the electron temperature and the electron density with the laser irradiance show that both these parameters increase with the increase in the laser irradiance [14].

4. Conclusions

In this paper were presented results regarding indium atom obtained by electron spectroscopy and laser induced breakdown spectroscopy. Possibility of detection of optically forbidden transitions is an advantage of electron spectroscopy. However, the resolution of our electron spectroscopy apparatus is lower than when using LIBS. The spectral lines obtained by electron spectroscopy can be combined with plasma emission measurements to select what emission lines could be used in the analysis. The possibilities of use of time resolved streak camera measurements may improve the analysis approach.

Analysis of streak images acquired using techniques explained here could be very useful for obtaining the important plasma parameters such as temperature, electron density as well as plasma-expansion velocity and plasma starting times, including their variation in time, or in the time frame of interest.

Acknowledgement

This work was financially supported within the Project of Ministry of Education, Science and Technological Development of the Republic of Serbia OI 171020

5. References

[1] Kuang Z, Perrie W, Liu D, Fitzsimons P, Edwardson S P, Fearon E, Dearden G and Watkins K G 2012 *Applied Surface Science* **258** 7601
[2] Fernandes S A Schoeps B Kowalick K, Nett R, Esen C, Pickshaus M and Ostendorf A 2013 *Physics Procedia* **41** 802
[3] Wang Y H Dumke R Liu T Stejskal A Zhao Y N Zhang J Lu Z H Wang L J Becker Th and Walther H 2007 *Optics Communications* **273** 526
[4] Hult J Burns I S Kaminski C F 2005 *Proceedings of the Combustion Institute* **30** 1535
[5] Chrystie R S M, Burns I S, Hult J and Kaminski C F 2009 *Optics Letters* **34** 2492
[6] Rabasovic M S Kelemen V I Tosic S D Sevic D Dovhanych M M Pejcev V Filipovic D M Remeta E Yu Marinkovic B P 2008 *Physical Review A* **77** 062713
[7] Rabasovic M S Tosic S D Sevic D Pejcev V Filipovic D M and Marinkovic B P 2009 *Nucl. Instrum. Meth. B* **267** 279
[8] D. Cremers and L. Radziemski 2006 *Handbook of Laser-Induced Breakdown Spectroscopy* John Wiley & Sons Ltd Chichester
[9] Vadillo J M and Laserna J J 2004 *Spectrochimica Acta Part B* **59** 147
[10] Milán M and Laserna J J 2001 *Spectrochimica Acta Part B* **56** 275
[11] Sevic D Rabasovic M S and Marinkovic B P 2011 *IEEE Transactions on plasma science* **39** 2782
[12] Das T Srivastava R Stauffer A D 2011 *Physics Letters A* **375** 568
[13] Miziolek A Palleschi V and Schechter I 2006 *Laser Induced Breakdown Spectroscopy* Cambridge University Press
[14] Shaikh N K Rashid B Hafeez S Jamil Y and Baig M A 2006 *J. Phys. D: Appl. Phys.* **39** 1384

The IST-LISBON database on LXCat

L L Alves

Instituto de Plasmas e Fusão Nuclear, Instituto Superior Técnico, Universidade de Lisboa, Av Rovisco Pais, 1049-001 Lisboa, Portugal

llalves@tecnico.ulisboa.pt

Abstract. LXCat is a web-based, community-wide project on the curation of data needed in the modelling of low-temperature plasmas. LXCat is organized in databases, contributed by members of the community around the world and indicated by the contributor's chosen title. This paper presents the status of the data available on the IST-LISBON database with LXCat. IST-LISBON contains up-to-date electron-neutral collisional data (together with the measured swarm parameters used to validate these data) resulting from the research effort of the Group of Gas Discharges and Gaseous Electronics with Instituto de Plasmas e Fusão Nuclear, Instituto Superior Técnico, Lisbon, Portugal. Presently, the IST-LISBON database includes complete and consistent sets of electron scattering cross sections for argon, helium, nitrogen, oxygen, hydrogen and methane.

1. Introduction. The LXCat project

Electrons are the prime species to convey energy into a gas, producing plasma, and thus knowledge of reliable electron–neutral scattering cross sections is of paramount importance to accurately describe the electron kinetics in low-temperature plasmas, bringing extra credibility to modelling results. The LXCat project (www.lxcat.net) [1] is an open-access website for collecting, displaying, and downloading electron and ion scattering cross sections for binary encounters (swarm-derived, measured or calculated using quantum mechanical theory), electron and ion swarm parameters (measured transport parameters and rate coefficients), interaction potentials, optical oscillator strengths and other data essential for modeling low-temperature plasmas. LXCat is organized in databases, contributed by members of the community around the world and indicated by the contributor's chosen title.

At present LXCat offers 21 databases (managed by approximately 30 contributors from 11 countries), 17 of which contain sets of cross sections for electron scattering from ground-state neutral atoms and molecules, covering a range of energies from thermal up to 100's of eV and higher. Some of the databases present complete and consistent sets of cross sections [2-4], with interest for modelling and/or diagnostic studies (e.g. based on optical measurements), corresponding to much more than a simple compilation of data. *Complete* cross section sets stand for arrangements able to describe the main electron momentum-loss, energy-loss and number-changing processes (ionization, attachment, recombination), either individually or grouped together. In this sense, cross sections for only one particular excitation process and/or for a limited energy range, as often encountered in the literature, are not relevant by themselves. *Consistent* sets are those able to reproduce measured values of swarm parameters when used as input data to evaluate the electron energy distribution function (eedf) from a Boltzmann solver or Monte Carlo/Particle-in-Cell codes.

The most popular strategy to determine a complete, consistent cross section set in gaseous electronics uses an iterative process to perform educated adjustments onto an initial compilation of data, as follows: (i) a set of cross sections is collected from the literature; (ii) the cross sections are used as input data to calculate eedfs and to evaluate electron transport parameters and rate coefficients; (iii) the magnitudes of the cross sections are adjusted (within experimental error or theoretical uncertainty, if possible), to improve the agreement between calculated and measured swarm data. Such an iterative procedure embeds a global consistency check, since it relies on reproducing measured electron parameters, but cannot be used to ensure either the uniqueness of the cross section set or the validation of the cross section for any individual process. Despite the poor resolution of the swarm method and the fact that it only allows adjusting integral cross sections, the procedure is highly meaningful for plasma modelling, since it provides a consistent estimation of transport parameters and macroscopic (excitation and ionization) rate coefficients. In fact, cross section compilations that lack this consistency check with swarm experiments are of limited interest in low-temperature simulations.

LXCat offers also a series of on-line tools enabling: (i) uploading / downloading data in ASCII format; (ii) selecting and displaying multiple data sets graphically; (iii) rescaling and downloading graphs; (iv) calculating electron swarm parameters from selected cross section data, in pure gases or in gas mixtures, using the Boltzmann equation solver BOLSIG+ [5]; and (v) displaying of calculated swarm parameters along with selected experimental values of these quantities obtained from the databases.

The LXCat team has been making systematic inter-comparisons of cross section data and comparisons of calculated and measured swarm parameters. These were published for noble gases [2-4] and reported in about 10 communications to conferences, also for common molecular gases (H_2, N_2 and O_2).

2. The IST-LISBON database

One of LXCat's databases is IST-LISBON, which contains up-to-date electron-neutral collisional data (together with the measured swarm parameters used to validate these data) resulting from the research effort of the Group of Gas Discharges and Gaseous Electronics with Instituto de Plasmas e Fusão Nuclear, Instituto Superior Técnico, Lisbon, Portugal.

Presently, IST-LISBON includes complete and consistent sets of electron scattering cross sections, and measured swarm parameters, for argon, helium, nitrogen (atomic and molecular, including also electron-impact rotational and vibrational excitation cross-sections), molecular oxygen (including also electron-impact rotational and vibrational excitation cross-sections), hydrogen (atomic and molecular, including also electron-impact rotational and vibrational excitation cross-sections) and methane. In general, our cross sections yield predictions for the electron transport parameters and rate coefficients, calculated with a two-term Boltzmann solver, which agree within less than 20% with measurements, for reduced electric fields $E/N \sim 10^{-4} - 500$ Td (where E is the electric field intensity and N is the neutral gas density; 1 Td = 10^{-17} V cm^2). This kind of agreement is acceptable in many plasma applications, although for highly accurate swarm experiments a minimum accuracy of 1% is required [6-8].

2.1. Argon

The compilation and adjustment of the complete set of electron–neutral scattering cross sections from ground-state argon is reported in [9]. The set includes 39 cross sections (elastic momentum-transfer, 37 excitations and ionization) up to 1 keV. Figure 1 summarizes the argon cross section set with IST-LISBON.

The elastic momentum-transfer cross section was obtained from [10]. Direct excitation cross sections up to 100 eV were obtained: from [11], for the 4s metastable and resonance levels (multiplied by a factor 0.5); from [12], for the 4p levels; from [13], for the 3d, 3d', 5s and 5s' levels; from [14], for the 5p levels; for the 4d, 6s and 4d' levels, only optically allowed transitions were considered using the expressions and data of [15-17]. The previous direct excitation cross sections were extended to 1

Journal of Physics: Conference Series **565** (2014) 012007 doi:10.1088/1742-6596/565/1/012007

keV, by using analytical asymptotic expressions similar to those proposed in [17-18]. Direct ionization adopted the cross section measured in [19].

When these cross sections are used as input data in a two-term Boltzmann solver, they yield swarm parameters (reduced mobility μN, characteristic energy u_k and Townsend ionization coefficient α/N) in general agreement with measurements, within the 2-25% experimental uncertainties reported. Note that calculations agree also with measurements (25-30% errors) of the total excitation coefficients (including cascade contributions coherent with the proposed cross section set) for the 4s-metastable, the 4s-resonance and the 4p levels. More details can be found in [2].

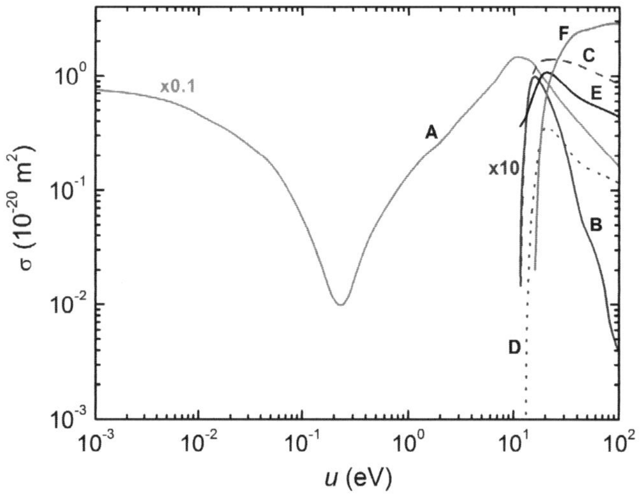

Figure 1. Summary of the argon cross section set with IST-LISBON, as a function of the electron kinetic energy. Elastic momentum-transfer (solid magenta line and A; multiplied by factor 0.1), excitation to 4s-metastable levels (solid blue and B; x10), excitation to 4s-resonance levels (dashed blue and C; x10), excitation to 2p levels (dotted blue and D), total excitation (solid black and E), and ionization (solid red and F).

2.2. Helium

The compilation and adjustment of the complete set of electron–neutral scattering cross sections from ground-state helium is reported in [20-21]. The set includes 44 cross sections (elastic momentum-transfer, 42 excitations and ionization) up to 100 eV. Figure 2 summarizes the helium cross section set with IST-LISBON.

The elastic momentum-transfer cross section was obtained from [22-23] at low energies, [24] at intermediate energies, and [25] at high energies. Total (direct plus cascade) excitation cross sections were obtained: for the 2^3S + 2^1S metastable levels (cross section multiplied by a factor 0.65), from [26] near threshold, from [27] at low energies and from [28] at high energies; for the 2^1P and 3^1P levels (x1.12), from [29]; for the 4^1P level (x1.12), from [30]; for the 2^3P level (x0.89) from [31]; for the 3^3P level (x0.89), from [32]. Direct cross sections for the $n^{1,3}$S,D (n = 3–7), n^1P (n = 5–7), n^3P (n = 4–7) and $n^{1,3}$FGHI (n < 7) levels were obtained from [33-34]. The direct cross sections for the low-lying levels $2^{1,3}$S, n^1P (n = 2–4) and n^3P (n = 2–3) were calculated from the corresponding total cross sections by correcting for the cascade transitions. The individual excitation cross sections for the 2^3S and 2^1S metastable levels were obtained from the corresponding lumped cross section by using the deconvolution parameter proposed in [33-34]. Direct ionization adopted the cross section measured in [19]. A review of this cross section set was recently published [35] (including also the extension of its maximum energy to 1 keV), and it will soon be used in updating the IST-LISBON database.

When these cross sections are used as input data in a two-term Boltzmann solver, they yield swarm parameters in general agreement with measurements, within the 5-10% experimental uncertainties reported and considering the dispersion observed in the measurements, reaching differences as high as factors of 4-5. More details can be found in [3].

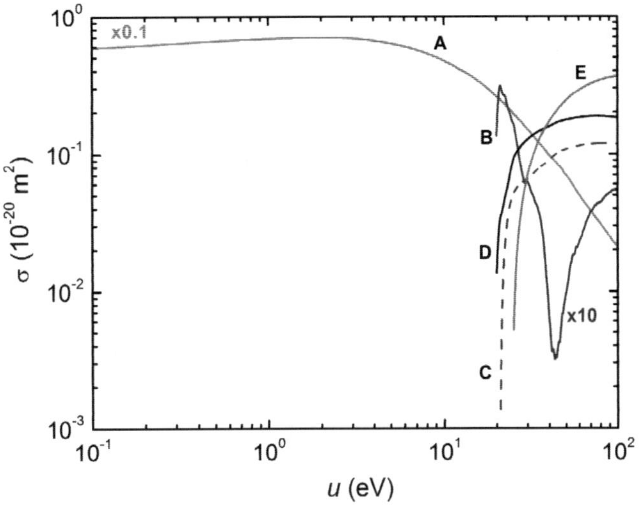

Figure 2. Summary of the helium cross section set with IST-LISBON, as a function of the electron kinetic energy. Elastic momentum-transfer (solid magenta line and A; multiplied by factor 0.1), excitation to metastable 2^3S+2^1S levels (solid blue and B; x10), excitation to 2^3P+2^1P levels (dashed blue and C), total excitation (solid black and D), and ionization (solid red and E).

2.3. Nitrogen

The complete set of electron–neutral scattering cross sections from ground-state molecular nitrogen was initially reported in [36]. The set includes 23 cross sections (effective; excitations to 10 vibrational levels and 11 electronic levels; and ionization), and it was compiled mostly from [37-38]. Originally, the set was limited to 40 eV kinetic energy, but prior to its publication on LXCat it was reviewed and extended to 1 keV, using information from the databases BIAGI-v8.9 and PHELPS. Figure 3 summarizes the N_2 cross section set with IST-LISBON.

The *effective* momentum-transfer cross section [37-38] corresponds to the sum of the elastic momentum-transfer, the total excitation and the ionization cross sections. This quantity appears in the collision term of the electron Boltzmann equation, after expansion in Legendre polynomials and if higher order anisotropies in the differential cross sections for inelastic scattering are neglected [9], as it is commonly assumed in two-term Boltzmann solvers.

Direct excitation cross sections (up to 40 eV) to electronic levels were obtained: from [37-38], for $A^3\Sigma_{u+}$ (v=0-4; v=5-9; v>9), $C^3\Pi_u$, $E^3\Sigma_{g+}$ and $a'^1\Sigma_{g+}$ levels; from [39], for $B^3\Pi_g$, $W^3\Delta_u$, $B'^3\Sigma_{u-}$, $a'^1\Sigma_{u-}$, $a^1\Pi_g$ and $w^1\Delta_u$ levels; from [40], for N_2 higher lying levels. The cross section for the N_2(v=0)-N_2(v=1) vibrational excitation was adjusted using also the results of [41-45]. The cross sections for the

N2(v=0)-N_2(v=2-8) excitations were adjusted using also the results of [36-37], and those for the N_2(v=0)-N_2(9-10) excitations, not included in PHELPS database, were adjusted using also the results of [36-37,46]. Direct ionization adopted the cross section measured in [19].

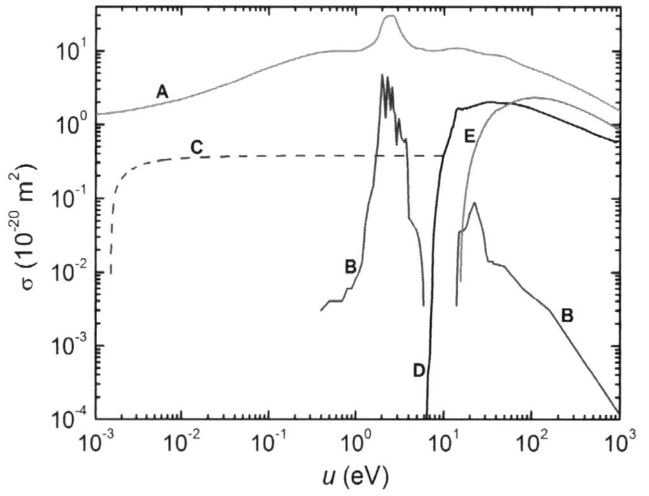

Figure 3. Summary of the N_2 cross section set with IST-LISBON, as a function of the electron kinetic energy. Effective momentum-transfer (solid magenta line and A), first vibrational excitation v=0→v=1 as an example (solid blue and B), first rotational excitation J=0→J=2 as an example (dashed blue and C), total excitation (solid black and D), and ionization (solid red and E).

When this complete set of cross sections (extended to 1 keV) is used as input data in a two-term Boltzmann solver, it yields calculated swarm parameters in agreement with measurements for E/N values between 10 and 1000 Td. Maximum deviations are of 27% for u_k and 22% for μN, between 300-400 Td, and of 45% for α/N around 150 Td. Below 10 Td the set must be further completed to include rotational excitation / de-excitation mechanisms, in order to reproduce measured swarm data. Here, we have adopted the rotational excitation cross sections of [47] for the $J \rightarrow J + 2$ transitions (J=0,...,30), taking Q = 1.09 and B = 2.5x10^{-4} eV [48] for the N_2 quadruple moment and rotational constant, respectively, assuming a Boltzmann distribution at gas temperature T_g for the rotational populations with the electronic ground-state. With the inclusion of the supplementary rotational mechanisms, the deviations between calculated and measured values of u_k and μN are well below 10%, in the low E/N region at T_g = 77, 300 K.

The IST-LISBON dataset for nitrogen includes also electron–neutral cross sections for the atomic system [49]. This set was compiled from [50] and [51], for the elastic momentum-transfer and the excitation of N(^{2}D) and N(^{2}P) levels from ground-state N(^{4}S), respectively.

2.4. *Oxygen*

The complete set of electron–neutral scattering cross sections from ground-state molecular oxygen was initially reported in [52]. The set was compiled mostly from [53], and it includes the following 14 cross sections: elastic momentum-transfer; excitations to 4 vibrational levels; excitations to 7 electronic levels ($a^1\Delta_g$; $b^1\Sigma_{g+}$; $A^3\Sigma_{u+}$, $C^3\Delta_u$ and $c^1\Sigma_{u-}$ bound states; $A^3\Sigma_{u+}$, $C^3\Delta_u$ and $c^1\Sigma_{u-}$ yielding dissociation into 2O(^{3}P) and continuum Herzberg bands; $B^3\Sigma_{u-}$ yielding dissociation into O(^{3}P)+O(^{1}D);

radiative levels with thresholds at 9.97 eV and 14.7 eV); dissociative attachment; and ionization. Originally, the set was limited to 40 eV kinetic energy, but prior to its publication on LXCat it was reviewed and extended to 1 keV, using information mainly from the databases PHELPS and BIAGI-v8.9. The exception is the cross section for the excitation of $O_2(a^1\Delta_g)$, which was extended up to 200 eV using the results of [54] and then extrapolated up to 1 keV. Figure 4 summarizes the O_2 cross section set with IST-LISBON.

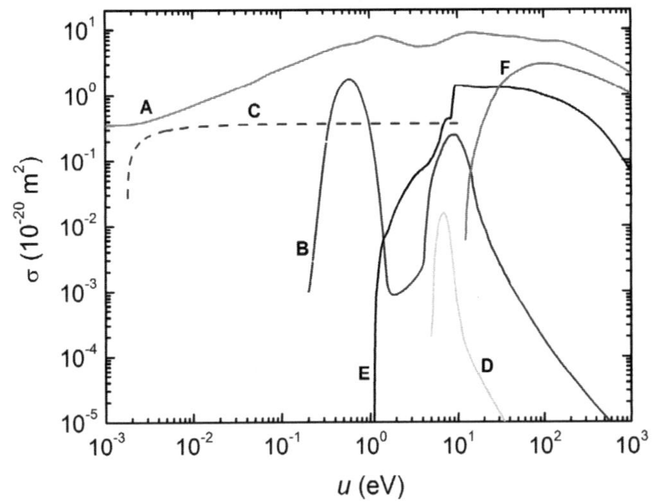

Figure 4. Summary of the O_2 cross section set with IST-LISBON, as a function of the electron kinetic energy. Effective momentum-transfer (solid magenta line and A), first vibrational excitation v=0→v=1 as an example (solid blue and B), first rotational excitation J=1→J=3 as an example (dashed blue and C), dissociative attachment (solid green and D), total excitation (solid black and E), and ionization (solid red and F).

When this complete set of cross sections (extended to 1 keV) is used as input data in a two-term Boltzmann solver, it yields calculated swarm parameters in agreement with measurements for E/N values between 10 and 1000 Td. The maximum deviations observed are the following: for u_k, 5% between 10-100 Td; for μN, 21% between 10-100 Td (where a 44% dispersion is found in the experimental values); for the reduced attachment Townsend coefficient, a factor of 2 difference in the region 10-100 Td, equivalent to the experimental dispersion; for the reduced ionization Townsend coefficient, a factor of 1.5 difference in the region 100-400 Td, also equivalent to the experimental dispersion. Below 10 Td the set must be further completed to include rotational excitation / de-excitation mechanisms, in order to reproduce measured swarm data. Here, we have adopted the rotational excitation cross sections of [47] for the $J \rightarrow J + 2$ transitions (J=1,3,5...,29), taking Q = 1.4 and $B = 1.8\text{x}10^{-4}$ eV [55], and assuming a Boltzmann distribution at T_g for the rotational populations with the electronic ground-state. With the inclusion of the supplementary rotational mechanisms, the deviations between calculated and measured values of u_k and μN in the low E/N region are found below the 20% experimental dispersion observed.

2.5. Hydrogen

The compilation of the complete set of electron–neutral scattering cross sections from ground-state hydrogen is reported in [56]. The set was compiled mostly from [57-58], and it includes the following

25 cross sections up to 1 keV: elastic momentum-transfer [59]; 4 rotational excitations (for J=0 → J=2, obtained from the swarm measurements of [60], at low energies, and from the theoretical expressions of [61], at intermediate energies; for $J \rightarrow J+2$ with J=1-3, obtained from the swarm measurements of [62], at low energies, and from the data proposed in [63], at intermediate energies); excitations to 3 vibrational levels [64]; excitation to 4 triplet dissociative levels ($b^3\Sigma_{u+}$; $c^3\Pi_u$; $a^3\Sigma_{g+}$ and $e^3\Sigma_{u+}$) and to 7 singlet partially dissociative Rydberg levels ($B^1\Sigma_{u+}$; $C^1\Pi_u$; $E^1\Sigma_{g+} - F^1\Sigma_{g+}$; $B'^1\Sigma_{u+}$; $D^1\Pi_u$; $B''^1\Sigma_{u+}$ and $D'^1\Pi_u$); dissociation into H(1s)+H(2s,2p,3,4,5) [65-67]; and ionization [68-70]. Figure 5 summarizes the H_2 cross section set with IST-LISBON.

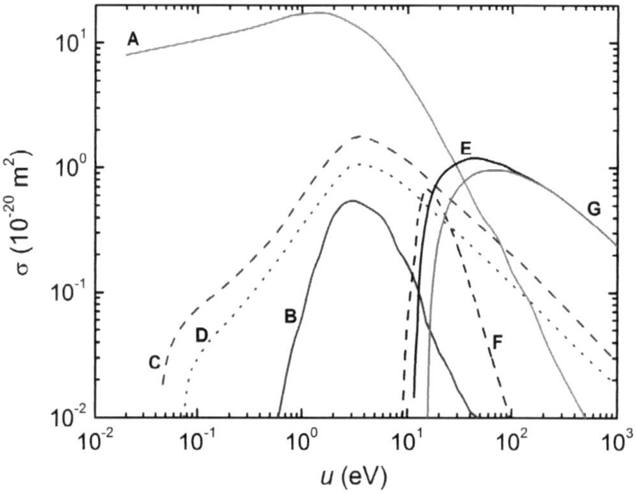

Figure 5. Summary of the H_2 cross section set with IST-LISBON, as a function of the electron kinetic energy. Elastic momentum-transfer (solid magenta line and A); total vibrational excitation (solid blue and B); first (J=0→J=2, dashed blue and C) and second (J=1→J=3, dotted blue and D) rotational excitations, as examples; total excitation to singlet levels (solid black and E) and to triplet levels (dashed black and F); and ionization (solid red and G).

When these cross sections are used as input data in a two-term Boltzmann solver, they yield swarm parameters in general agreement with measurements, for reduced electric fields in the range 10^{-3} to 1000 Td and gas temperatures of 300, 77 K. At low E/N, the agreement can only be obtained by including inelastic stepwise and superelastic transitions between the rotational levels considered, assuming a Boltzmann distribution at T_g for their populations. The maximum deviations observed are the following: for u_k, 8% in the region 10^{-2}-1 Td and 19% at 150 Td (for a 30% dispersion in the experimental values), this deviation increasing with E/N; for μN, 6% in the region 10^{-2}-1 Td and 16% at 180 Td; for α/N, ~50% in the region 150-850 Td. A similar agreement is also observed between calculated and measured values of the total dissociation coefficient and the excitation coefficient for all singlet levels.

The IST-LISBON dataset for hydrogen includes also electron–neutral cross sections for the atomic system [56]. This set was compiled from [71-72] for the elastic momentum-transfer, and adjusted from [15] for the excitation of H(2s,2p,3,4,5) levels and for the ionization from ground-state H(1s).

Journal of Physics: Conference Series **565** (2014) 012007 doi:10.1088/1742-6596/565/1/012007

2.6. Methane

The complete set of swarm-derived electron–neutral cross sections for methane was initially reported in [73], for kinetic energies up to 40 eV. Prior to its publication on LXCat, this set was substantially reviewed and extended up to energies of at least 400 eV. The present set, mainly based on the works [74-77], includes the following 9 cross sections: elastic momentum-transfer [74]; excitation of 2 vibrational modes (1,3) and (2,4) [74-75], obtained from [78-79] at low energies and from [80] at high energies; dissociation into CH_3 and CH_2 [75], obtained by subtracting the dissociative ionization cross section proposed in [77] to the total dissociation cross section reported in [81], and estimating the CH_3/CH_2 branching ratio from [76]; dissociative attachment into CH_2^- and H^- [74] (not shown in figure 6, due to the small value of the corresponding cross section); and ionization into CH_4^+ and CH_3^+ [77], yielding a total ionization cross section that agrees with the results of [19,75].

When this complete set of cross sections is used as input data in a two-term Boltzmann solver, it yields calculated swarm parameters in agreement with measurements, for E/N values in the range 10^{-1} to 400 Td. The maximum deviations observed are the following: for u_k, 12% around 30 Td and 30-40% below 1 Td (similar to the dispersion in the experimental values); for μN, 10-15% in the region 10-300 Td and 28-40% below 1 Td (again, similar to the dispersion in the experimental values); for α/N, 25% at 150 Td.

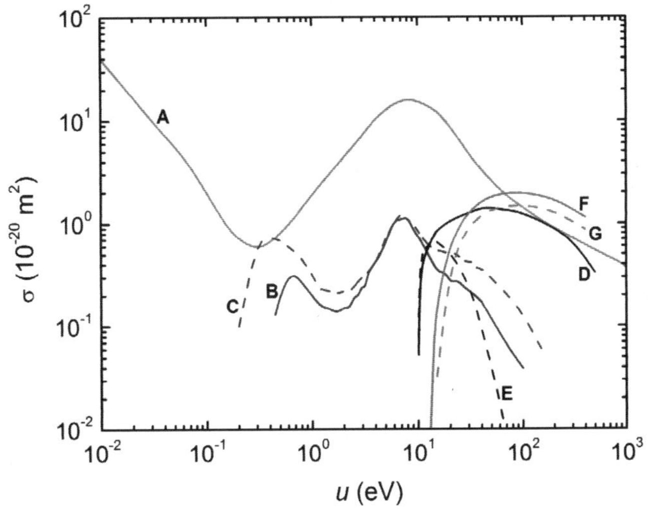

Figure 6. Summary of the methane cross section set with IST-LISBON, as a function of the electron kinetic energy. Elastic momentum-transfer (solid magenta line and A); vibrational excitations (1,3) (solid blue and B) and (2,4) (dashed blue and C); dissociation into CH_3 (solid black and D) and CH_2 (dashed black and E); and ionization into CH_4^+ (solid red and F) and CH_3^+ (dashed red and G).

3. Final remarks

This paper has presented the current status of the IST-LISBON database with LXCat, summarizing its complete and consistent sets of electron-neutral scattering cross sections for argon, helium, nitrogen, oxygen, hydrogen and methane. We have also reported on the quality of the calculated swarm parameters (characteristic energy, reduced mobility and excitation / ionization coefficients) obtained with these cross sections, when used as input data in a two-term Boltzmann solver, giving indication of the maximum deviations observed with respect to experimental values of these parameters, also available on LXCat. The IST-LISBON database will continue to be updated with the most recent data on cross sections and swarm parameters compiled, adjusted and checked by the Lisbon's group. Future work involves the update of helium cross sections and the upload of data for CO_2.

This paper has also briefly presented the web-based, open-access LXCat project. Our future plans include continuing the enlargement of LXCat databases (attracting new contributors and/or introducing new data types) and pursuing the evaluation / checking of the data published on-line. We are also committed to give each contributor a proper recognition for his/her work, by setting a consistent format for referencing among editors of journals (eventually, including citations of the original papers where the data was obtained from) and/or by introducing DOI numbers for the individual databases.

Acknowledgments

This work was partially supported by Fundação para a Ciência e a Tecnologia, under Project Pest-OE/SADG/LA0 010/2013.

References

[1] Pancheshnyi S, Biagi S F, Bordage M C, Hagelaar G J L, Morgan W L, Phelps A V and Pitchford L C 2012 *Chem. Phys.* **398** 148
[2] Pitchford L C *et al* 2013 *J. Phys. D: Appl. Phys.* **46** 334001
[3] Alves L L *et al* 2013 *J. Phys. D: Appl. Phys.* **46** 334002
[4] Bordage M C *et al* 2013 *J. Phys. D: Appl. Phys.* **46** 334003
[5] Bolsig 2005 www.bolsig.laplace.univ-tlse.fr/
[6] Petrović *et al* 2009 *J. Phys. D: Appl. Phys.* **42** 194002
[7] Petrović *et al* 2007 *Plasma Sources Sci. Technol.* **16** S1
[8] White *et al* 2003 *J. Phys. D: Appl. Phys.* **36** 3125
[9] Yanguas-Gil A, Cotrino J and Alves L L 2005 *J. Phys. D: Appl. Phys.* **38** 1588
[10] Yamabe C, Buckman S J and Phelps A V 1983 *Phys. Rev.* A **27** 1345
[11] Khakoo M A *et al* 2004 *J. Phys. B: At. Mol. Opt. Phys.* **37** 247
[12] Chilton J E, Boffard J B, Schappe R S and Lin C C 1998 *Phys. Rev.* A **57** 267
[13] Hayashi M 2003 *Report NIFS-DAT-72*, National Institute for Fusion Science of Japan
[14] Weber T, Boffard J B and Lin C C 2003 *Phys. Rev. A* **68** 032719
[15] Drawin H W 1967 *Report No EUR-CEA-FC-383*, Fontenay-aux-Roses
[16] Lee C M and Lu K T 1973 *Phys. Rev.* A **8** 1241
[17] Bretagne J, Callède G, Legentil M and Puech V 1986 *J. Phys. D: Appl. Phys.* **19** 761
[18] Bretagne J, Godart J and Puech V 1982 *J. Phys. D: Appl. Phys.* **15** 2205
[19] Rapp D and Englander-Golden P 1965 *J. Chem. Phys.* **43** 1464
[20] Alves L L and Ferreira C M 1991 *J. Phys. D: Appl. Phys.* **24** 581
[21] Alves L L, Gousset G and Ferreira C M 1992 *J. Phys. D: Appl. Phys.* **25** 1713
[22] Crompton R W, Elford M T and Jory R L 1967 *Aust. J. Phys.* **20** 369
[23] Crompton R W, Elford M T and Robertson A G 1970 *Aust. J. Phys.* **23** 667
[24] Milloy H B and Crompton R W 1977 *Phys. Rev.* A **15** 1847
[25] Register D F, Trajmar S and Srivastava S K 1980 *Phys. Rev.* A **21** 1134
[26] Schulz G J and Fox R E 1957 *Phys. Rev.* **106** 1179
[27] Schulz G J and Philbrick J W 1964 *Phys. Rev. Lett.* **13** 477
[28] Mason N J and Newell W R 1987 *J. Phys. B: At. Mol. Phys.* **20** 1357
[29] Westerveld W B, Heideman H G M and Van Eck J 1979 *J. Phys. B: At. Mol. Phys.* **12** 115
[30] Donaldson F G, Hender M A and McConkey J W 1972 *J. Phys. B: At. Mol. Phys.* **5** 1192
[31] Jobe J D and St John R M 1967 *Phys. Rev.* **164** 117
[32] de Heer F J and Jansen R H J 1977 *J. Phys. B: At. Mol. Phys.* **10** 3741
[33] Gousset G and Boulmer-Leborgne C 1984 *J. Physique* **45** 689

[34] Daviaud S 1989 *PhD Thesis* University Paris-Sud, Orsay, France
[35] Santos M, Noël C, Belmonte T and Alves L L 2014 *J. Phys. D: Appl. Phys.* **47** 265201
[36] Loureiro J and Ferreira C M 1986 *J. Phys. D: Appl. Phys.* **19** 17
[37] Pitchford L C and Phelps A V 1982 *Bull. Am. Phys. Soc.* **27** 109
[38] Tachibana K and Phelps A V 1979 *J. Chem. Phys.* **71** 3544
[39] Cartwright D C, Trajmar S, Chutjian A and Williams W 1977 *Phys. Rev.* A **16** 1041
[40] Zipf E C and McLaughlin R W 1978 *Planet. Space Sci.* **26** 449
[41] Schulz G J 1962 *Phys. Rev.* **125** 229
[42] Schulz G J 1964 *Phys. Rev.* **135** A988
[43] Schulz G J 1973 *Rev. Mod. Phys.* **45** 423
[44] Engelhardt A G, Phelps A V and Risk C G 1964 *Phys. Rev.* **135** A1566
[45] Pavlovic Z, Boness M J W, Herzenberg A and Schulz G J 1972 *Phys. Rev.* A **6** 676
[46] Boness M J W and Schulz G J 1973 *Phys. Rev.* A **8** 2883
[47] Gerjouy E and Stein S 1955 *Phys. Rev.* **97** 1671
[48] Huot J and Bose T K 1991 *J. Chem. Phys.* **94** 3849
[49] Guerra V and Loureiro J 1995 *J. Phys. D: Appl. Phys.* **28** 1903
[50] Geltman S 1973 *J. Quantum Spectrosc. Radiat. Transfer* **13** 601
[51] Berrington K A, Burke P G and Robb W D 1975 *J. Phys. B: At. Mol. Phys.* **8** 2500
[52] Gousset G, Ferreira C M, Pinheiro M, Sá P A, Touzeau M, Vialle M and Loureiro J 1991 *J. Phys. D: Appl. Phys.* **24** 290.
[53] Phelps A V 1985 *Technical Report 28*, JILA Information Center Report, University of Colorado, Boulder, Colorado, USA
[54] Wakiya K J 1978 *J. Phys. B: At. Mol. Phys.* **11** 3931
[55] Huber K P and Herzberg G 1979 *Molecular Spectra and Molecular Structure. IV. Constants of Diatomic Molecules* (D. van Nostrand Reinhold Co., Princeton:New Jersey)
[56] Marques L, Jolly J and Alves L L 2007 *J. Appl. Phys.* **102** 063305
[57] Phelps A V, private communication
[58] Tawara H, Itikawa Y, Nishimura H and Yoshino M 1990 *J. Phys. Chem. Ref. Data* **19** 617
[59] Wingerden B, Weigold E, de Heer F and Nygaard K J 1977 *J. Phys. B: At. Mol. Phys.* **10** 1345
[60] Crompton R W, Gibson D K and McIntosh A I 1969 *Aust. J. Phys.* **22** 715
[61] Henry R J W and Lane N F 1969 *Phys. Rev.* **183** 221
[62] Gibson D K 1970 *Aust. J. Phys.* **23** 683
[63] Heaps M G and Green A E S 1975 *J. Appl. Phys.* **46** 4718
[64] Nishimura H, Danjo A and Sugahara H 1985 *J. Phys. Soc. Jpn.* **54** 1757
[65] Fujimoto T, Sawada K and Takahata K 1989 *J. Appl. Phys.* **66** 2315
[66] Sawada K, Eriguchi K and Fujimoto T 1993 *J. Appl. Phys.* **73** 8122
[67] Lavrov B P 1977 *Opt. Spectrosc.* **42** 250
[68] Gorse C, Capitelli M, Bacal M, Bretagne J and Laganà A 1987 *Chem. Phys.* **117** 177
[69] Simko T, Martisovits M, Bretagne J and Gousset G 1997 *Phys. Rev.* E **56** 5908
[70] Rapp D, Englander-Golden P and Braglia D 1965 *J. Chem. Phys.* **42** 4081
[71] Mukherjee M, Basu M and Gosh A 1987 *Phys. Rev.* A **36** 937
[72] Bray I, Konovalov D A and McCarthy I E 1991 *Phys. Rev.* A **43** 5874
[73] Pintassilgo C D 1995 *MSc Thesis* Universidade Técnica Lisboa, Lisbon, Portugal
[74] Davies D K, Kline L E and Bies W E 1989 *J. Appl. Phys.* **65** 3311
[75] Ohmori Y, Kitamori K, Shimozuna M and Tagashira H 1986 *J. Phys. D: Appl. Phys.* **19** 437
[76] Nakano T, Toyoda H and Sugai H 1991 *Jpn J. Appl. Phys.* **30** 2912
[77] Chatham H, Hils D, Robertson R and Gallagher A 1984 *J. Chem Phys.* **81** 1770
[78] Rohr K 1980 *J. Phys. B: At. Mol. Phys.* **13** 4897
[79] Sohn W, Jung K and Ehrhardt H 1983 *J. Phys. B: At. Mol. Phys.* **16** 891
[80] Tanaka H, Kubo M, Onodera N and Suzuki A 1983 *J. Phys. B: At. Mol. Phys.* **16** 2861
[81] Winters H F 1975 *J. Chem. Phys.* **63** 3462

Journal of Physics: Conference Series **565** (2014) 012008 doi:10.1088/1742-6596/565/1/012008

Resistive Plate Chambers: electron transport and modeling

D Bošnjaković, Z Lj Petrović and S Dujko

Institute of Physics, University of Belgrade, Pregrevica 118, 11070 Belgrade, Serbia

E-mail: dbosnjak@ipb.ac.rs

Abstract. We study the electron transport in gas mixtures used by Resistive Plate Chambers (RPCs) in high energy physics experiments at CERN. Calculations are performed using a multi term theory for solving the Boltzmann equation. We identify the effects induced by non-conservative nature of electron attachment, including attachment heating of electrons and negative differential conductivity (NDC). NDC was observed only in the bulk component of drift velocity. Using our Monte Carlo technique, we calculate the spatially resolved transport properties in order to investigate the origin of these effects. We also present our microscopic approach to modeling of RPCs which is based on Monte Carlo method. Calculated results for a timing RPC show good agreement with an analytical model and experimental data. Different cross section sets for electron scattering in $C_2H_2F_4$ are used for comparison and analysis.

1. Introduction

Resistive Plate Chambers (RPCs) emerged in 1980s as a practical alternative to spark counters with localized discharge [1, 2]. Their reliability comes from electrodes of high resistivity, such as glass or bakelite, which are used to suppress the destructive discharges. Owing to their good efficiency and spatial resolution in conjunction with outstanding timing resolution [3, 4], they are mainly used for timing and triggering in high energy physics experiments [5–7], but they also found their way in other fields such as medical imaging [8, 9] and geophysics [10].

There were many approaches to RPC modeling, mostly analytical [11, 12], fluid [13–15] or Monte Carlo based [16]. The common thread among all these models is that they require electron transport data as input parameters. Electron transport data are usually calculated from the knowledge of cross sections for electrons scattering using Boltzmann equation analysis and/or Monte Carlo technique. A Monte Carlo tool for such a task — MAGBOLTZ [17] has cross sections imbedded in the code and is often used by the particle detector community. The question of reliability was raised in case of cross sections for $C_2H_2F_4$ [18] which is the main component in gas mixtures for RPCs operated in avalanche mode. Another issue is the correct implementation of transport data in RPC modeling when non-conservative collisions such as attachment or ionization are present. In this regard, the duality of transport coefficients is often neglected which in some cases may lead to incorrect results for RPCs performances, i.e. efficiency, timing resolution and charge spectra.

In this paper, we review our recent efforts and outline some of our key results in studying of electron transport and modeling of RPCs. Results are obtained using our multi term theory for solving Boltzmann's equation and Monte Carlo method which are briefly described in Section 2.

In Section 3.1, we investigate the electron transport in RPC gas mixtures used in ALICE, ATLAS and CMS experiment at CERN. Finally, our new microscopic model of RPCs based on Monte Carlo simulation technique is described in Section 3.2. For more details the reader is referred to our recent papers about the electron transport in RPCs [19] and our microscopic approach to RPC modeling [20].

2. Theoretical methods

2.1. Boltzmann equation analysis

Transport coefficients for electrons drifting and diffusing through a background gas under the influence of spatially homogeneous electric field $\mathbf{E}$, can be evaluated by solving Boltzmann's equation:

$$\frac{\partial f}{\partial t} + \mathbf{c} \cdot \frac{\partial f}{\partial \mathbf{r}} + \frac{e}{m}\mathbf{E} \cdot \frac{\partial f}{\partial \mathbf{c}} = -J(f, f_0)\,, \tag{1}$$

where $f(\mathbf{r}, \mathbf{c}, t)$ is the phase-space distribution function and e and m are the charge and mass of the electron. $J(f, f_0)$ is the collision integral for electron-gas interactions, including elastic, inelastic and non-conservative (e.g. electron attachment and/or ionization) collisions. f_0 is the Maxwellian velocity distribution function of the neutral gas at $T = 293$ K.

We solve the Boltzmann equation (1) using a numerical technique initially developed by Robson and Ness [21] and later extended by White et al. [22] and Dujko et al. [23, 24]. Among many important aspects, we emphasize the following important steps in solving Boltzmann's equation:

(i) The angular dependence of f is represented in terms of an expansion in spherical harmonics. In contrast to classical two-term theory, our analysis is 'multi term', and we take as many spherical harmonics as are necessary to achieve the full convergence of transport coefficients.

(ii) The speed (energy) dependence of f is represented by an expansion about a Maxwellian at an arbitrary temperature, in terms of Sonine polynomials.

(iii) Transport coefficients are calculated assuming the hydrodynamic conditions. Thus, the spatial dependence of f is treated by the density gradient expansion.

Substitution of these expansions into the Boltzmann equation (1) leads to a hierarchy of coupled differential equations for the moments of f. These equations are numerically solved and all transport coefficients of interest, including the mean energy, drift velocity and diffusion tensor as well as rate coefficients are expressed in terms of these moments [19].

2.2. Monte Carlo simulation technique

Here we outline only the basic concepts behind of our Monte Carlo technique. A thorough description can be found in our recent reviews [25–27].

We assume that the background gas molecules are stationary (zero gas temperature). Each electron is tracked by the analytical solution of the equation of electron motion in electric (and magnetic) field. The moment of collision is determined using the integration method where an electron is moved in very small time steps (compared to collision frequency) while the cumulative probability for a collision is less than a generated random number. When the collision occurs, the type of collision is determined using relative probabilities of different processes determined by the cross sections. The scattering is assumed to be isotropic. In an elastic collision, the electron loses energy of about $2m/M$ where m and M are the electron and molecule mass, respectively. When an inelastic collision or ionization takes place, the electron energy is reduced by the energy loss of that process. After ionization, each division of the remaining energy between the primary and secondary electron is considered to be equally probable. If attachment occurs, the electron is removed from the simulation.

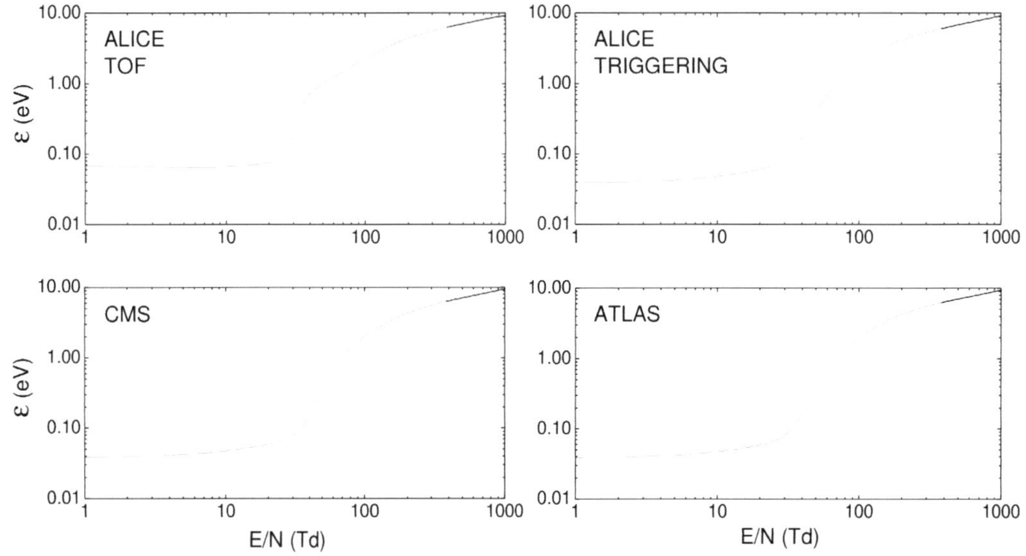

Figure 1. Variation of the mean energy with E/N for RPCs used in ALICE, CMS and ATLAS experiments at CERN. Taken from [19]. © IOP Publishing. Reproduced by permission of IOP Publishing. All rights reserved.

Our Monte Carlo technique can also be used to calculate the spatially resolved transport data using the following procedure. First we calculate the mean position (centre of mass) and the standard deviation (σ) of the electron swarm, for each coordinate axis. The interval $(-3\sigma, +3\sigma)$ around the mean position is then divided into 100 cells along each axis. Finally, the data including the electron average energy and average velocity, and attachment rate coefficient are calculated for every cell.

3. Results and discussion

3.1. A study of electron transport in RPC gases

First, we present the results for electron transport calculated using Boltzmann equation analysis. The results are calculated for the RPC gas mixtures of $C_2H_2F_4$/iso-C_4H_{10}/SF_6 employed at CERN LHC: (1) ALICE timing (TOF) RPC, 90/5/5; (2) ALICE triggering RPC, 89.7/10/0.3; (3) CMS triggering RPC, 96.2/3.5/0.3; and (4) ATLAS triggering RPC, 94.7/5/0.3. The following is assumed: the cross section set for $C_2H_2F_4$ was developed by our group [28], cross sections for iso-C_4H_{10} are taken from MAGBOLTZ 2.7.1 code and cross sections for SF_6 are taken from Itoh et al. [29]; the temperature and pressure of the background gas is 293 K and 1 atm, respectively; the reduced electric field strength E/N is given in Td (1 Td = 1×10^{-21} Vm^2).

Mean electron energies as a function of E/N are shown in Figure 1. One can observe that the mean energy is nearly constant up to about 30 Td where the energy begins to rise noticeably. This cooling effect is due to rising collision frequency for vibrational excitation of $C_2H_2F_4$. Above 30 Td the cross sections for vibrational excitations begin to drop and therefore the energy increases. Another interesting effect can be observed in case of ALICE TOF mixture with the highest SF_6 content: for lower electric fields, the mean energy is significantly higher than the thermal value of 0.038 eV. This effect is known as *attachment heating* [21] since low energy electrons are consumed in attachment and consequently the mean energy increases.

Figure 2 shows the calculated drift velocities. In case of ALICE TOF mixture, bulk component

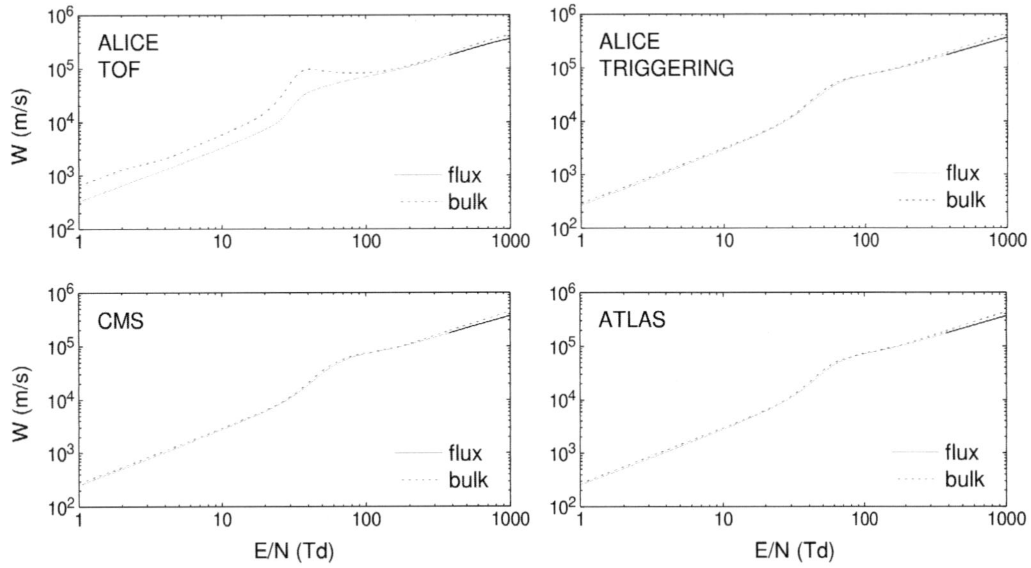

Figure 2. Variation of the bulk and flux drift velocities with E/N for RPCs used in ALICE, CMS and ATLAS experiments at CERN. Taken from [19].

of drift velocity exhibits a strong negative differential conductivity (NDC). NDC is a kinetic phenomenon defined as a decrease of drift velocity with increasing electric field [30]. It can arise due to a certain combination of elastic and inelastic cross sections and is usually present in both flux and bulk components of drift velocity. In this case however, the NDC is only present in the bulk drift velocity and the question of whether this is possible was already raised [31]. The effect can be understood by analyzing the spatial profiles of the electron swarm. For that purpose we use our Monte Carlo method to calculate the spatially resolved data including the mean energy and attachment rate coefficient. The Monte Carlo calculations are performed using cold gas approximation ($T = 0$) and therefore the calculated values are shifted with respect to those obtained using Boltzmann equation. Figure 3a shows the spatial profile and mean energy along the swarm for $E/N = 10$ Td which corresponds to the peak of bulk drift velocity (the onset of NDC). It can be seen that the mean energy is not constant along the swarm. Electrons at front of the swarm have more energy than those at the back and thus the attachment rate is higher at the back than at the front having in mind the energy dependence of cross section for attachment. As a consequence, the center of mass of the swarm shifts forward which is observed as an increase of bulk drift velocity over the flux component. The difference between flux and bulk component is roughly proportional to the spatial gradient of the attachment rate coefficient (Figure 3b). Indeed, one can see that the highest gradient corresponds to the peak of bulk drift velocity for $E/N = 10$ Td. As electric field increases, the gradient drops and the NDC gradually diminishes.

3.2. Our Monte Carlo model of RPCs

In this section, we present the results for timing distributions of electron avalanches and also the performance characteristics of a timing RPC with a gas mixture of 85% $C_2H_2F_4$ + 5% iso-C_4H_{10} + 10% SF_6. Results are calculated using our Monte Carlo method described in Section 2.2.

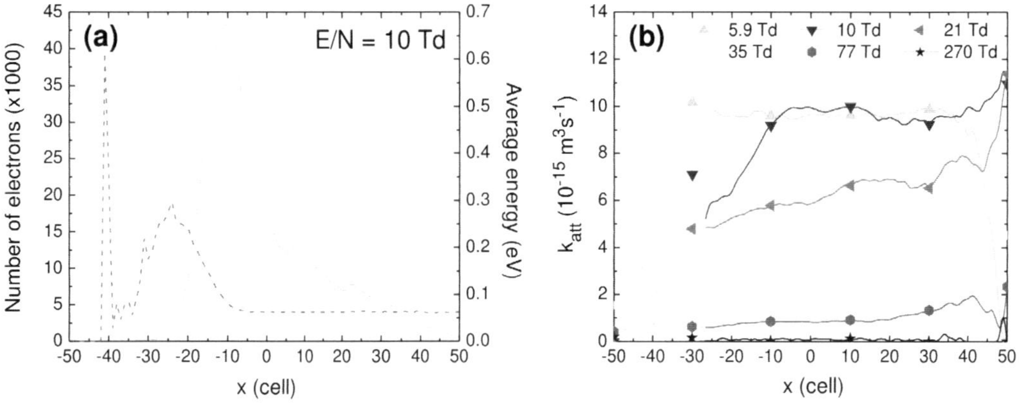

Figure 3. Spatially resolved transport properties for ALICE timing RPC gas mixture ($t = 1$ ns). The x axis is in the direction of the electric field. (a) Spatial profile of electrons (solid line) and spatially resolved averaged energy (dashed line) at $E/N = 10$ Td. (b) Spatially resolved attachment rate coefficient for a range of E/N. Taken from [19]. © IOP Publishing. Reproduced by permission of IOP Publishing. All rights reserved.

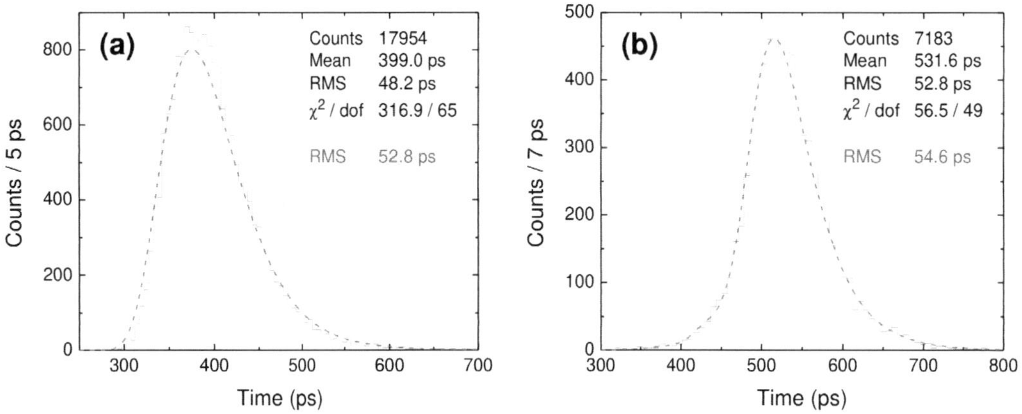

Figure 4. Timing distributions compared to analytical model [11] (dashed lines) for $E/N = 439$ Td. Our cross sections for $C_2H_2F_4$ [28] are assumed. (a) Single electron avalanches: the threshold is set to 10000 electrons. (b) Realistic RPC events: electron avalanches started by primary ionization generated using HEED cluster size distribution; gas gap 0.3 mm; the threshold is set to 2 fC. Taken from [20]. © SISSA Medialab Srl. Reproduced by permission of IOP Publishing. All rights reserved.

Cross sections for electron scattering are the same as in Section 3.1.

Figure 4a shows the timing distribution for single electron avalanches in an infinite space. The distribution agrees well with an analytical model [11]. This analytical model is based on Legler's model [32] of avalanche fluctuations and thus the slight difference between the timing distributions can be attributed to the approximation of constant probability for an ionization which is assumed by Legler's model.

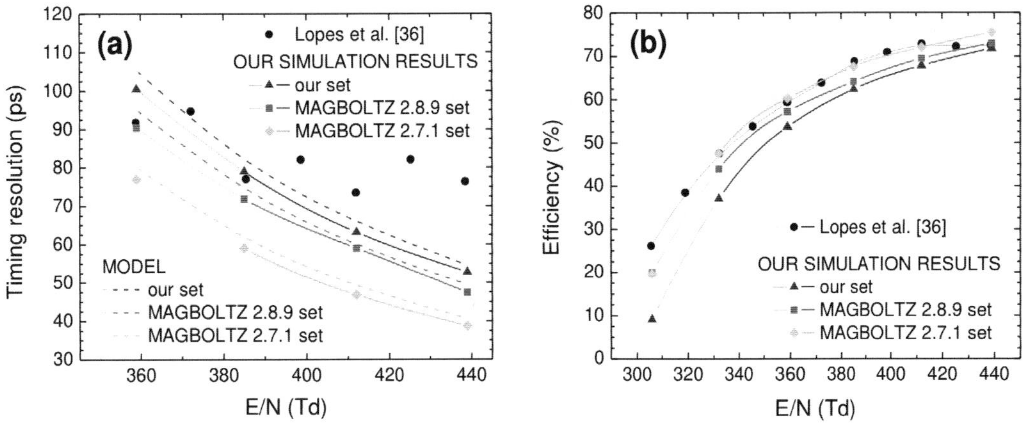

Figure 5. (a) Timing resolutions and (b) efficiencies calculated for a specific timing RPC using different $C_2H_2F_4$ cross section sets and HEED cluster size distribution. Comparison with experimental values by Lopes et al. [36] and analytical model [11] (dashed lines). Taken from [20].

The realistic RPC model, in addition to electron avalanche fluctuations, includes the effects of primary ionization, boundaries (i.e. gas gap) and the threshold given by induced charge. Primary ionization is implemented using a simple model which assumes that the primary electrons are grouped in clusters while the distance between the neighboring clusters follows the exponential distribution. Mean distance between the clusters and cluster size distribution are calculated using HEED [33] assuming minimum ionizing particles. The boundaries are determined by the 0.3 mm gas gap with "absorbing" anode. The threshold is set to 2 fC of induced charge which corresponds to about 10^6 electrons in the gas gap. The induced charge is calculated as an integral of the induced current which is given by Ramo's theorem [34].

A timing distribution calculated using the described RPC model is shown on Figure 4b. The distribution has a typical Gaussian-like shape. The left tail of the distribution represents the fastest events which most probably come from large primary ionization while the right tail is mainly determined by the electron avalanche fluctuations.

Finally, we show the results for the timing resolution and detection efficiency. For comparison, the results are made using different cross sections sets for $C_2H_2F_4$. In addition to our set, we use a set from MAGBOLTZ 2.8.9 (which is updated concerning the latest measurements [35]) and the older version from MAGBOLTZ 2.7.1. The timing resolution (Figure 5a) and detection efficiency (Figure 5b) both noticeably depend on the cross sections used. The experimentally measured values [36] are also shown for comparison.

4. Conclusion

We have presented some of our recent results in studying of electron transport in RPC gases and modeling of RPCs. Studies of electron transport reveal some interesting phenomena, induced by the explicit and implicit effects of electron attachment. In particular, the NDC effect is present only in the bulk component of drift velocity while unusually strong attachment heating is observed in the profile of the electron mean energy. Both of these phenomena are observed in case of ALICE TOF gas mixture. Using our Monte Carlo technique to evaluate the spatial profiles of the electron swarm we are able to illustrate and prove that the NDC is induced by strong attachment due to higher concentration of SF_6. In addition, our microscopic approach to

RPC modeling shows how to calculate the RPC performance characteristic of a timing RPC by following the individual electrons in a Monte Carlo fashion. The results for timing distributions agree well with an analytical model which uses the approximation of constant probability for ionization. The calculated timing resolution and efficiency agree well with experimental data and they also show sensitivity with respect to cross section sets for electron scattering used for $C_2H_2F_4$.

Acknowledgments

This work is supported by MPNTRRS Projects OI171037 and III41011.

References

[1] Santonico R and Cardarelli R 1981 *Nucl. Instrum. Meth.* **187** 377–380
[2] Cardarelli R, Santonico R, Di Biagio A and Lucci A 1988 *Nucl. Instrum. Meth. A* **263** 20–25
[3] Santonico R 2012 *Nucl. Instrum. Meth. A* **661** S2–S5
[4] Fonte P 2002 *IEEE Trans. Nucl. Sci.* **49** 881–887
[5] The ATLAS Collaboration 2008 *J. Instrum.* **3** S08003
[6] The ALICE Collaboration 2008 *J. Instrum.* **3** S08002
[7] The CMS Collaboration 2008 *J. Instrum.* **3** S08004
[8] Blanco A, Couceiro M, Crespo P, Ferreira N C, Ferreira Marques R, Fonte P, Lopes L and Neves J A 2009 *Nucl. Instrum. Meth. A* **602** 780–783
[9] Couceiro M, Crespo P, Mendes L, Ferreira N, Ferreira Marques R and Fonte P 2012 *Nucl. Instrum. Meth. A* **661** S156–S158
[10] Cârloganu C, Niess V, Béné S, Busato E, Dupieux P, Fehr F, Gay P, Miallier D, Vulpescu B, Boivin P, Combaret C, Labazuy P, Laktineh I, Lénat J-F, Mirabito L and Portal A 2013 *Geosci. Instrum. Method. Data Syst.* **2** 55–60
[11] Riegler W 2009 *Nucl. Instrum. Meth. A* **602** 377–390
[12] Mangiarotti A, Fonte P and Gobbi A 2004 *Nucl. Instrum. Meth. A* **533** 16–21
[13] Khosravi Khorashad L, Moshaii A and Hosseini S 2011 *EPL (Europhys. Lett.)* **96** 45002
[14] Khosravi Khorashad L, Eskandari M and Moshaii A 2011 *Nucl. Instrum. Meth. A* **628** 470–473
[15] Moshaii A, Khosravi Khorashad L, Eskandari M and Hosseini S 2012 *Nucl. Instrum. Meth. A* **661** S168–S171
[16] Riegler W and Lippmann C 2004 *Nucl. Instrum. Meth. A* **518** 86–90
[17] Biagi S F 1999 *Nucl. Instrum. Meth. A* **421** 234–240
[18] Gonzalez-Diaz D and Sharma A 2012 Current Challenges and Perspectives in Resistive Gaseous Detectors: a manifesto from RPC 2012 *in proceedings of the "XI workshop on Resistive Plate Chambers and Related Detectors (RPC2012)", PoS (RPC2012) 084*
[19] Bošnjaković D, Petrović Z Lj, White R D and Dujko S 2014 *J. Phys. D: Appl. Phys.* **47** 435203
[20] Bošnjaković D, Petrović Z Lj and Dujko S 2014 *J. Instrum.* **9** P09012
[21] Robson R E and Ness K F 1986 *Phys. Rev. A* **33** 2068–2077
[22] White R D, Robson R E, Dujko S, Nicoletopoulos P and Li B 2009 *J. Phys. D: Appl. Phys.* **42** 194001
[23] Dujko S, White R D, Petrović Z Lj and Robson R E 2010 *Phys. Rev. E* **81** 046403
[24] Dujko S, White R D, Petrović Z Lj and Robson R E 2011 *Plasma Sources Sci. Technol.* **20** 024013
[25] Dujko S, Raspopović Z M and Petrović Z Lj 2005 *J. Phys. D: Appl. Phys.* **38** 2952–2966
[26] Dujko S, White R D, Ness K F, Petrović Z Lj and Robson R E 2006 *J. Phys. D: Appl. Phys.* **39** 4788–4798
[27] Dujko S, White R D and Petrović Z Lj 2008 *J. Phys. D: Appl. Phys.* **41** 245205
[28] Šašić O, Dupljanin S, de Urquijo J and Petrović Z Lj 2013 *J. Phys. D: Appl. Phys.* **46** 325201
[29] Itoh H, Matsumura T, Satoh K, Date H, Nakao Y and Tagashira H 1993 *J. Phys. D: Appl. Phys.* **26** 1975–1979
[30] Petrović Z Lj, Crompton R W and Haddad G N 1984 *Aust. J. Phys.* **37** 23
[31] Vrhovac S B and Petrović Z Lj 1996 *Phys. Rev. E* **53** 4012–4025
[32] Legler W 1961 *Z. Naturforsch.* **16a** 253
[33] Smirnov I 2005 *Nucl. Instrum. Meth. A* **554** 474–493
[34] Ramo S 1939 *Proc. I.R.E.* **27** 584
[35] de Urquijo J, Juárez A M, Basurto E and Hernández-Ávila J L 2009 *Eur. Phys. J. D* **51** 241–246
[36] Lopes L, Fonte P and Mangiarotti A 2012 *Nucl. Instrum. Meth. A* **661** S194–S197

Journal of Physics: Conference Series **565** (2014) 012009 doi:10.1088/1742-6596/565/1/012009

Wake effect in graphene due to moving charged particles

I Radović[1], D Borka[1] and Z L Mišković[2]

[1]VINČA Institute of Nuclear Sciences, University of Belgrade, P.O. Box 522, 11001, Belgrade, Serbia

[2]Department of Applied Mathematics, and Waterloo Institute for Nanotechnology, University of Waterloo, Waterloo, Ontario, Canada N2L 3G1

E-mail: iradovic@vin.bg.ac.rs

Abstract. We study the wake effect in a supported graphene layer induced by external charged particles moving parallel to it by using the dynamic polarization function of graphene within the random phase approximation for its π electrons described as Dirac's fermions. We explore the effects of a substrate assuming that graphene is supported by an insulating substrate, such as SiO_2, and a strongly polar substrate, such as SiC, under the gating conditions. Strong effects are observed in the wake pattern in the induced density of charge carriers in supported graphene due to finite size of the graphene-substrate gap, as well as due to strong coupling effects, and plasmon damping of graphene's π electrons. We find that the excitation of surface phonons in the substrate may exert quite strong influences on the wake effect in the total electrostatic potential in the graphene plane at low particle speeds.

1. Introduction

Graphene is a flat monolayer of carbon atoms tightly packed into a two-dimensional (2D) honeycomb lattice [1]. It is a basic building block for graphitic materials of all other dimensionalities: highly oriented pyrolytic graphite (HOPG, a stack of graphene layers), carbon nanotubes (rolled-up cylinders of graphene), and fullerene molecules (consisting of wrapped graphene by the introduction of pentagons on the hexagonal lattice) [2, 3].

The ability to screen an external electric field is an important property of any nanostructured material, including graphene. Depending on the speed of the external charge, the screening mechanism changes its character dramatically, going from a Debye-like screened potential of a static charge to a dynamic regime characterized by an oscillatory potential contained in a cone trailing a moving charge, which is commonly known as the wake effect [4]. It is characterized by the onset of oscillations in the polarization of the medium, which arise from resonances due to excitations of collective modes in the medium and often provide an effective mechanism of energy loss for an external charge. While the wake effect in three-dimensional (3D) plasmas has been known for almost sixty years [5], its current significance encompasses diverse new areas, such as dust-crystal formation in complex plasmas [6], Coulomb explosion of large clusters, such as C_{60}, in thin solid foils [7] and plasmas [8], channelling of fast ions through nano-capillaries in solids [9] and carbon nanotubes [4, 10], as well as interactions of charged particles with 1D electron gas (1DEG) [11], one [12] and two [13, 14] layers of a 2D quantum electron gas (2DQEG), supported thin metal films [15], nanosphere [16], and with magnetized two-component plasmas [17, 18].

We have studied, in our previous publications [19, 20], the wake effect due to fast charged particles that move at speeds in excess of the Bohr's speed over a supported 2DEG characterized by a single energy band with parabolic dispersion. We have used a one-fluid hydrodynamic model with the parameters characteristic of graphene [19], which treats all four carbon's valence electrons as a single 2DEG, to present the oscillatory wake effect which develops in the induced number density when the particle speed exceeds a threshold value for the collective excitations in a 2DEG corresponding to the four valence electrons in graphene. We have also used a two-fluid hydrodynamic model with the parameters characteristic of graphene [20], which makes distinction between the contributions of carbon's σ and π electrons, to calculate the induced number density per unit area of electrons in the 2DEG and compare it with the induced number density in the one-fluid model. Our results showed that, when the particle speed matches the phase velocity of the quasi-acoustic π plasmon, the induced number density shows the usual wake oscillations. In addition, we have presented calculations of the total electrostatic potential in the 2DEG plane, induced by a fast point charge moving parallel to it. Those results indicated a possibility of realizing the so-called wake riding effect in 2D [4, 21], whereby other charged particles may be captured in a potential well, or their state manipulated in the presence of the wake potential induced by a fast external charge.

In order to study the wake effect due to slow charged particles moving parallel to a graphene layer under the gating conditions, the dielectric-response theory for surfaces and layered structures [22] is a convenient way to proceed, given that the dielectric function for graphene is available within the random phase approximation (RPA) based on a linear approximation for the π electron bands [23, 24, 25]. As regards the applicability of the latter approximation in describing the elementary excitations in the wavenumber-frequency, (k,ω), domain pertinent to graphene, we note that conditions $k < 2k_c$ and $\omega < 2\varepsilon_c/\hbar$ should be satisfied, where $k_c \approx a^{-1}$ is a high-momentum cutoff (with $a \approx 2.46$ Å being the lattice constant), and $\varepsilon_c \approx 1$ eV is a high-frequency cutoff [2, 23, 25]. For particles moving parallel to graphene at a fixed distance z_0 and constant speed v, the former condition will be satisfied for distances $z_0 > a$ (thereby neglecting the size of the π electron orbitals in graphene). The latter condition can be transformed into a restriction on the particle speed by invoking the Bohr's adiabatic criterion and requiring that $v < 2z_0\varepsilon_c/\hbar$. When the projectile speed is normalized by the Fermi speed of graphene, v_F ($v_F \approx c/300$, where c is the speed of light in free space), the latter condition amounts to $v/v_F \leq 0.3z_0$ with z_0 expressed in angstroms.

Although we consider the RPA dielectric function to be a basic, parameter-free model that provides an adequate description of both the inter-band and intra-band single-particle excitations (SPEs), as well as plasmon excitations, of graphene's π electrons, the model nevertheless has its shortcomings. For example, it ignores the local-field effects (LFE) due to electron-electron correlations [26, 27] and assigns an infinitely long lifetime to the electron excitations. Only one of these shortcomings can be qualitatively corrected in the RPA dielectric function at a time, e.g. by using either the Hubbard approximation (HA) for the LFE in the static limit [28] or by introducing finite relaxation time, or decay (damping) rate, γ, using Mermin's procedure [29, 30, 31, 32].

We present the oscillatory wake effect in the charged carrier density and in the total electrostatic potential in the graphene plane in a supported graphene layer, induced by a slowly moving charged projectile. We study the effects of a substrate assuming that graphene is supported by an insulating substrate, such as SiO_2, and a strongly polar substrate, such as SiC, under the gating conditions, taking into account the influence of: the equilibrium charge carrier density n due to doping of graphene, the distance z_0 and the speed v of the projectile, the size of the graphene-substrate gap h, and the damping rate γ of elementary excitations in graphene.

The equilibrium charge carrier density is a particularly important parameter because it determines the Fermi wavenumber of graphene's π-electrons, $k_F = \sqrt{\pi n}$ (we shall assume $n > 0$, i.e., graphene doped by electrons, without loss of generality), and the corresponding Fermi energy, $E_F = \hbar k_F v_F$.

Note that we use Gaussian electrostatic units and denote the charge of a proton by $e > 0$.

2. Basic theory

We use a Cartesian coordinate system with coordinates $\{\vec{R}, z\}$ and assume that single-layer graphene is located in the plane $z = 0$, where $\vec{R} = \{x, y\}$ is position in the plane and z distance from it. A semi-infinite substrate is assumed to occupy the region $z \le -h < 0$ underneath graphene, whereas the region $z > -h$ is assumed to be vacuum or air. It can be shown that, for a point charge Ze moving parallel to graphene along the x axis with speed v at a fixed distance $z_0 > 0$, the induced number density per unit area of electrons in graphene and the total electrostatic potential in the graphene plane are given by, respectively

$$n_{gr}(x, y, t) = \frac{Z}{\pi^2} \int_0^\infty \int_0^\infty e^{-kz_0} \operatorname{Re}\left\{\left[1 - \frac{\varepsilon_{bg}(k, k_x v)}{\varepsilon(k, k_x v)}\right] e^{i[k_x(x-vt)+k_y y]}\right\} dk_x dk_y\,, \tag{1}$$

$$\Phi_{tot}(x, y, z, t)\big|_{z=0} = \frac{2Ze}{\pi} \int_0^\infty \int_0^\infty \frac{e^{-kz_0}}{k} \operatorname{Re}\left\{\frac{e^{i[k_x(x-vt)+k_y y]}}{\varepsilon(k, k_x v)}\right\} dk_x dk_y\,, \tag{2}$$

where we have used the symmetry properties of the real and imaginary parts of the dielectric function of the combined graphene-substrate system

$$\varepsilon(k, \omega) = \varepsilon_{bg}(k, \omega) + \frac{2\pi e^2}{k} \chi(k, \omega)\,, \tag{3}$$

with $\chi(k, \omega)$ being the polarization function of non-interacting π electrons in free graphene [23, 24, 25], and

$$\varepsilon_{bg}(k, \omega) = \frac{\varepsilon_s(\omega) + 1}{2} \frac{1 + \coth(kh)}{\varepsilon_s(\omega) + \coth(kh)}\,, \tag{4}$$

is the background dielectric function which quantifies the effects of substrate on the response of graphene [33, 34]. Note that $\varepsilon_{bg}(k, \omega)$ takes the values in the range between 1 and $[\varepsilon_s(\omega) + 1]/2$, characterizing, respectively, the case of a free-standing graphene ($h \to \infty$) and the case of a zero gap ($h = 0$) between graphene and a substrate.

In equation (4), $\varepsilon_s(\omega)$ is the bulk dielectric function of the substrate, which is given in the local approximation for a material supporting only one transverse optical (TO) phonon of frequency ω_{TO} by

$$\varepsilon_s(\omega) = \varepsilon_\infty + (\varepsilon_0 - \varepsilon_\infty) \frac{\omega_{TO}^2}{\omega_{TO}^2 - \omega(\omega + i\gamma_{TO})}\,, \tag{5}$$

where ε_∞ and ε_0 are the dielectric constants at high and low frequencies, respectively, and γ_{TO} is the damping rate of the TO phonon. Note that, when substrate phonons are ignored in graphene studies, it is customary to take the static limit of the dielectric function in equation (5), $\varepsilon_s(0) = \varepsilon_0$ [2].

In order to estimate the effects of strong coupling, we go beyond the RPA regime by using the aproach outlined in reference [35] for interactions of slow charged particles with 2DEG, whence the RPA polarization function $\chi(k, \omega)$ is to be replaced with

$$\chi_{LFE}(k,\omega)=\frac{\chi(k,\omega)}{1-G(k)V(k)\chi(k,\omega)}, \tag{6}$$

where $V(k)=2\pi e^2/k$ is the Coulomb interaction in 2D.

For the sake of simplicity, we use static limit of the LFE correction function, $G(k)$, which is given in the Hubbard approximation (HA) by $G(k)=(1/4)k(k^2+k_F^2)^{-1/2}$ [28, 36]. On the other hand, the finite lifetime of the excitation modes of charge carriers in graphene is treated by introducing a finite damping rate, γ, in the RPA polarization function through Mermin's procedure [29, 30, 31, 32], whereby one replaces $\chi(k,\omega)$ with

$$\chi_M(k,\omega,\gamma)=\frac{\chi(k,\omega+i\gamma)}{1-\frac{i\gamma}{\omega+i\gamma}\left[1-\frac{\chi(k,\omega+i\gamma)}{\chi_s(k)}\right]}, \tag{7}$$

where the static limit of the RPA polarization, $\chi_s(k)=\lim_{\omega\to 0}\chi(k,\omega)$, is given elsewhere [23, 24, 29].

3. Results and discussion

We choose the external particle to be proton ($Z=1$) moving along the x axis with speed v at a fixed distance $z_0=20$ Å above graphene with the equilibrium charge carrier density n.

Firstly, we use equation (1) to calculate the induced number density for free-standing graphene ($h\to\infty$), without the LFE, and with vanishing damping rate ($\gamma\to 0$). Our calculations of the induced number density for $y=0$ show that the wake effect generally takes place for particle speeds satisfying $v>v_F$, and is manifested as oscillations trailing the proton ($x-vt<0$). In figure 1, we explore the spatial dependence of wake oscillations in the induced number density n_{gr} (multiplied by z_0^2) as a function of both $x-vt$ and y (both normalized by z_0), for a proton moving along the x axis with a speed $v=4v_F$ at a distance $z_0=20$ Å above graphene with an equilibrium charge carrier density of $n=10^{13}\text{cm}^{-2}$. One can observe that the V shape wake-field appears apparently behind the particle, along with multiple oscillatory lateral wakes. One notices, away from the symmetry axis $y=0$, that the amplitudes of the lateral wakes are damped which is a consequence of plasmon damping due to the inter-band SPEs.

Next, we calculate the induced number density for graphene supported by a SiO_2 substrate in the static mode ($\varepsilon_0=3.9$), without the LFE, and with vanishing damping rate ($\gamma\to 0$). We explore the effects of a variable gap height and consider the gap heights: $h=0$ for the zero gap commonly considered in the literature, $h=4$ Å for a realistic value [37] and $h\to\infty$ for free graphene. In figure 2, we compare the spatial distribution of the induced number density n_{gr} (normalized by n) as a function of $x-vt$ (multiplied by $k_F=\sqrt{\pi n}$) with $y=0$, for a proton moving at a distance $z_0=20$ Å above graphene, for several gap heights and densities, with vanishing damping ($\gamma\to 0$), at two speeds: (a) $v=2v_F$ and (b) $v=4v_F$. One can observe in figure 2(a) that the oscillation period of the induced number density decreases when the size of the gap increases, as was also shown in our previous publication [19] in the case of the wake effect in interactions of fast ions with a supported 2DEG. This effect of the gap size on the oscillation periods seems to be weaker at higher particle speeds, as shown in figure 2(b) for $v=4v_F$, but the amplitude of these oscillations in the normalized number density n_{gr}/n appears to be suppressed by the close proximity of a substrate. Such a strong influence of the size of the graphene-substrate gap on the wake effect is in accord with the conclusions found in our previous publications [29, 30, 38] about the importance of the gap size on the stopping and image forces on external charges moving near graphene.

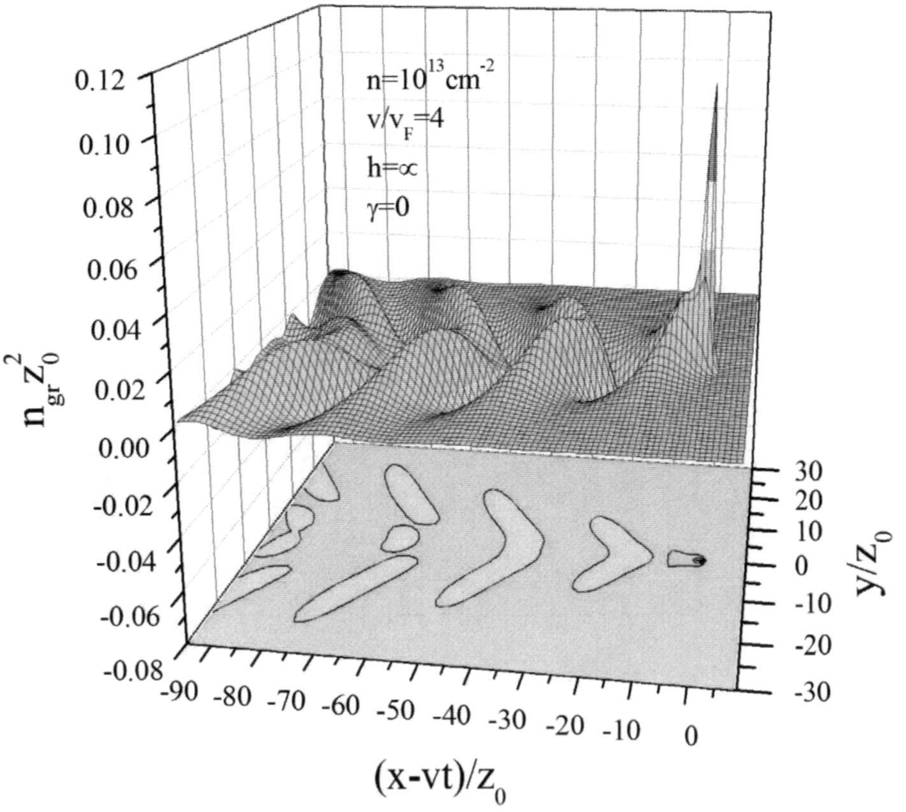

Figure 1. (Color online.) The induced number density n_{gr} (multiplied by the square of the particle distance z_0^2) as a function of $x - vt$ and y (both normalized by z_0) in the wake-field region created by a proton moving along the x axis with a speed $v = 4v_F$ at a distance $z_0 = 20\,\text{Å}$ above free-standing graphene ($h \to \infty$) with an equilibrium charge carrier density of $n = 10^{13}\,\text{cm}^{-2}$, and with vanishing damping ($\gamma \to 0$).

We also discuss the local-field effects (LFE) and the effects of finite damping rate on the induced number density. While the LFE is an inherent feature of graphene's electronic response that is particularly important at low particle speeds [35], the damping rate γ depends on graphene's environment and hence may vary quite strongly from sample to sample because γ arises from various mechanisms for charge carrier scattering, including charged impurities, local defects, and ripples in graphene [28]. For example, it was found for epitaxial graphene that a reasonable value for the damping rate γ is on the order of several hundred meV [29]. Thus, we compare in figure 3 several cases: graphene with $\gamma = 0$ and $\gamma = 100\,\text{meV}$ (without LFE), and graphene with the LFE included (whence $\gamma = 0$). Figure 3 shows the spatial distribution of the induced number density n_{gr}

(normalized by n) as a function of $x-vt$ (multiplied by $k_F=\sqrt{\pi n}$) with $y=0$, for two equilibrium densities n of 10^{11} cm^{-2} to 10^{13} cm^{-2}, for a proton moving at a distance $z_0=20$ Å above graphene on a SiO_2 substrate ($\varepsilon_0=3.9$ and $h=4$ Å) at two speeds: (a) $v=2v_F$ and (b) $v=4v_F$. One notices that the main result of including the LFE is an increase in the period of the wake oscillations, which is similar to the effect of decreasing the graphene-substrate gap in figure 2. As expected, the LFE is seen to be more pronounced at the lower speed in figure 3 [35]. On the other hand, one notices that finite damping rate exerts quite strong influence in reducing the amplitudes of the wake oscillations at both speeds shown in figure 3. The observed suppression of the wake oscillations with increasing γ occurs due to the broadening of the plasmon resonance in the (k,ω) plane.

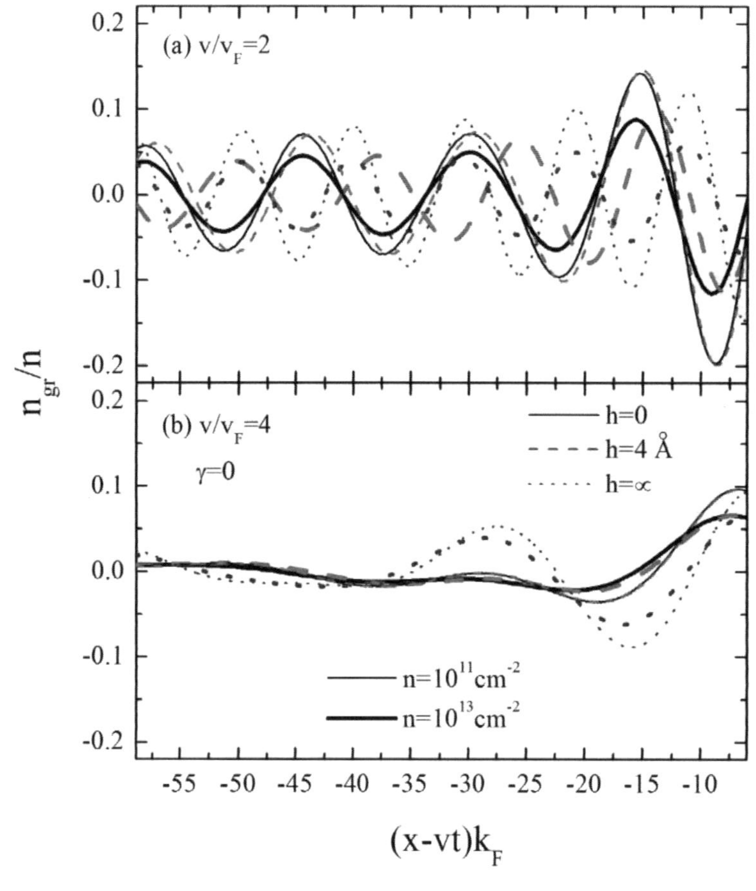

Figure 2. (Color online.) The spatial distribution of the induced number density n_{gr} (normalized by the equilibrium charge carrier density n) as a function of $x-vt$ (multiplied by the Fermi momentum $k_F=\sqrt{\pi n}$) with $y=0$, for a proton moving at a distance $z_0=20$ Å above graphene on a SiO_2 substrate ($\varepsilon_0=3.9$), with vanishing damping ($\gamma\to 0$), at two speeds: (a) $v=2v_F$ and (b) $v=4v_F$. Results are shown for three values of the gap heights: $h=0$ (black solid lines), $h=4$ Å (red dashed lines), and $h\to\infty$ (blue dotted lines). Thin and thick lines represent the cases of $n=10^{11}$ cm^{-2} and $n=10^{13}$ cm^{-2}, respectively.

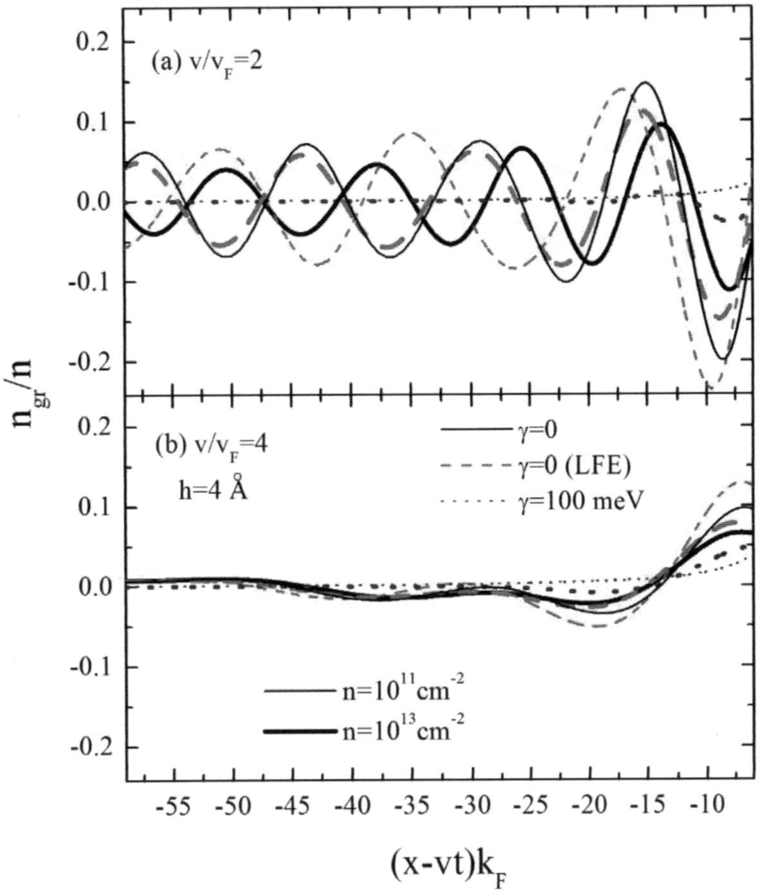

Figure 3. (Color online.) The spatial distribution of the induced number density n_{gr} (normalized by the equilibrium charge carrier density n) as a function of $x-vt$ (multiplied by the Fermi momentum $k_F=\sqrt{\pi n}$) with $y=0$, for a proton moving at a distance $z_0=20\,\text{Å}$ above graphene on a SiO_2 substrate ($\varepsilon_0=3.9$ and $h=4\,\text{Å}$), at two speeds: (a) $v=2v_F$ and (b) $v=4v_F$. Results are shown for three values of the damping rates γ: $\gamma=0$ (black solid lines), $\gamma=0$ with the LFE (red dashed lines), and $\gamma=100\,\text{meV}$ (blue dotted lines). Thin and thick lines represent the cases of $n=10^{11}\,\text{cm}^{-2}$ and $n=10^{13}\,\text{cm}^{-2}$, respectively.

Finally, we use equation (2) to calculate the total electrostatic potential in the plane of graphene with an equilibrium charge carrier density of $n=10^{13}\,\text{cm}^{-2}$, without the LFE, and with vanishing damping rate ($\gamma\to 0$). We explore the effects of a strongly polar substrate, such as SiC. Note that we use the following values in equation (5): $\varepsilon_\infty\approx 6.7$, $\varepsilon_0\approx 9.7$, $\omega_{TO}\approx 97\,\text{meV}$, and $\gamma_{TO}\approx 10\,\text{meV}$ [30]. For the sake of simplicity, we set $h=0$ noting that the effects of finite h are considered in figure 2. In figure 4, we show the total electrostatic potential in the plane of graphene Φ_{tot} (normalized by $\Phi_0=Ze/z_0$) as a function of $x-vt$ (normalized by z_0) with $y=0$, for four model systems: graphene on a substrate with phonon (g-SiC-phonon), graphene on a substrate in the static mode (g-SiC-static), bare substrate with phonon (SiC-phonon), and free graphene (g-free). Since the wake effect is critically dependent on the particle speed, we consider three cases: $v/v_F = 0.5$, 1 and 2, in the panels (a), (b) and (c) of figure 4, respectively. Numerical results show that, when the projectile speed exceeds a threshold value on the order of graphene's Fermi speed v_F, the oscillatory wake effect develops in the total electrostatic potential in the graphene plane trailing the particle, as shown in figure 4(c) for the speed $v = 2v_F$. However, the potential for graphene on a substrate that supports a phonon mode also exhibits spatial oscillations at sub-threshold speeds, $v \leq v_F$, as shown in figure 4(a) for $v = 0.5v_F$ and in figure 4(b) for the speed $v = v_F$. This may be due to hybridization of the plasmon in graphene and the substrate phonon.

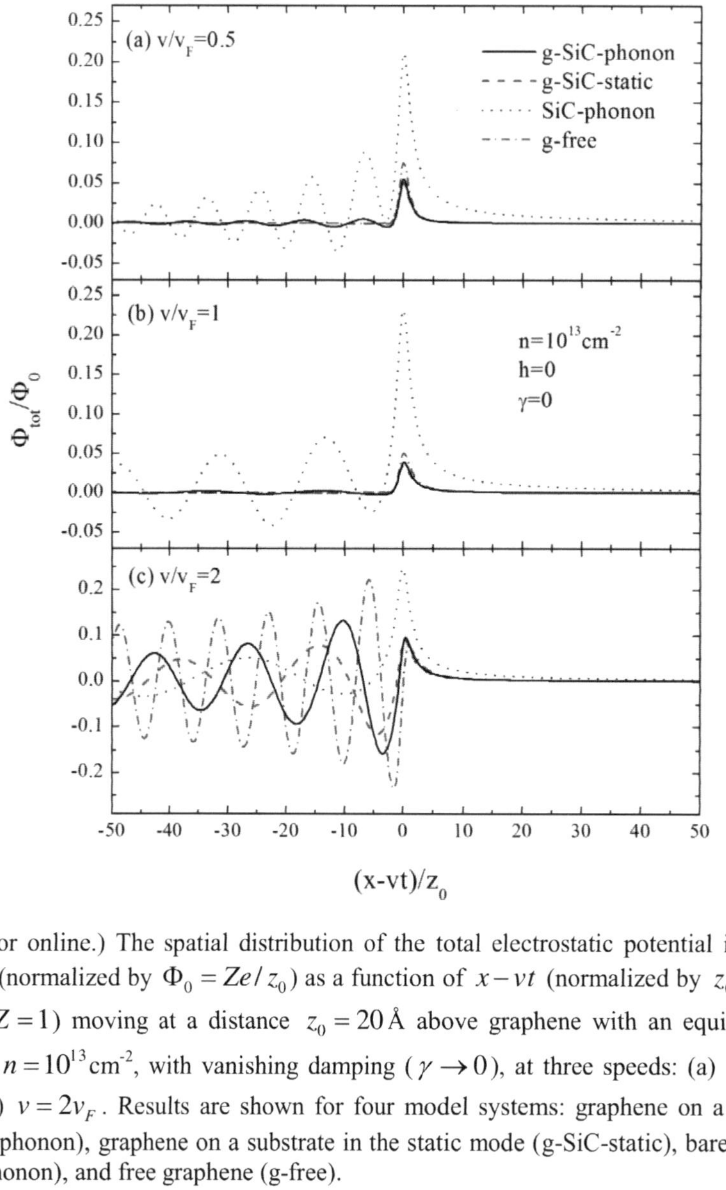

Figure 4. (Color online.) The spatial distribution of the total electrostatic potential in the plane of graphene Φ_{tot} (normalized by $\Phi_0 = Ze/z_0$) as a function of $x - vt$ (normalized by z_0) with $y = 0$, for a proton ($Z = 1$) moving at a distance $z_0 = 20$ Å above graphene with an equilibrium charge carrier density $n = 10^{13}$ cm^{-2}, with vanishing damping ($\gamma \to 0$), at three speeds: (a) $v = 0.5v_F$, (b) $v = v_F$, and (c) $v = 2v_F$. Results are shown for four model systems: graphene on a substrate with phonon (g-SiC-phonon), graphene on a substrate in the static mode (g-SiC-static), bare substrate with phonon (SiC-phonon), and free graphene (g-free).

4. Conclusions

We have presented the wake effect in interactions of moving external charges with supported graphene under the gating conditions. Calculations of the induced number density per unit area of electrons in graphene, as well as of the total electrostatic potential in the graphene plane as functions of the speed of the projectile moving parallel to graphene and the particle distance, were performed within the random phase approximation (RPA) based on a linear approximation for the π electron bands.

We have found that the presence of a substrate can exert strong influence on the wake effect as well, e.g., via the dependence of the oscillation period of the induced number density trailing the charged particle on the size of the graphene-substrate gap (the oscillation period decreases when the size of the graphene-substrate gap increases). Similarly, strong influence was also observed at lower particle speeds when local-field effects were included in graphene's electronic response. In addition, we have found that the amplitudes of oscillations are heavily suppressed by increasing the damping rates of plasmons in graphene.

The total electrostatic potential in the plane of graphene exhibits the usual wake effect in both free graphene and graphene on a substrate in the static limit, showing spatial oscillations that lag the particle, which moves at a speed v exceeding the threshold value of v_F, the Fermi speed in graphene. Surprisingly, the potential for graphene on a substrate that supports a phonon mode also exhibits spatial oscillations at sub-threshold speeds, $v \leq v_F$, that trail a moving charge. While this may be due to hybridization of the plasmon in graphene and the substrate phonon, such an unusual manifestation of the wake effect warrants further investigation.

Acknowledgments

This work has been supported by the Ministry of Education, Science and Technological Development of the Republic of Serbia (Project No. 45005). Z.L.M. also acknowledges support from the Natural Sciences and Engineering Research Council of Canada.

References

[1] Geim A K and Novoselov K S 2007 *Nature Mater.* **6** 183
[2] Castro Neto A H, Guinea F, Peres N M R, Novoselov K S and Geim A K 2009 *Rev. Mod. Phys.* **81** 109
[3] Geim A K and Kim P 2008 *Sci. Am.* **298** 90
[4] Mowbray D J, Mišković Z L, Goodman F O and Wang Y N 2004 *Phys. Lett.* A **329** 94
[5] Ritchie R H and Neufeld J 1955 *Phys. Rev.* **98** 1632
[6] Hou L J, Wang Y N and Mišković Z L 2003 *Phys. Rev.* E **68** 016410
[7] Wang Y N, Qiu H T and Mišković Z L 2000 *Phys. Rev. Lett.* **85** 1448
[8] Wang G Q *et al.* 2012 *Phys. Rev.* A **86** 043201
[9] Arista N R 2001 *Phys. Rev.* A **64** 032901
[10] Mowbray D J, Mišković Z L, Goodman F O and Wang Y N 2004 *Phys. Rev.* B **70** 195418
[11] Zhang Y, Song Y H and Wang Y N 2011 *Phys. Plasmas* **18** 112705
[12] Li C Z, Song Y H and Wang Y N 2008 *Phys. Lett.* A **372** 4500
[13] Li C Z, Song Y H and Wang Y N 2008 *Chin. Phys. Lett.* **25** 2981
[14] Li C Z, Wang Y N, Song Y H and Mišković Z L 2014 *Phys. Lett.* A **378** 1626
[15] Li C Z, Song Y H and Wang Y N 2009 *Phys. Rev.* A **79** 062903
[16] An S B, Wang Y X, Song Y H and Wang Y N 2012 *Phys. Lett.* A **376** 763
[17] Wang Q, Song Y H and Wang Y N 2010 *Phys. Lett.* A **374** 4678
[18] Hu Z H, Song Y H and Wang Y N 2010 *Phys. Rev.* E **82** 026404
[19] Radović I and Borka D 2010 *Phys. Lett.* A **374** 1527
[20] Radović I and Borka D 2010 *Nucl. Instr. Meth.* B **268** 2649
[21] Burgdörfer J 1992 *Nucl. Instr. Meth.* B **67** 1
[22] Pitarke J M, Silkin V M, Chulkov E V and Echenique P M 2007 *Rep. Prog. Phys.* **70** 1

[23] Wunsch B, Stauber T, Sols F and Guinea F 2006 *New J. Phys.* **8** 318
[24] Hwang E H and Das Sarma S 2007 *Phys. Rev.* B **75** 205418
[25] Barlas Y, Pereg-Barnea T, Polini M, Asgari R and Mac-Donald A H 2007 *Phys. Rev. Lett.* **98** 236601
[26] Kramberger C *et al.* 2008 *Phys. Rev. Lett.* **100** 196803
[27] Trevisanutto P E, Giorgetti C, Reining L, Ladisa M and Olevano V 2008 *Phys. Rev. Lett.* **101** 226405
[28] Adam S, Hwang E H, Rossi E and Das Sarma S 2009 *Solid State Commun.* **149** 1072
[29] Allison K F, Borka D, Radović I, Hadžievski Lj and Mišković Z L 2009 *Phys. Rev.* B **80** 195405
[30] Allison K F and Mišković Z L 2010 *Nanotechnology* **21** 134017
[31] Mermin N D 1970 *Phys. Rev.* B **1** 2362
[32] Asgari R, Vazifeh M M, Ramezanali M R, Davoudi E and Tanatar B 2008 *Phys. Rev.* B **77** 125432
[33] Radović I, Borka D and Mišković Z L 2012 *Phys. Rev.* B **86** 125442
[34] Radović I, Borka D and Mišković Z L 2014 *Phys. Lett.* A **378** 2206
[35] Wang Y N and Ma T C 1995 *Phys. Rev.* B **52** 16395
[36] Jonson M 1976 *J. Phys.* C **9** 3055
[37] Ishigami M, Chen J H, Cullen W G, Fuhrer M S and Williams E D 2007 *Nano Lett.* **7** 1643
[38] Radović I, Hadžievski Lj and Mišković Z L 2008 *Phys. Rev.* B **77** 075428

Liquid-phase reactions induced by atmospheric pressure glow discharge with liquid electrode

Fumiyoshi Tochikubo, Naoki Shirai, Satoshi Uchida

Department of Electrical and Electronic Engineering, Tokyo Metropolitan University, 1-1, Minami-Osawa, Hachioji, Tokyo 192-0397, Japan

E-mail: tochi@tmu.ac.jp

Abstract. We experimentally investigated some of the initial reactions in a liquid induced by electron or positive-ion irradiation from an atmospheric-pressure dc glow discharge in contact with the liquid. We used an H-shaped glass reactor to observe the effects of electron irradiation and positive-ion irradiation on the liquid-phase reaction separately and simultaneously. Aqueous solutions of NaCl, $AgNO_3$, $HAuCl_4$, and $FeCl_2$ are used as the electrolyte. Solutions of $AgNO_3$ and $HAuCl_4$ are used for the generation of Ag and Au nanoparticles, respectively. Solution of $FeCl_2$ is used for the generation of ferromagnetic particles. Experimental results showed that electron irradiation of the liquid surface generates OH^- in water and that positive-ion irradiation of the liquid surface generates H^+ in water even without the dissolution of gas-phase nitrogen oxide. A possible reaction process is qualitatively discussed. We also showed that the control of reductive and oxidative environment in the liquid is possible not only by the gas composition for the plasma generation but also by the liquid composition.

1. Introduction

Nonthermal atmospheric-pressure plasmas in contact with liquids have been widely studied for realizing a variety of plasma applications such as material processing, water treatment including sterilization, analytical chemistry, and so forth[1, 2]. DC glow discharge with a liquid electrode is one of the simplest methods to obtain a stable gas discharge plasma in contact with a liquid. In our previous paper, we reported the characteristics of atmospheric-pressure dc glow discharge with a liquid cathode[3] and liquid anode[4]. A dc glow discharge with a liquid electrode was studied to analyze the composition of the electrolyte by spectroscopic investigation[5, 6]. The surface treatment of polypropylene and polyethylene in a liquid was demonstrated using an atmospheric-pressure air glow discharge with an electrolyte cathode[7]. A dc glow discharge with a liquid electrode has also been applied for the synthesis of metallic nanoparticles in a liquid[8, 9, 10, 11, 12, 13, 14]. A dc glow discharge with a liquid electrode is considered as electrolysis with a plasma electrode. The plasma electrode supplies electrons or positive ions to the liquid surface via a different chemical reaction from that in conventional electrolysis with metal electrodes. Although electrolysis with a plasma electrode is not a new technology[15], the interfacial phenomena between the plasma and the liquid are not yet understood well.

In this work, we experimentally investigated some of the initial reactions in a liquid induced by electron or positive-ion irradiation from an atmospheric-pressure dc glow discharge in contact with a liquid electrode by using Hoffman electrolysis apparatus. A possible reaction process is qualitatively discussed on the basis of the experimental results.

2. Experimental method

Figure 1 shows the experimental setup used in this work. An H-shaped Hoffman electrolysis reactor is used to separate the anodic and cathodic reactions. The reactor is filled with an electrolyte, the solution surface of which is roughly 1 cm below the top mouth of the H-shaped reactor. Aqueous solutions of sodium chloride (NaCl), silver nitrate ($AgNO_3$), chlorauric acid ($HAuCl_4$), and ferrous chloride ($FeCl_2$) are used as the electrolyte. Solutions of $AgNO_3$ and $HAuCl_4$ are used for the generation of Ag and Au nanoparticles, respectively[14]. Solution of $FeCl_2$ is used for the generation of ferromagnetic particles.

A nozzle electrode made of stainless steel with inner and outer diameters of 500 and 800 μm, respectively, is set about 2 mm above the solution surface in each mouth of the H-shaped reactor. Helium gas with a flow rate of 200 sccm is injected from the nozzle electrode. We refer to this electrode configuration as the normal-type electrode [see Fig. 1(b)]. We also use a sheath-flow-type electrode, in which the core helium flow is surrounded by a sheath flow of nitrogen or oxygen [see Fig. 1(c)]. The nozzle electrode for the core helium flow is the same as that of the normal-type electrode. The typical flow rate of the sheath flow is 1000 sccm while that of the helium core flow is kept at 200 sccm. We expected to change the reductive and oxidative environment by changing the sheath-flow gas. By applying a dc voltage between the two nozzle electrodes, two glow discharges as an anode and a cathode for the plasma-assisted electrolysis are formed in contact with the solution along the helium flows. Let us use the terms "plasma anode" and "plasma cathode" in this paper. Positive ions enter the solution at the plasma anode while electrons enter the solution at the plasma cathode. For comparison, conventional electrolysis was also observed with two platinum electrodes immersed in the solution.

We measure the pH value of the solution at approximately 10 mm below the solution surface using a pH meter. To visualize the local changes in the pH value, a small amount of bromothymol blue (BTB) is added as a pH indicator. The color of BTB changes to yellow at a pH value of less than 6.0 while it becomes blue at a pH value of larger than 7.6. A gas detector tube is used to detect Cl_2, HCl, and H_2 above the liquid surface resulting from the liquid-phase reaction. The detection limit of the gas detector tube is 0.1 ppm for Cl_2, 2 ppm for HCl, and 0.05% for H_2. We also measure HClO concentration in the liquid by DPD color comparison method. DPD stands for N,N-diethyl-p-phenylenediamine, a chemical reagent. The DPD color changes into pink when DPD is oxidized by HClO. Nanoparticles generated in liquid are analyzed by transmission electron microscopy (TEM) and an energy dispersive X-ray spectroscopy (EDX).

3. Experimental results

3.1. Comparison between conventional and plasma-assisted electrolysis

Figure 2 shows photographs of conventional and plasma-assisted electrolysis in NaCl solution with a concentration of 0.5 M taken 15 min after starting electrolysis at a constant current of 3 mA. A small amount of BTB was added to the solution. The initial pH of the solution was 7.4. In the conventional electrolysis in Fig. 2(a), the liquid color around the cathode became blue, alkalified, and that around the anode was decolorized. The pH values measured by the pH meter were 10.3 near the Pt cathode and 3.0 near the Pt anode. We detected Cl_2 and HCl from the anode side and H_2 from the cathode side using the gas tube detector although we could not measure their generation yields quantitatively. The reaction of conventional electrolysis in NaCl solution is well known to be

$$\text{anode} : 2Cl^- \rightarrow Cl_2 + 2e, \quad (1)$$

$$\text{cathode} : 2H_2O + 2e \rightarrow H_2 + 2OH^-. \quad (2)$$

Thus, Cl_2 gas is generated at the anode while H_2 gas is generated at the cathode. Hydroxide ion (OH^-) generation causes the alkalification of the liquid around the cathode. The Cl_2 gas

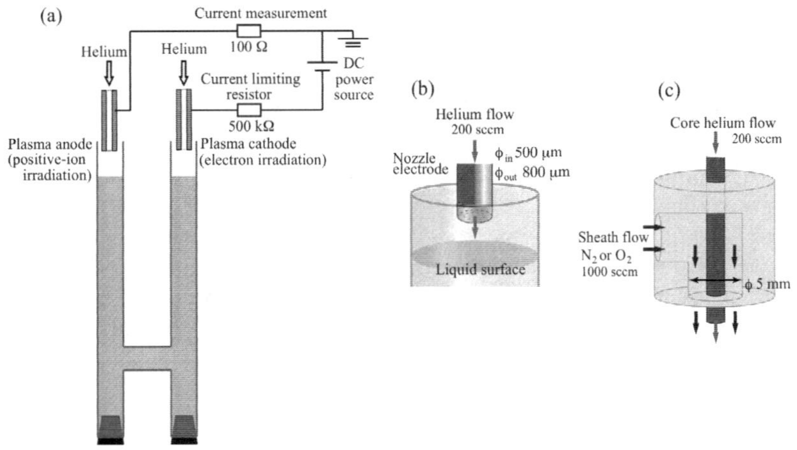

Figure 1. Experimental setup for plasma-assisted electrolysis using H-shaped Hoffman electrolysis apparatus, (a) overall arrangement of apparatus, (b) normal-type electrode, and (c) sheath-flow-type electrode.

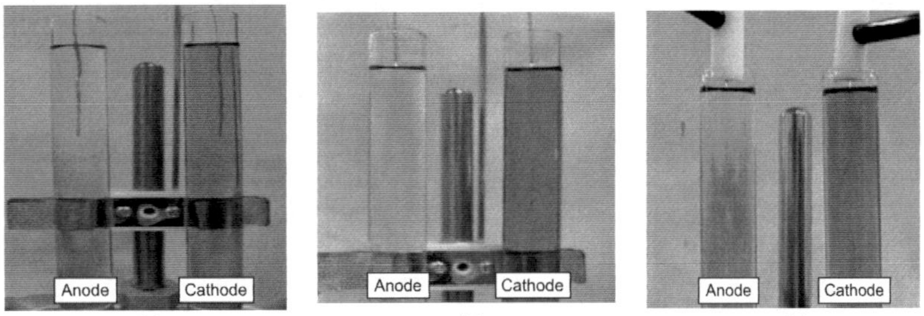

(a) (b) (c)

Figure 2. Photograph of electrolysis in 0.5 M NaCl solution with added BTB with a constant current of 3 mA at 15 min after starting electrolysis, (a) conventional electrolysis, (b) plasma-assisted electrolysis with the normal-type electrode, (c) plasma-assisted electrolysis with the sheath-flow-type electrode with O_2 sheath flow.

partly reacts with water molecule as

$$H_2O + Cl_2 \rightarrow HCl + HClO. \tag{3}$$

Therefore, the liquid around the anode is acidified. Since HClO has a high oxidation potential, it easily decomposes BTB, and the solution around the anode is decolorized. We confirmed the generation of HClO in liquid by DPD color comparison method.

In the plasma-assisted electrolysis with the normal-type electrode in Fig. 2(b), the liquid color below the plasma cathode became blue, and that below the plasma anode became yellow without the decoloration of BTB. The pH values measured by the pH meter were 9.0 below the plasma cathode and 3.0 below the plasma anode. We attempted to measure the generation of H_2, Cl_2, and HCl gases using the gas detector tube. These gases were below the detection limit although they were detected in conventional electrolysis. The concentration of HClO in the liquid was also below the decetction limit of the DPD color comparison method. Therefore, we deduce that

Cl_2 gas generation is less active in plasma electrolysis than that in conventional electrolysis. It is reasonable to consider that the acidification of the solution below the plasma anode results from the dissolution of gaseous nitrogen oxides generated in the plasma. To evaluate this effect, same experiment was carried out using the sheath-flow-type electrode with O_2 sheath flow, which avoid contact between the discharge plasma and the surrounding air to reduce the generation of nitrogen oxides. The photograph of the plasma-assisted electrolysis with the sheath-flow-type electrode is shown in Fig. 2(c). The pH values measured by the pH meter were 9.3 below the plasma cathode and 3.4 below the plasma anode. These results imply that electron irradiation of the liquid surface generates OH^- in water and that positive-ion irradiation of the liquid surface generates H^+ in water even without nitrogen oxide generation, although we did not measure the nitrate ion concentration in the liquid. A detailed discussion of the reaction process is given in section 4.

3.2. Plasma-assisted electrolysis using aqueous solutions of $AgNO_3$ and $HAuCl_4$

Plasma-assisted electrolysis was carried out using aqueous solutions of $AgNO_3$ for Ag nanoparticles generation and $HAuCl_4$ for Au nanopatricles generation. The concentrations of $AgNO_3$ and $HAuCl_4$ were typically on the order of 10^{-4} M. Sodium dodecyl sulfate (SDS) with a concentration of 0.05% (1.7 mM) was added to the solution as a surface active agent. Specimen photographs of the plasma-assisted electrolysis are shown in Fig. 3 with TEM images of nanoparticles generated in the liquid. In the plasma-assisted electrolysis with $AgNO_3$ solution, the solution color below the plasma cathode becomes yellow, the color of which results from the surface plasmon resonance of Ag nanoparticles. That is, Ag nanoparticles were only synthesized on the plasma cathode side via the reduction of Ag^+. In the case of using $HAuCl_4$ solution, the solution colors below both the plasma anode and the plasma cathode become red as a result of surface plasmon resonance of Au nanoparticles. That is, Au nanoparticles were synthesized by both positive-ion irradiation and electron irradiation of the solution surface. Positive-ion irradiation with the plasma anode is more effective than electron irradiation for the synthesis of Au nanoparticles. The detail of the nanoparticle generation is shown in our previous paper[14].

3.3. Plasma-assisted electrolysis using aqueous solutions of $FeCl_2$

$FeCl_2$ is ionized into Fe^{2+} and Cl^- in the liquid. When Fe^{2+} is reduced by reductant such as hydrated electrons, ferromagnetic Fe particles can be generated[13]. When oxidative reaction is dominant, Fe^{2+} is oxidized to Fe^{3+}, which shows yellow color in the liquid. If Fe^{2+} and Fe^{3+} are simultaneously oxidized in the liquid with reasonable molar ratio, ferromagnetic particles of magnetige, Fe_3O_4, will be generated, otherwise paramagnetic particles of hematite, Fe_2O_3, will be mainly generated. Figure 4(a) shows the photograph of the plasma-assisted electrolysis in aqueous solution of 0.01 M $FeCl_2$ with 1.7 mM SDS using normal-type electrode for both plasma anode and plasma cathode. Brown red particles, which are not attracted by a magnet, are generated below both plasma anode and plasma cathode. The particle generation is more active by the positive-ion irradiation below plasma anode. We confirmed that the generated particles consist of Fe and O by the EDX analysis. Therefore, the generated particles are deduced to be hematite. This result suggests that Fe^{2+} is considerably oxidized to Fe^{3+} even in the reductive environment by electron irradiation of the liquid surface below plasma cathode. To achieve further reductive environment at the liquid surface, the sheath-flow-type electrode with N_2 sheath flow is used as the plasma cathode, and its result is shown in Fig. 4(b). Black particles, which are easily attracted by a magnet, are generated in addition to the brown red particles below plasma cathode. The magnetic property of the paricles generated below plasma cathode was analyzed by a superconducting quantum interference device (SQUID), and is shown in Fig. 5. The magnetization curve has hysteresis characteristics with saturated magnetization value of 23 emu/g. The saturated magnetization of magnetite is 80 emu/g. Therefore, roughly

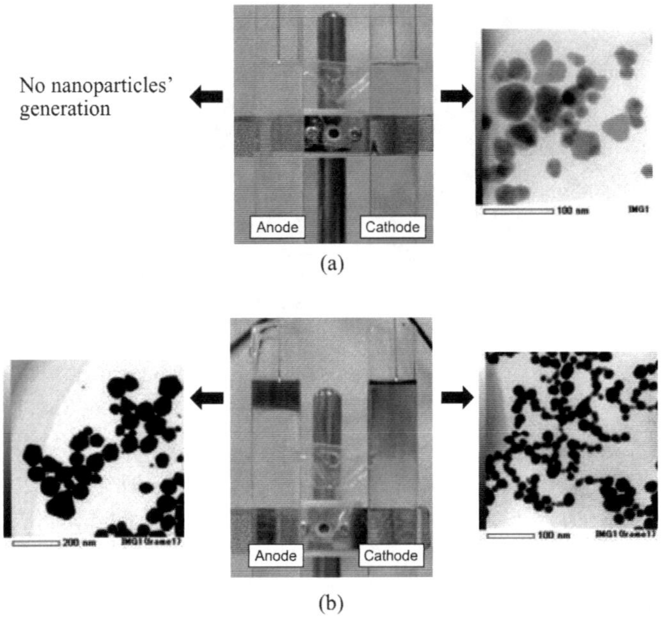

Figure 3. Photographs of plasma-assisted electrolysis taken 5 min after starting electrolysis at a constant current of 3 mA. TEM images of the generated nanoparticles are also shown. (a) 0.1 mM $AgNO_3$ solution with 1.7 mM SDS, and (b) 0.24 mM $HAuCl_4$ with 1.7 mM SDS.

a quarter of the generated particle is considered to be magnetite. It will be possible to increase the ratio of magnetite by optimizing the experimental condition such as $FeCl_2$ concentration and current density.

Surface active agent such as SDS is added in the solution to prevent the agglomeration of particles. Therefore, larger particles are often generated without adding the surface active agent. However, magnetite particles were not generated but hematite particles without adding the surface active agent. This result suggests that the main role of SDS in the present experiment is not to prevent the agglomeration of particles but to suppress the excess oxidation of Fe^{2+} to Fe^{3+}. To verify this hypothesis, ethanol is added in the solution of $FeCl_2$ instead of SDS. The amount of the added ethanol is the same as that of SDS. Ethanol is known to trap OH radical as

$$OH + C_2H_5OH \rightarrow H_2O + CH_3CHOH, \quad k = 2.2 \times 10^9\ M^{-1}s^{-1}[16]. \qquad (4)$$

Magnetite particles were generated below plasma cathode by the plasma-assisted electrolysis using $FeCl_2$ solution with adding ethanol. It is confirmed that the role of SDS and ethanol is to suppress the excess oxidation of Fe^{2+} in liquid. That is, the control of reduction and oxidation in the liquid is possible not only by the gas composition for plasma generation but also by the liquid composition in the plasma-assisted electrolysis.

4. Discussion

In this work, we investigated the liquid-phase reactions induced by electron and positive-ion irradiation separately from atmospheric-pressure glow discharges. The results are summarized in Table 1. In this section, we discuss the reactions in a liquid induced by electron irradiation and positive-ion irradiation.

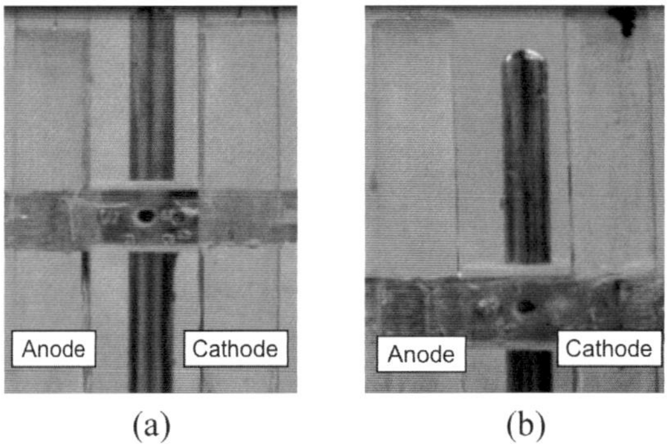

Figure 4. Photographs of plasma-assisted electrolysis using 0.01 M $FeCl_2$ solution with 1.7 mM SDS taken 5 min after starting electrolysis at a constant current of 2 mA. (a) normal-type electrode is used for both plasma anode and plasma cathode, and (b) sheath-flow-type electrode with N_2 sheath flow is used for plasma cathode while normal-type electrode is used for plasma anode.

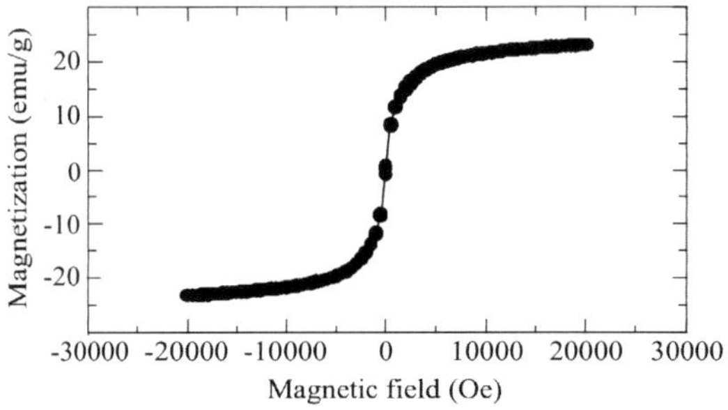

Figure 5. Magnetic property of the particles generated in $FeCl_2$ solution below plasma cathode using sheath-flow-type electrode with N_2 sheath flow.

4.1. Reactions in liquid induced by electron irradiation

In the present case, electrons in the positive column of the atmospheric-pressure glow discharges are exposed to the liquid surface. Thus, the electron impact dissociation or ionization of water molecules at the liquid surface is not very effective judging from electron impact cross-section data[17]. The electrons exposed to the solution surface will be stabilized as hydrated electrons. The important reactions involving hydrated electrons are listed in Table 2. The reaction path of the hydrated electrons depends on the concentration of the reaction partners. In reactions (R1) and (R2) in Table 2, hydrated electrons react with water molecules to form OH^-. Reaction (R3) is also one of the dominant reactions, the frequency of which depends on the pH of the solution. At a solution pH of 3.0, the reaction time constant of hydrated electrons for (R3) is 42 ns, and even at a neutral pH of 7.0, it is 0.42 ms. Since reaction (R3) neutralizes H^+ to form H atoms in the solution, the OH^- concentration is increased by the ionization equilibrium of water molecules. Therefore, it is reasonable to consider that electron irradiation of the water surface alkalifies the solution.

In $AgNO_3$ solution, reaction (R4) is considered to be the dominant reaction causing the trapping of hydrated electrons. A high concentration of neutral Ag in the liquid leads the

Table 1. Summary of experimental results. NP denotes a nanoparticle.

Solution (typical concentration)	Conventional electrolysis		Plasma-assisted electrolysis	
	Anode	Cathode	Plasma anode positive-ion irradiation	Plasma cathode electron irradiation
NaCl (0.5 M)	Acidified Cl_2, HCl gas decoloration	Alkalified H_2 gas	Acidified H^+ generation	Alkalified OH^- generation
$AgNO_3$ (0.1 mM)	No visible change	No visible change	Ag-NPs not observed	Ag-NP generation
$HAuCl_4$ (0.24 mM)	No visible change	No visible change	Au-NP generation (more effective)	Au-NP generation
$FeCl_2$ (0.01 M)	Cl_2 gas	Fe deposition	Fe_2O_3 generation	Fe_2O_3 generation, With N_2 sheath flow Fe_3O_4 generation

Table 2. Important reactions induced by electron irradiation. e_{aq} denotes a hydrated electron.

No.	Reaction	Rate coefficient	Ref.
(R1)	$e_{aq} + H_2O \rightarrow OH^- + H$	10^3 s^{-1}	[18]
(R2)	$e_{aq} + e_{aq} \rightarrow 2OH^- + H_2$	6.0×10^9 M^{-1}s^{-1}	[19]
(R3)	$e_{aq} + H^+ \rightarrow H$	2.4×10^{10} M^{-1}s^{-1}	[20]
(R4)	$e_{aq} + Ag^+ \rightarrow Ag$	4.8×10^{10} M^{-1}s^{-1}	[21]
(R5)	$H + Ag^+ \rightarrow Ag + H^+$	2.0×10^{10} M^{-1}s^{-1}	[22]
(R6)	$H + AuCl_4^- \rightarrow AuCl_4^{2-} + H^+$	5.7×10^9 M^{-1}s^{-1}	[23]
(R7)	$e_{aq} + Fe^{2+} \rightarrow$ products	1.6×10^8 M^{-1}s^{-1}	[24]
(R8)	$e_{aq} + Fe^{3+} \rightarrow Fe^{2+}$	6.0×10^{10} M^{-1}s^{-1}	[25]

nuclear formation of Ag nanoparticles. H atoms in the liquid also reduce Ag^+ via reaction (R5).

Since $HAuCl_4$ is a strong acid, the initial pH of the solution under the present experimental conditions was approximately 3.0. Thus, the dominant reaction path for hydrated electrons will be reaction (R3). H atoms in a liquid will reduce $AuCl_4^-$ (Au^{3+}) via reaction (R6). Although the elementary process for the reduction of Au ions in $HAuCl_4$ solution has not been clearly shown to the best of our knowledge, it is reasonable to consider that H atoms or hydrated electrons contribute to the successive reduction from Au^{3+} to Au by electron irradiation of the solution surface by the plasma.

Hydrated electrons have high reaction rate coefficient with Fe^{2+} and Fe^{3+} as shown in Table 2. However, the experimental results suggested that the oxidation of Fe^{2+} was more active than its reduction. It is difficult to identify what species is responsible for the oxidation of Fe^{2+}. One possible answer is the oxidation of Fe^{2+} at plasma-liquid interface by OH radicals generated in the plasma. Adding SDS or ethanol in the liquid suppress the excess oxidation of Fe^{2+} for generating magnetite particles.

4.2. Reactions induced by positive-ion irradiation

Our experimental results strongly suggested that positive-ion irradiation from the plasma to the liquid surface acidifies the solution even without the dissolution of nitrogen oxides from the gas phase. That is, positive-ion irradiation directly or indirectly results in H^+ generation. One hypothesis is the ionization of water molecules by energetic ions at the liquid surface as

$$\oplus + H_2O \rightarrow H_2O^* \rightarrow e_{aq} + H_2O^+, \quad (5)$$

$$H_2O^+ + H_2O \rightarrow H_3O^+ + OH. \quad (6)$$

However, the kinetic energy of positive ions impinging on the cathode is estimated to be of eV order in atmospheric-pressure glow discharge in helium because of the frequent energy loss due to collisions[26]. Although the ionization energy of a water cluster is lower than that of an isolated water molecule, it is still difficult for impinging low-energy positive ions to initiate reaction (5). If the same amounts of hydrated electrons and H^+ are generated via reactions (5) and (6), the solution pH does not change. In addition, Ag nanoparticles might be obtained by positive-ion irradiation to the surface of $AgNO_3$ solution if reaction (5) is dominant. Therefore, an other explanation is necessary for the generation of H^+. A strong candidate for this reaction is charge transfer collision between an impinging ion X^+ and H_2O.

$$X^+ + H_2O \rightarrow X + H_2O^+ \quad (7)$$

Gas-phase charge transfer collisions with water molecules have been experimentally investigated for Ar^+ and N_2^+[27], for example. These ions have large collision cross sections for reaction (7) at low energies; the cross section is roughly 10^{-14} cm^{-2} at 0.1 eV and decreases with increasing ion energy. Similar reaction is expected between He^+ and H_2O, and it is reasonable to consider that the irradiation of low-energy positive ions from the plasma anode to the liquid surface effectively generates H^+ and OH via charge transfer collision.

Since positive-ion irradiation of the liquid surface essentially enhances the oxidation in the liquid, metallic particle generation by the reduction of metallic ion is difficult. However, Au nanoparticles are generated by positive-ion irradiation more effectively than by electron irradiation. We do not have clear answer for this problem at this moment.

5. Conclusion

We experimentally investigated the initial liquid-phase reaction induced by electron or positive-ion irradiation from an atmospheric-pressure dc glow discharge. We used an H-shaped glass reactor as an electrolysis apparatus with two atmospheric-pressure glow discharges as electrodes for plasma-assisted electrolysis. This apparatus enabled us to observe the effects of electron irradiation and positive-ion irradiation on the liquid-phase reaction separately and simultaneously. Aqueous solutions of NaCl, $AgNO_3$, $HAuCl_4$, and $FeCl_2$ are used as the electrolyte. Possible reaction processes induced in the liquid were qualitatively discussed. Following results are obtained.

(i) At the plasma cathode, where electrons are exposed to the solution surface, the pH of the solution increased roughly from the initial value of 7 to 10 in NaCl solution. This result implies that electron irradiation of the liquid surface generates OH^- in water via neutralization of H^+ to form H atom or disocciative electron attachment of H_2O. Ag nanoparticles were synthesized with $AgNO_3$ solution and Au nanoparticles were synthesized with $HAuCl_4$ solution via the reduction of metallic ion by hydrated electron or H atom.

(ii) At the plasma anode, where positive ions are exposed to the solution surface, the pH of the solution decreased roughy from 7 to 3 in NaCl solution even under the condition of preventing the dissolution of nitrogen oxides in the liquid. The generation of Cl_2 gas by the plasma anode is negligible compared with that by the metal electrode in the conventional electrolysis. This result implies that positive-ion irradiation of the liquid surface generates H^+ in water. The contribution of charge transfer collision between low-energy incident positive ion and water molecule at liquid surface is introduced to explain the overall tendency of the experimental results.

(iii) Although electron irradiation of the liquid surface essentially enhances the reductive reaction in the liquid, Fe^{2+} was not reasonably reduced but even oxidized in $FeCl_2$ solution, and hematite particles were generated. Ferromagnetic magnetite particles were synthesized in further reductive environment by adding chemical agent into the liquid to trap oxidizing agent and by changing the gas composition for plasma generation. In other words, the control of reductive and oxidative environment in the liquid is possible not only by the gas composition for plasma generation but also by the liquid composition in the plasma-assisted electrolysis.

Acknowledgments
This work was supported financially in part by Kakenhi (No 21110007), Japan. Authors thank Prof. Osuke Miura at Tokyo Metropolitan University for his useful suggestion on the property of the genrated magnetic particles. Authors also thank Mr. Y. Shimokawa, Mr. T. Ishida, and Mr. A. Nito in our laboratory for their help in performing a part of the experimental work.

References

[1] Locke B R, Sato M, Sunka P, Hoffman M R and Chang J S 2006 *Ind. Eng. Chem. Res.* **45** 882
[2] Bruggeman P and Leys C 2009 *J. Phys.* D **42** 053001
[3] Shirai N, Ichinose K, Uchida S and Tochikubo F 2011 *Plasma Sources Sci. Technol.* **20** 034013
[4] Shirai N, Uchida S, Tochikubo F and Ishii S 2011 *IEEE Trans. Plasma Sci.* **39** 2652.
[5] Cserfalvi T, Mezei P and Apai P 1993 *J. Phys.* D **26** 2184
[6] Cserfalvi T and Mezei P 1996 *Fresenius J. Anal. Chem.* **355** 813
[7] Titov V A, Rybkin V V, Shikova T G, Ageeva T A, Golubchikov O A and Choi H S 2005 *Surf. Coat. Technol.* **199** 231
[8] Richmonds C and Sankaran R M 2008 *Appl. Phys. Lett.* **93** 131501
[9] Kaneko T, Baba K, Harada T and Hatakeyama R 2009 *Plasma Process. Polym.* **6** 713
[10] Mariotti D and Sankaran R M 2010 *J. Phys.* D **43** 323001
[11] Chang F -C, Richmonds C and Sankaran R M 2010 *J. Vac. Sci. Technol.* A **28** L5
[12] Chiang W H, Richmonds C and Sankaran R M 2010 *Plasma Sources Sci. Technol.* **19** 034011
[13] Tokushige M, Yamanaka T, Matsuura A, Nishikiori T and Ito Y 2009 *IEEE Trans. Plasma Sci.* **37** 1156
[14] Shirai N, Uchida S and Tochikubo F 2014 *Jpn. J. Appl. Phys.* **53** 046202
[15] Davies R A and Hickling A 1952 *J. Chem. Soc.* 3595
[16] Motohashi N and Saito Y 1993 *Chem. Pharm. Bull.* **41** 1842
[17] Itikawa Y and Mason N 2005 *J. Phys. Chem. Ref. Data* **34** 1
[18] Schwarz H A 1992 *J. Phys. Chem.* **96** 8937
[19] Schmidt K H and Bartels D M 1995 *Chem. Phys.* **190** 145
[20] Shiraishi H, Sunaryo G R and Ishigure K 1994 *J. Phys. Chem.* **98** 5164
[21] Ershov B G, Janata E, Henglein A and Fojtik A 1993 *J. Phys. Chem.* **97** 4589
[22] Buxton G V, Greenstock C L, Helman W P and Ross A B 1988 *J. Phys. Chem. Ref. Data* **17** 513
[23] Ghosh-Mazumdar A S and Hart E J 1968 *Adv. Chem. Ser.* **81** 193
[24] Baxendale J H, Fielden E M and Keene J P 1965 *Proc. Roy. Soc. Ser.* A **286** 320
[25] Jonah C D, Miller J R and Matheson M S 1977 *J. Phys. Chem.* **81** 1618
[26] Choi J, Iza F, Lee J K and Ryu C 2007 *IEEE Trans. Plasma Sci.* **35** 1274
[27] Dressler R A, Salter R H and Murad E 1993 *J. Chem. Phys.* **99** 1159

3D Mapping of plasma effective areas via detection of cancer cell damage induced by atmospheric pressure plasma jets

Xu Han[1,2], Yueing Liu[3], M. Sharon Stack[3,4] and Sylwia Ptasinska[1,2,3]*

[1]Radiation Laboratory, [2]Department of Physics, [3]Harper Cancer Research Institute, [4]Department of Chemistry and Biochemistry, University of Notre Dame, Notre Dame, IN, 46556, USA

E-mail: Sylwia.Ptasinska.1@nd.edu

Abstract. In the present study, a nitrogen atmospheric pressure plasma jet (APPJ) was used for irradiation of oral cancer cells. Since cancer cells are very susceptible to plasma treatment, they can be used as a tool for detection of APPJ-effective areas, which extended much further than the visible part of the APPJ. An immunofluorescence assay was used for DNA damage identification, visualization and quantification. Thus, the effective damage area and damage level were determined and plotted as 3D images.

1. Introduction

Atmospheric pressure plasma jets (APPJs) offer many opportunities for applications at the intersection of plasma physics, radiation chemistry and medicine [1]. Currently, much scientific effort is focused on ascertaining the effects of atmospheric pressure plasmas interacting with cancer cells (oral [2], prostate [3], skin [4], brain [5], pancreatic [6,7], melanoma [8], human breast [9], lung [10], and liver [11]), understanding the fundamental mechanisms that lead to cell death, and finding the optimal conditions for cancer treatment. These studies aim to explore the pros and cons of plasma radiation as an alternative to antitumor therapy. Our long-term goal is to elucidate whether the APPJ, which produces a cocktail of plasma species upon reactions with air and liquid, including charged particles (ions and electrons), photons, radicals and neutral species (excited atoms and molecules) can be controlled and targeted to selectively destroy cancer cells while causing little damage to the healthy tissue.

Our previous studies on damage induced to isolated DNA extracted from *E. coli* bacteria indicated that DNA is very sensitive to short-term APPJ exposure [12]. Extremely rapid degradation of DNA was observed, yielding 60% damage within the first 10 s of APPJ treatment [12]. We also observed that the influence of the medium (e.g., water [13,14] or amino acid solution [13]) has some effect on the level of DNA damage, but high percentages of DNA damage were still reached under these experimental conditions. Our studies on isolated DNA showed mainly the formation of single strand breaks, which can easily be repaired by the cell, as opposed to double strand breaks within the DNA molecule. Therefore, the question arose as to whether APPJ can induce damage to DNA in the cell, which is a much more complex biological system than the above mentioned media. The answer to this question would provide more insight into the possibility of using APPJ as a tool for cancer treatment.

* To whom any correspondence should be addressed.

Therefore, we recently performed a study to detect damage to DNA in cancer cells [2]. We observed that even short irradiation times by APPJ (< 30 s) caused destruction of DNA in 60% of cells [2]. This indicates a relatively fast cascade of biochemical reactions leading to multiple strand breaks in DNA initiated by the reactive species from the APPJ. For longer irradiation times, the damage level increases, reaching nearly 80% after only 2 min of APPJ treatment. At such prolonged irradiation times, the effective area of the APPJ is much larger than the area in which the APPJ is in direct contact with a sample [2]. This finding can have significant consequences for applications of APPJs, where the plasma has to be used for the treatment of localized areas (e.g., tumors) of a specific size or shape.

In this work, we discuss effective areas measured for our APPJ source, compare them with other effective areas obtained from various APPJ sources, and suggest a way to plot these areas as a 3D map.

2. Experimental set-up

Recently, we successfully constructed and developed various APPJ sources, which we used for the treatment of different materials such as semiconductors (e.g., Si) [15] and bio-macromolecules (e.g., DNA) [14]. The APPJ source used in this study operates based on a dielectric barrier discharge and consists of a dielectric tube (inner diameter of 2 mm, outer diameter of 3 mm) with two thin copper electrodes alternately wrapped around it in a spiral configuration [2,16], with one electrode connected to a high voltage (HV) power supply and the other to the ground (Fig. 1). Nitrogen gas (99.99% from Airgas, USA) was flowed through the dielectric tube at a flow rate of 1.5 slm. The gas discharge was ignited inside the tube by a relatively low-frequency (28 kHz) AC sinusoidal input with a peak-to-peak voltage of 22.4 kV and peak-to-peak current of 59 mA measured by voltage (Tektronix TCP A300) and current (Tektronix P6015A) probes, respectively. Once a plasma jet was launched outside the source, it formed an almost cylindrical jet.

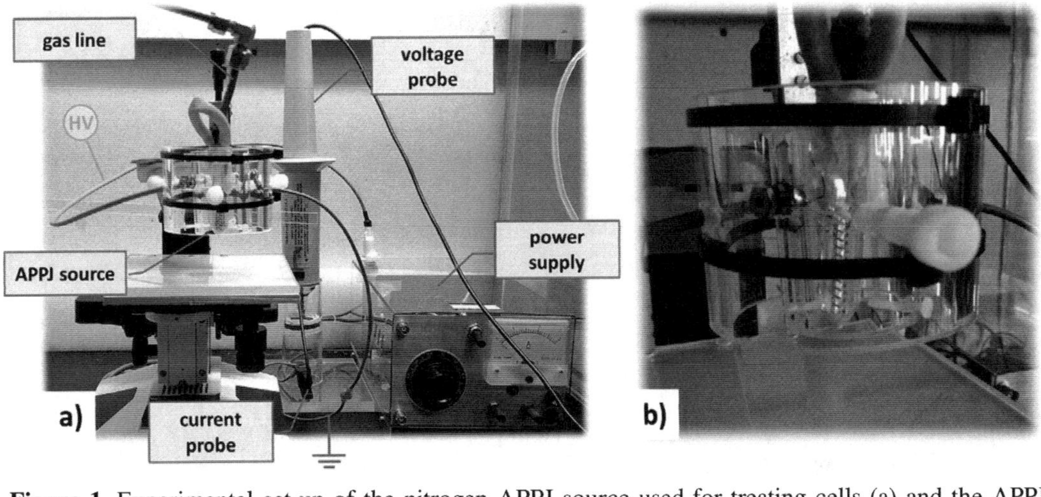

Figure 1. Experimental set-up of the nitrogen APPJ source used for treating cells (a) and the APPJ source in an operational mode (b).

Oral cancer cells (SCC-25) grown on photoetched coverslips were irradiated by the nitrogen APPJ to determine its effectiveness in inducing DNA damage [2]. Damaged DNA molecules served as surrogate markers for high-level damage in cells. To visualize damaged DNA in the nuclei, both treated and control cells were probed with H2AX antibody, followed by a fluorescently tagged secondary antibody (green fluorescence protein, GFP), and then imaged by fluorescence microscopy. Phosphorylated H2AX is a common marker used for DNA damage resulting from direct radiation and/or correlated to apoptosis. Therefore, in the present study, we are unable to determine the type of

DNA and cell alteration induced by APPJ treatment. By comparing the number of cells with GFP to the total number of nuclei stained by DAPI (4', 6-diamidino-2-phenylindole), the ratio of cells with DNA destroyed on the coverslip was obtained. This ratio of cells with DNA strand breaks over the total number of cells hereafter is called the damage ratio.

3. Results and discussion

Our previous studies showed that isolated DNA molecules that have been extracted from *E. coli* can be damaged by plasma radiation [12–14,17]. Damage levels in DNA were evaluated in terms of various distances from the source orifice and plasma exposure times. In the present study, we investigated damage to cellular DNA. It is worth noting that treatment with the plasma can affect cells on multiple levels, causing damage to cell walls (due to etching), the membrane (local disruption and lipid peroxidation), DNA (oxidative DNA damage, base modification, thymine dimer formations, and strand breaks), RNA (strand breaks) and proteins (oxidative protein damage, unfolding, and amino acid modifications) [18].

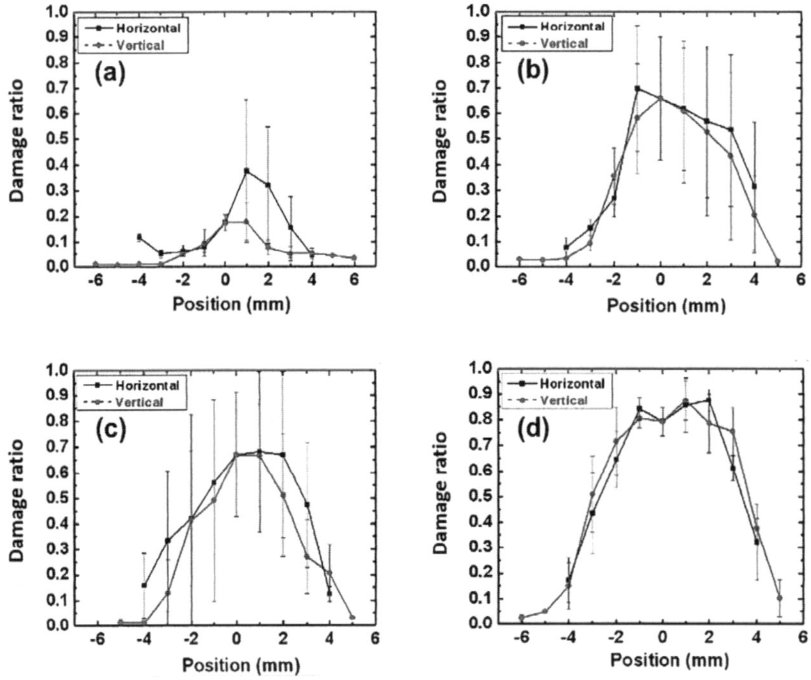

Figure 2. The damage ratio of cancer cells at different positions on the coverslip, which was placed at a distance of 2 cm from the source orifice, for various plasma treatment exposures: (a) 10 s, (b) 30 s, (c) 1 min and (d) 2 min. An error bar for each point was estimated as an average of two trials. Adapted from [2] with the permission of AIP Publishing LLC.

Since the diameter of our plasma jet was relatively small (~2 mm), we were interested in exploring how large an area can be affected by plasma irradiation. The obtained finding could provide significant information about the precision of plasma treatment when only a tumor area has to be targeted, while areas with healthy cells are intact. In this case, the ability to visualize damaged cells and distinguish them from intact cells is the key reason for choosing the technique of immunofluorescence microscopy for analysis of treated areas after plasma irradiation. In order to explore the effective area of plasma irradiation, the damage ratio was plotted in terms of spatial parameters for four different plasma treatment durations. Figure 2 shows damage ratios at different positions on the coverslip detected in two perpendicular directions (labelled as "vertical" and "horizontal") from the point of direct APPJ contact with the sample. The damaged cell ratio was observed to increase with plasma irradiation time, reaching 40% after 10 s and 90% after 2 min. Moreover, the increase of plasma treatment time contributed to an enlargement in spatial peak widths, indicating larger effective areas for the APPJ. After a treatment of 10 s, the width of the plotted peaks corresponded to that of the diameter of the visible part of APPJ; however, after 2 min treatment, the width was five times larger than the diameter of the APPJ.

In order to obtain detailed information on the effective area of the APPJ, a 3D map was plotted for two coverslips with cancer cells treated by APPJ radiation over 10 s and 2 min (Fig. 3). The data in Figure 3 represent one set of multiple experimental trials performed under the same conditions as those in Figure 2. The damage level increased significantly with APPJ exposure time as is shown also in Figure 2, as did the area in which damaged cells were detected. For these two particular trials, the damage ratios in Figures 3a and 3b were estimated to be 20% and 66%, respectively. The estimated effective area from Figure 3b was ~ 50 mm^2, while the area of direct contact of the visible part of the APPJ with the sample was ~5 mm^2. Since the diameter of the plasma jet is very small, the damaging effects acting on locations distant from the center are not likely to be due to direct interactions of the APPJ with the cells. Rather, these damaging effects are likely due to secondary interactions as a result of the diffusion of reactive radical species and electrons generated in the APPJ, followed by complex chemical reactions activating cells to cause DNA strand breaks. This assumption is also supported by the work of the Laroussi group [19] and other groups discussed below, who found a similar effect of plasma acting on bacteria; in their studies, a larger inactivation area of bacteria was observed with a longer plasma treatment.

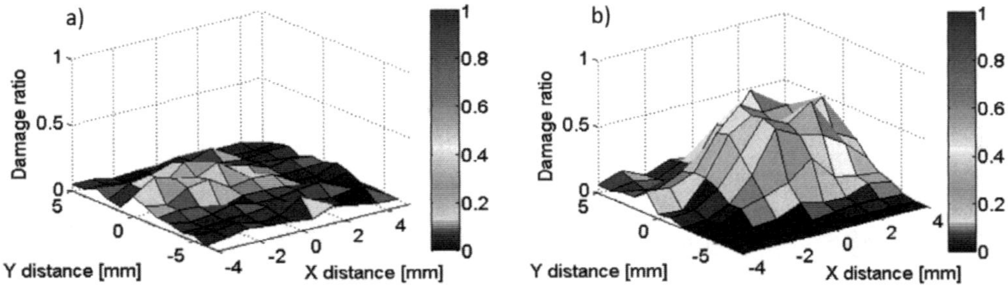

Figure 3. Spatial distributions of damage ratio of cancer cells placed at a distance of 2 cm from the source orifice and irradiated by the APPJ for 30 s (a) and 120 s (b).

Additionally, in our recent study, we investigated the extent of the APPJ effective area by exposing samples of dry plasmid DNA to APPJ [17]. The radial and axial lengths of the visible part of the APPJ were 4 mm and 55 mm, respectively; however, DNA damage was detected at distances of 20 mm radially and 250 mm axially from the plasma source orifice. We found that the volume of the physical extent of the APPJ is about 50 times greater than that of the visible part. While the chemical composition of the visible part of the APPJ was identified by an optical emission spectroscopy, the detection of reactive species beyond this part could not be measured due to their significantly lower emission signals. Nevertheless, we proposed that the DNA damage at long distances is caused by reactive oxygen species [17].

As mentioned above, other groups have also observed enlarged effective areas of APPJs in comparison with the visible jet diameter. Most such investigations were performed with an APPJ used for bacterial inactivation [19–22]. In a study conducted by Deng et al. [20], the diameter of the dielectric tube was about 6 mm; thus, the area of plasma treatment was estimated to be < 1 cm^2. However, the effective area of inactivation was much larger, particularly for longer APPJ exposure (~5 min). The diameters of the inhibition region were 1 and 3.5 cm after 0.5 and 5 min treatments, respectively. These authors suggested that the diffusion of reactive plasma species and also the presence of UV photons were the main factors for an enlarged sterilized zone.

Another group reported studies of atmospheric pressure plasma effects on different types of bacteria (*B. cereus*, *S. aureus*, *E. coli* and *P. aeruginosa*), which suggest possible applications of plasmas for the decontamination of surfaces [21]. All the tested bacteria treated by plasma exhibited the presence of clear zones, where growth was inhibited. The measured diameters of these zones increased with treatment times between 30 s and 2 min, and in all cases were extended beyond the visible plasma part that was < 5 mm in diameter at the point of direct contact with the sample. Additionally, the effective area varied for different types of treated bacteria with *P. aeruginosa* being the least susceptible to inactivation by APPJ exposure [21].

Colagar et al. [22] also concluded that an increase in APPJ exposure time led to more efficient and larger areas of *E. coli* inactivation. They reported that the area of the inactivation zone was increased 1.3% at 1.5 min compared with 1 min of APPJ exposure, while the area of the inactivation zone after 5.5 min was approximately three times greater than the area of the inactivation zone after 1 min irradiation.

In contrast to our studies with cancer cells [2] and studies performed by other groups on bacterial cells [19–22], in which the effective plasma areas expanded with increasing exposure time, a study on cell permeabilization using an APPJ showed that the effective areas are limited to the area at which the plasma comes in direct contact with the sample [23]; however, only short exposure treatments (1-30 s) were used for this investigation [23].

Other techniques have also been used to investigate the effective size of APPJs, mainly water contact angle measurements before and after surface treatment [24]. The contact angle on a clean glass surface before the plasma treatment was between 40-50°. After exposure to the APPJ, the treated surface became more hydrophilic with the contact angle < 10°, and the width of the treated area was ~25 mm. In order to eliminate ambient air diffusion and entrainment, an additional dielectric tube was placed around the APPJ. In this case, the treated area showed a similar change in hydrophilicity, but the effective area width was ~60 mm. This observation clearly indicates the possibility of controlling the effective area of plasma treatment [24].

4. Conclusions

In the present study, cancer cells grown on coverslips and then irradiated by an APPJ have been used to detect cells with damaged DNA and to measure the spatial distribution of plasma effective areas. The trends of all measured distributions suggested a lower number of cells with damaged DNA located farther from the irradiation center, where the APPJ was in direct contact with the sample. A comparison of the peak widths indicated a direct correlation between plasma effective area and treatment time. Mapping of damaged cells can be used as a method for determining the extent of any kind of plasma jet; however, the mechanism underlying the observed spread in damage needs to be further investigated.

Moreover, these enhancements in our knowledge of the effective plasma area can contribute to the further development and improvement of plasma therapeutic techniques.

Acknowledgments

This material is based upon work supported by the U.S. Department of Energy Office of Science, Office of Basic Energy Sciences under Award Number DE-FC02-04ER15533. This is contribution number NDRL 5035 from the Notre Dame Radiation Laboratory.

References

[1] Von Woedtke T, Reuter S, Masur K and Weltmann K-D 2013 *Phys. Rep.* **530** 291–320

[2] Han X, Klas M, Liu Y, Stack M S and Ptasinska S 2013 *Appl. Phys. Lett.* **102** 233703
[3] Hirst A M, Frame F M, Maitland N J and O'Connell D 2014 *Biomed Res. Int.* **2014** 878319
[4] Kim P Y, Kim Y-S, Koo I G, Jung J C, Kim G J, Choi M Y, Yu Z and Collins G J 2011 *PLoS One* **6** e24104
[5] Tanaka H, Mizuno M, Ishikawa K, Nakamura K, Kajiyama H, Kano H, Kikkawa F and Hori M 2011 *Plasma Med.* **1** 265–77
[6] Partecke L I, Evert K, Haugk J, Doering F, Normann L, Diedrich S, Weiss F-U, Evert M, Huebner N O, Guenther C, Heidecke C D, Kramer A, Bussiahn R, Weltmann K-D, Pati O, Bender C and von Bernstorff W 2012 *BMC Cancer* **12** 473
[7] Brullé L, Vandamme M, Riès D, Martel E, Robert E, Lerondel S, Trichet V, Richard S, Pouvesle J-M and Le Pape A 2012 *PLoS One* **7** e52653
[8] Sensenig R, Kalghatgi S, Cerchar E, Fridman G, Shereshevsky A, Torabi B, Arjunan K P, Podolsky E, Fridman A, Friedman G, Azizkhan-Clifford J and Brooks A D 2011 *Ann. Biomed. Eng.* **39** 674–87
[9] Kim S J, Chung T H, Bae S H, Leem S H 2010 *Appl. Phys. Lett.* **97** 023702
[10] Kim J Y, Ballato J, Foy P, Hawkins T, Wei Y, Li J and Kim S-O 2011 *Biosens. Bioelectron.* **28** 333–8
[11] Zhang X, Li M, Zhou R, Feng K and Yang S 2008 *Appl. Phys. Lett.* **93** 021502
[12] Ptasińska S, Bahnev B, Stypczyńska A, Bowden M, Mason N J and Braithwaite N S J 2010 *Phys. Chem. Chem. Phys.* **12** 7779–81
[13] Stypczyńska A, Ptasińska S, Bahnev B, Bowden M, Braithwaite N S J and Mason N J 2010 *Chem. Phys. Lett.* **500** 313–7
[14] Han X, Cantrell W A, Escobar E E and Ptasinska S 2014 *Eur. Phys. J. D* **68** 46
[15] Zhang X and Ptasinska S 2014 *J. Phys. D. Appl. Phys.* **47** 145202
[16] Klas M and Ptasinska S 2013 *Plasma Sources Sci. Technol.* **22** 025013
[17] Bahnev B, Bowden M D, Stypczyńska A, Ptasińska S, Mason N J and Braithwaite N S J 2014 *Eur. Phys. J. D* **68** 140
[18] Lackmann J-W and Bandow J E 2014 *Appl. Microbiol. Biotechnol.* **98** 6205–13
[19] Laroussi M, Tendero C, Lu X, Alla S and Hynes W L 2006 *Plasma Process. Polym.* **3** 470–3
[20] Deng S, Cheng C, Ni G, Meng Y and Chen H 2008 *Jpn. J. Appl. Phys.* **47** 7009–12
[21] Alkawareek M Y, Algwari Q T, Gorman S P, Graham W G, O'Connell D and Gilmore B F 2012 *FEMS Immunol. Med. Microbiol.* **65** 381–4
[22] Colagar A H, Memariani H, Sohbatzadeh F and Omran A V 2013 *Appl. Biochem. Biotechnol.* **171** 1617–29
[23] Leduc M, Guay D, Leask R L and Coulombe S 2009 *New J. Phys.* **11** 115021
[24] Hsu C and Yang Y 2010 *IEEE Trans. Plasma Sci.* **38** 496–9

Journal of Physics: Conference Series **565** (2014) 012012 doi:10.1088/1742-6596/565/1/012012

Application of low temperature plasmas for restoration/conservation of archaeological objects

F Krčma[1,4], L Blahová[1], P Fojtíková[1], W G Graham[2], H. Grossmannová[1,3], L Hlochová[1], J Horák[1], D Janová[1], C P Kelsey[2], Z Kozáková[1], V Mazánková[1], M Procházka[1], R Přikryl[1], L Řádková[1], V Sázavská[1], M Vašíček[1], R Veverková[1], M Zmrzlý[1]

[1] Faculty of Chemistry, Brno University of Technology, Purkyňova 118, 612 00 Brno, Czech Republic

[2] Queens University Belfast, Center Plasma Physics, Belfast BT7 1NN, Antrim, North Ireland

[3] Technical Museum Brno, Purkyňova 105, 612 00 Brno, Czech Republic

E-mail: krcma@fch.vutbr.cz

Abstract. The low-temperature low-pressure hydrogen based plasmas were used to study the influence of processes and discharge conditions on corrosion removal. The capacitive coupled RF discharge in the continuous or pulsed regime was used at operating pressure of 100–200 Pa. Plasma treatment was monitored by optical emission spectroscopy. To be able to study influence of various process parameters, the model corroded samples with and without sandy incrustation were prepared. The SEM-EDX analyzes were carried out to verify corrosion removal efficiency. Experimental conditions were optimized for the selected most frequent materials of original metallic archaeological objects (iron, bronze, copper, and brass). Chlorides removal is based on hydrogen ion reactions while oxides are removed mainly by neutral species interactions. A special focus was kept for the samples temperature because it was necessary to avoid any metallographic changes in the material structure. The application of higher power pulsed regime with low duty cycle seems be the best treatment regime. The low pressure hydrogen plasma is not applicable for objects with a very broken structure or for non-metallic objects due to the non-uniform heat stress. Due to this fact, the new developed plasmas generated in liquids were applied on selected original archaeological glass materials.

1. Introduction

The conservation/restoration operations are applied widely in practice for various historical objects. Conservation means a human activity (direct or indirect) which is aimed at increasing their expected life or damaged objects or their collections [1]. The father of modern archaeological conservation is Friedrich Rathgen [2]. Restoration means only a direct human activity. It is aimed at ensuring that the damaged object in a collection gets its aesthetic (sometimes original) or historic look [1].

The conservation/restoration of excavated archaeological objects is a very important procedure to keep these unique historical documents for future generations. The main conservation goals are

[4] To whom any correspondence should be addressed.

elimination of corrosive agents and stimulators of corrosion (mainly chlorine ions), removal or reduction of corrosions layers (i.e. discovering the original objects surface) and finally, protection against further corrosion during the object storage [3].

The excavated ancient objects are made of various materials depending on the historical epoch. The majority of objects is made of metals (mainly iron and bronze), glass or ceramics. The objects are commonly affected by various corrosion kinds with different intensity. The given state of any objects is dependent on many facts, mainly on object material, artefact manufacturing technology (strongly dependent on the epoch and locality), storing duration between manufacturing and excavation, and composition of corrosive surrounding. The storage conditions between object excavation and conservation as well as precedent conservation procedures must be included, too, if object is not conserved immediately after excavation. Thus every object, even excavated at the same locality, is original with its own history.

Cleaning of archaeological artefacts surface is an irreversible process. Its main task is revealing the original surface or its remains if the object is too damaged. Classical cleaning methods used in practice are surface mechanical methods, solvent-based methods, and chemical methods.

The mechanical cleaning methods include a manual removal of dust usually by fine cloth, an air cleaning (vacuuming and blowing air), a mechanical cleaning with handy instruments (scalpel, saber, dental drill, etc.), a removal of non-functional surface coating systems of organic origin using heat, vibration and ultrasonic cleaning, abrasive methods, blasting, grinding and polishing, scrubbing, and laser. Application of nearly all these techniques induces a great mechanical and sometimes also local thermal overheating that can lead to non-wished object damage [4].

Cleaning with solvents and solutions transform impurities and corrosion compounds into the solution and it leads to their washing out. Distilled water is used very often because it is cheap and its polar character wets metals and corrosion products very well. Various surfactants are added into the cleaning water solution or pure organic solvents are applied if impurities are of organic origin (varnishes, oils, waxes, etc.).

The principle of chemical cleaning procedures is the conversion of chemical impurities using specific chemical reactions. Alkaline solutions (like sodium hydroxide or ammonia water solutions) remove organic as well as inorganic impurities. Although many of corrosion products are soluble in acidic solutions very well, secondary corrosion initiated by these solutions could be the serious problem. The electrochemical cleaning of metals is another possible surface cleaning method. During the electrolytic cleaning, the metal core of the treated object is connected to the negative pole of an external source, and it becomes cathode. The anode can be carbon, stainless steel or other electrically conductive material [5]. The last chemical cleaning method is the use of complex forming agents. These are water-soluble chemical compounds that form soluble complexes with the metal ions. For example, Chelaton III, polyphosphates (removal of lime crusts on the surface of archaeological artefacts) or thiourea (cleaning silver corrosion products) are often used in practice [6].

All these classical conservation procedures are typically men powered (mechanical conservation) or time (other) consuming. Thus there is a task to develop some other or additional procedures that will allow reducing these demands.

One of the alternative modern methods of cleaning is a plasma chemical treatment of archaeological artefacts. The great advantage of the plasma treatment is the fact that it is a "dry way" process that can be simultaneously applied on more objects depending on their kind and device dimensions. This prevents further corrosion, which may be caused by the presence of an aqueous environment. Also, there is no dissolution of the metal itself, as for example in aqueous solutions of Chelaton III. The complete removal of corrosion layers using the hydrogen based low pressure plasma is not possible from all materials. It was confirmed for silver made objects, only [7]. In other cases, the corrosion layers are not removed fully but they start to be much brittle and thus significantly softer mechanical or shorter chemical cleaning can be applied. Of course, some disadvantages of plasma chemical treatment exist. If the treatment is carried out at high power and continuous mode, object could be damaged by high temperature itself or indirectly by non-uniform thermal stress. The surface layer called patina presented at the copper based materials surface (bronze or brass) can be damaged, too [5], and thus some historical information can be loosed.

One of the first mentions of the reduction by the low temperature plasma comes from Daniels, who was able to reduce polluted layers of silver back to silver. He treated Daguerreotypes, too [8].

At mid-80s, Stanislav Veprek from the Institute of Inorganic Chemistry, University of Zurich successfully applied plasma chemical reduction on various ancient metallic objects [9]. The plasma chemical equipment consisted of the discharge Pyrex glass tube (diameter of 15 cm, length of 45 cm) with two outer water cooled electrodes which were connected to the RF generator (80 MHz, 2 kW). Hydrogen and some hydrogen based gas mixtures were used in his studies. The temperature was measured by a mercury thermometer placed inside the reactor. The treated objects were hung on a thin iron wire or lying on a quartz glass grid. Veprek´s group treated mainly iron objects and temperature during the process could not exceed 400°C [10]. Later experiments confirmed that such temperature is too high to prevent local crystallographic changes in iron based materials, and thus contemporary museums workers require lower temperatures, in case of iron not exceeding 200°C. This temperature, as was confirmed recently, should be too high for some other metals, mainly for bronze.

Plasma reduction technique has been applied at Swiss National Museum since 1994. Radiofrequency generator (27 MHz) is used, and the gas pressure (hydrogen and argon) is 15–40 Pa. Iron and silver artefacts were treated [11].

In the Central Bohemian Museum in Roztoky u Prahy (Czech Republic), the plasma chemical equipment was used at the Department of Conservation Collections since 1996. Unfortunately, this equipment was damaged by floods in 2002 and was not rebuild yet.

Our research group has a good collaboration with Technical Museum in Brno where a big plasma chemical experimental device (diameter of 40 cm and length of 150 cm) is installed under the Methodical Centre of Conservation (MCK). Thus our results presented later are immediately applied in the conservators practice.

However, the low-pressure low-temperature hydrogen based plasma is non-applicable on objects with broken structure as well as on glass and ceramics. The application of plasma jets penetrating into liquid was tested for the corrosion removal from glass samples [12]. Recent progress in the field of electrical discharges generated directly in liquids allowed application of devices used in plasma based surgery for the surface treatment of ancient objects.

The presented paper gives our recent results and progress in the field of low-temperature low-pressure plasma application on metallic model samples. The second part gives the first results of discharges in liquids application on the corrosion removal from ancient glass objects.

2. Experimental set up

2.1 Low-temperature low- pressure RF device

The plasma treatment was carried out using the experimental set up depicted in figure 1. The Quartz cylindrical reactor (90 cm long and 9.5 cm in the diameter) was surrounded by two external copper electrodes supplied by a radio-frequency generator (13.56 MHz) through an automatic matching network. The sample was placed onto a glass grid holder (consisting of two parallel Pyrex tubes, with outer diameter of 6 mm, fixed at the ends of reactor) at the rector central part. The experiment took place in pure hydrogen (99.9 %) or various hydrogen-argon mixtures at pressure around 160 Pa and constant total gas flow rate of 50 sccm. Samples were treated at different applied RF plasma power of 100 W, 200 W, 300 W and 400 W in a pulsed mode with a selected duty cycle of 25–100% with frequency of 1 kHz. The treatment duration was up to 120 minutes.

During the plasma chemical reduction, hydrogen and argon ions, radicals and excited molecules react with oxygen and chlorine that are the main elements of the corrosion layers (see in [13, 14]). OH radicals and HCl molecules are formed by these reactions. The OH radicals were monitored continuously by the optical emission spectroscopy using the Ocean Optics HR4000 spectrometer with 2400 gr/mm grating. An optical fiber was placed at a quartz window mounted at the reactor axis. Data obtained from this method were used to calculate integral intensities of OH radicals that were used for the process monitoring according to the procedure described in [15]. The HCl presence was directly confirmed by IR absorption spectroscopy because no HCl bands were found in the emission spectra. HCl presence was also confirmed indirectly by a consequent corrosion of aluminum parts downstream of the discharge. To avoid the pump oil damage by the HCl molecules, a liquid nitrogen trap with aluminum wires was added into the pumping system.

Temperature of the sample during the treatment was measured by a K-type thermocouple installed 5 mm inside the sample metallic body. To avoid electromagnetic induction influence on the measured temperature, the temperature values were read out at the end of the 5 s discharge break.

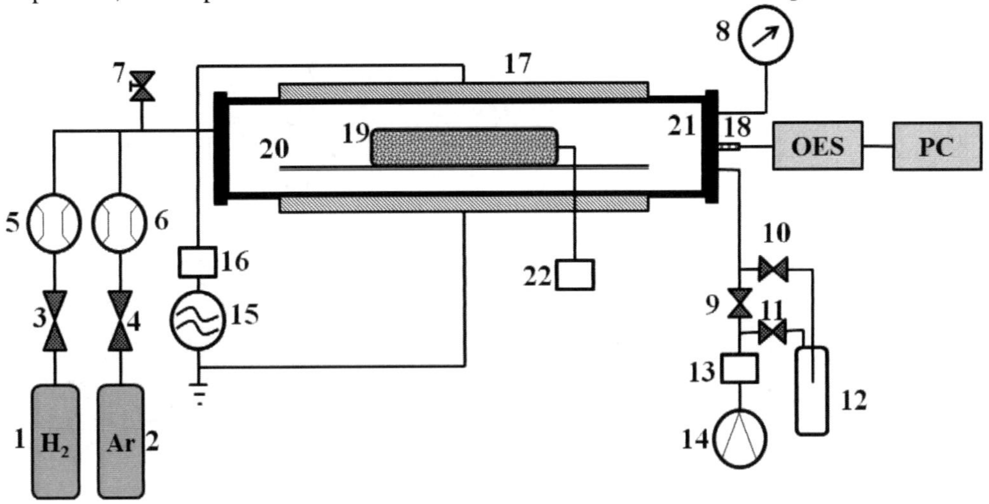

Figure 1. Scheme of the low-temperature low-pressure RF plasma device:
1, 2 – high pressure bottles; 3,4 – on/off valves; 5, 6 – mass flow controllers; 7 – needle valve; 8 – capacitance gauge; 9, 10, 11 – on/off valves; 12 – liquid nitrogen trap filled by aluminum wires; 13 – butterfly valve, 14 – rotary oil pump, 15 – RF generator; 16 – matching network; 17 – copper electrodes; 18 – quartz optical fiber; 19 – treated sample; 20 – glass grid holder; 21 – Quartz reactor; 22 – thermocouple.

Model samples used in the presented study were prepared from the most common archaeological metals using hydrochloric acid vapor (for iron, copper, bronze and brass samples) and ammonia solution (for brass samples, only). Two different sets were prepared: The first one with a pure corrosion layer, the second one with a sand incrustation for better simulation of the real situation of an excavated archaeological artifact. The samples preparation was described in details in [16].

The composition of the corrosion layers is typically very complex [3]. Various oxides and chlorides (both also hydrogenated or with crystalline water) were detected by XRD analysis. The elementary abundance was obtained by SEM-EDX analysis.

2.2 Plasma operating in liquid

Plasma was created by a special multielectrode device which principal scheme is shown in figure 2. It is a high-performance instrument which is commonly used in biomedicine at the present time. This system even allows a direct contact with the targeted sample. This system operates as a multi-electrode setup with a driving circuit producing 100 kHz RF bipolar square wave at relatively low voltage signals (up to 600 V). Current was determined via voltage drop at known resistance. The main factor for the treatment was obviously the solution conductivity which depended on temperature and added salt concentration and the distance between the treated sample and the electrode. Further, the presence of other dissolved or suspended particles or nearby physical structures had a inconsiderable influence.

The discharge chemistry in each specific instance was also affected by the applied current. The optimal experimental conditions for the archaeological artefacts treatment were obtained using 0.9% water solution of sodium chloride (solution conductivity of 13 mS·cm^{-1}). The best distance between the treated surface and the front of the multi-electrode head was about 1 mm, and the optimal applied power was about 5 W.

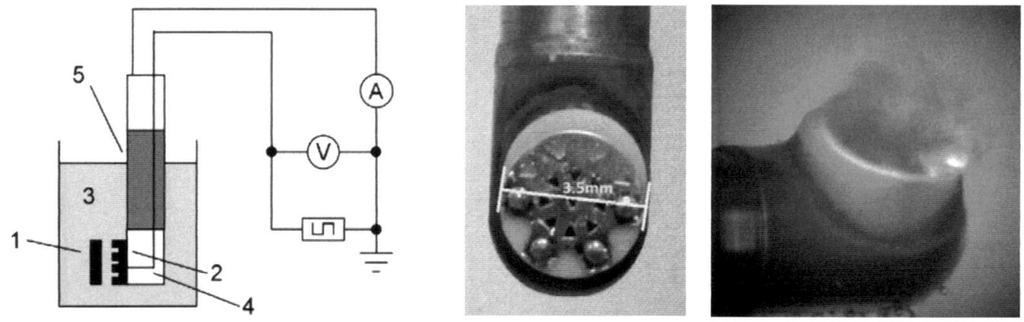

Figure 2. Left: The scheme of the experimental set-up: 1 – treated sample; 2 – powered electrode; 3 – water solution; 4 – insulator; 5 – grounded electrode; right – electrode head and electrode head under operation.

The underwater discharge was used for the glass samples treatment. Presented results were obtained using the real archaeological artefacts which have laid in the soil since 17th–18th century. The beads and the ceramics were from different locations (Mistřín, Moravičany and Brno, Czech Republic) and the waste from blowing glass, glass from vessels and ingots from glass were found in the pond Huťský, Josefov, Czech Republic.

3. Results

3.1 Treatment of metallic samples by low pressure device

The example of the sample temperature and OH radical intensity that reflects corrosion removal process is given by figure 3 for iron corroded samples without and with incrustation.

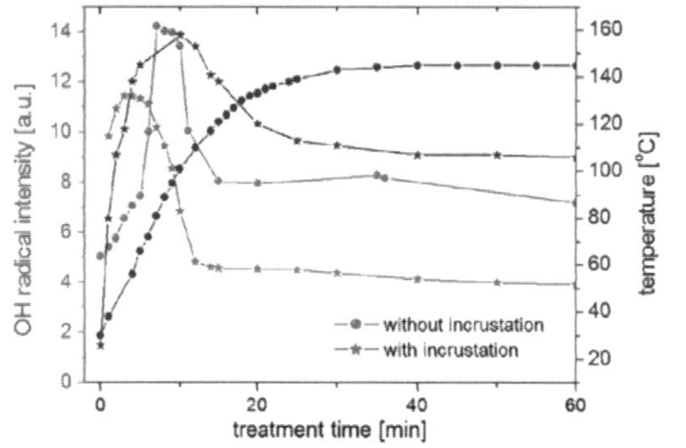

Figure 3. Temperature and OH radical intensity during the plasma treatment of iron samples at applied power of 400 W.

It can be seen that the OH intensity increases at the beginning of the process at both cases. This intensity increase is faster in case of the incrusted sample. When the slightly bounded oxygen at incrustation layers is removed, the rest of incrustation plays a shielding role and thus the plasma active particles do not interact directly with the corroded surface. Consequently, the OH radical intensity drops down. This is the probable reason why OH radical intensity is lower after the initial peak in case of sample covered by corrosion with incrustation.

The temperature of the non-incrusted samples increased during the first 30 minutes of the plasma treatment, and after that it remained constant. During the treatment of the incrusted samples, much faster but only a temporal increase of temperature was observed. After that, the temperature was significantly lower in contrary to the non-incrusted samples as it can be seen from figure 3. In this case, the maximal temperature was even higher than the maximal temperature of the non-incrusted sample.

Temperature dependences on applied power showed that the maximal temperatures were significantly lower at the lower applied powers. Due to the fact, no temperature peak at the process beginning is observed. The similar results have been obtained for all used metals [16].

To verify the influence of sample size and shape on the maximal temperature during the plasma treatment, two sets of non-corroded samples of different weight were prepared for each studied material. The first set contained samples of almost cubic shape (in graphs they are denoted as volume samples) and the planar samples of 2 mm thickness were in the second one. These samples were treated in plasma under the standard conditions described above (pressure and hydrogen flow) at applied power of 300 W in the continuous regime. Applied power of 150 W was used for the bronze samples treatment due to higher temperature sensitivity of this material. An example of temperature evolution of selected brass samples was shown recently [16]. The temperature increase is slower if a bigger sample is treated, and the thermal stabilization takes longer time. The maximal temperature increases with the sample mass (see figure 4). After this measurement, the same samples were corroded in HCl vapor in sandy environment to prepare corrosion with incrustation, and the experiment was repeated. The temperature time evolution was according to combination of results described above in text and in figure 3. The maximal temperatures for copper and brass samples are given in figure 4, the other materials showed similar behavior.

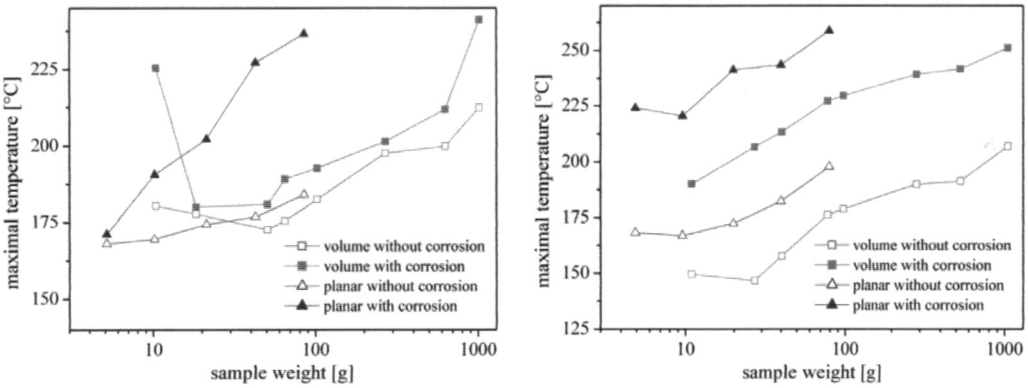

Figure 4. Maximal sample temperatures reached during the plasma treatment of non-corroded and corroded with incrustation samples of different shape; left – copper; right – brass.

This figure clearly shows that maximal temperature is exponentially proportional to the sample mass. This means that the sample heating is dominantly due to the direct interaction with plasma active particles (note that neutral gas temperature calculated from OH radicals is 300–350 °C [17]) and the inductive heating by RF field generating the plasma is negligible. Therefore, planar samples are heated to a slightly higher temperature. Corrosion layers removal leads to a significant increase of the maximal temperatures, mainly in the case of the planar samples. The corrosion layers influence is much higher in case of brass. The process is also dependent on the plasma treatment conditions, mainly on the applied power. Nearly no difference between samples with and without corrosion was observed for bronze samples that were treated at half applied power (not shown here).

The previous results showed importance of the applied power with respect to the treated object maximal temperature. Particles generated in plasma have some lifetime, and thus we have tried to verify the role of the post-discharge period on the corrosion removal process. A lot of experiments have been carried out with different applied power and the duty cycle. An example is shown in figure 5 left. It is clearly visible that the corrosion removal is enhanced by the increasing applied power and it is slower at shorter duty cycles. By other words, it is strongly dependent on the mean energy. If mean energy is too low, the corrosion removal is too slow or even inefficient. On the other hand, too high mean applied power leads to too high maximal temperature (see figure 5 right) during the processing, and thus the crystallographic changes can be induced. Adjustment of the treatment conditions is more complicated in case of bronze due to the local overheating in the surrounding of ion impact to the surface that can lead to an unwanted evaporation of tin from the sample surface. It can be pointed that the maximal temperature is lower at higher applied power during the pulse with the lower duty cycle at the same mean power as it can be seen from figure 5 right.

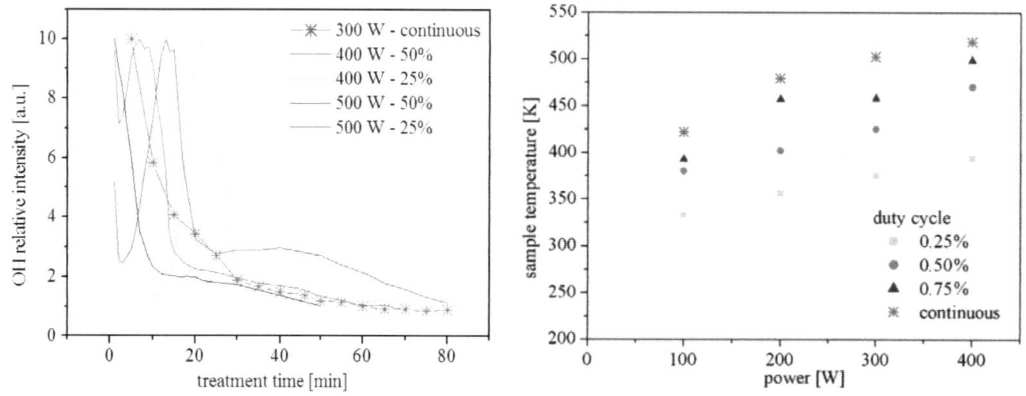

Figure 5. Left – influence of applied power and duty cycle on the evolution of OH radicals during the plasma treatment of iron samples; right – maximal brass sample temperatures as a function of applied power and duty cycle.

The elementary analysis of surface layers showed a significant difference of corrosive elements abundance depending on the treatment conditions as it is shown in figure 6. The comparison of the pulsed and the continuous regime of plasma treatment is also included. It can be seen that the amount of oxygen and chlorine decreases with increasing applied power in the continual regime. The same dependence of oxygen concentration can be seen in the pulsed regime. The concentration of chlorine is more or less independent on the duty cycle parameter. This means that oxygen compounds removal is more dependent on the applied mean power or sample temperature. Chlorides removal is dependent on the absolute value of the applied power during the active part of the duty cycle. Thus, it can be concluded that chlorides removal is mainly due to the reaction of hydrogen ions with surface compounds, but oxygen compounds are decomposed by hydrogen ions as well as by neutral hydrogen atoms.

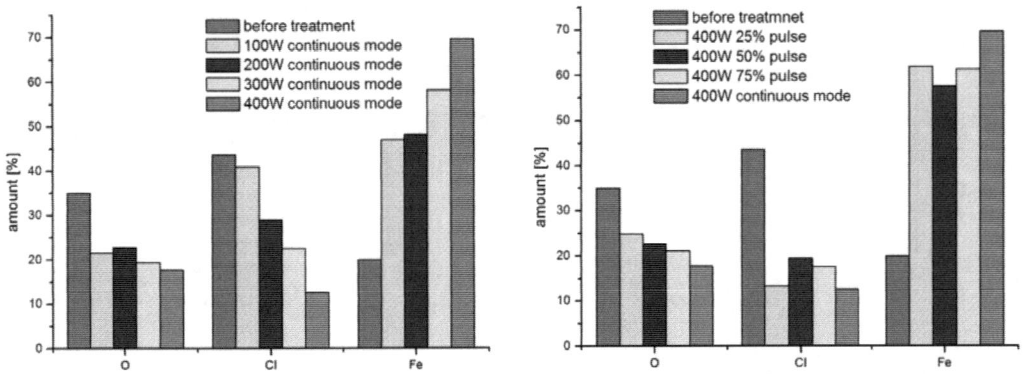

Figure 6. Material analyzes of HCl vapor corroded iron samples after plasma treatment at different powers in continuous mode (left) and at 400W in pulsed regime (right).

3.2 Treatment of non-metallic samples by plasma generated in liquids

An example of the glass vessel treated by the discharge in liquid is given in figure 7.

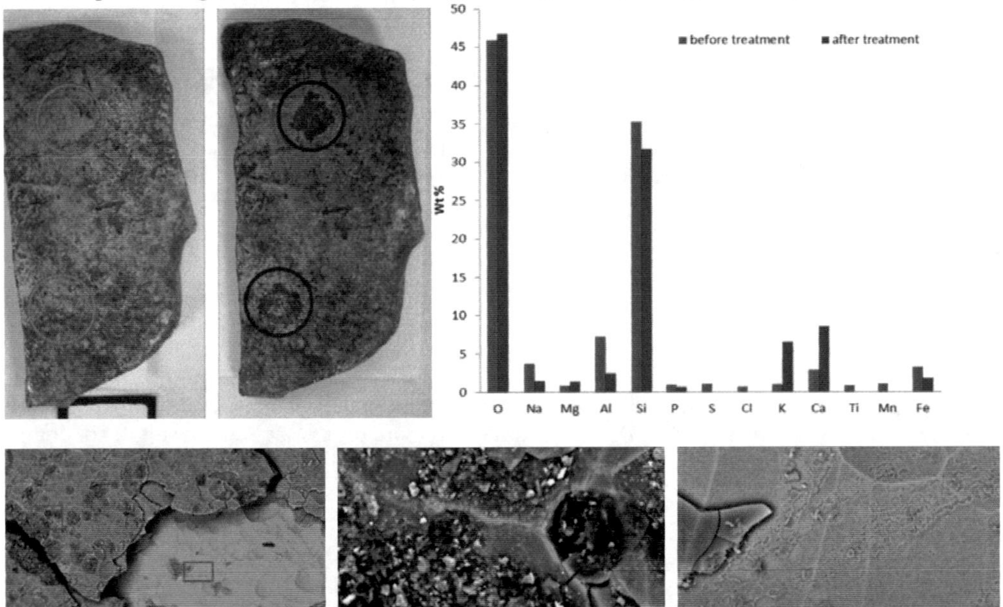

Figure 7. Plasma treatment of glass vessel: top left – vessel before and after treatment at marked areas; top right – comparison of elementary composition of the surface layers; bottom left – boundary between non treated and treated areas; bottom middle – high resolution SEM of non-treated area; bottom right – high resolution SEM of plasma treated area.

The treated areas are marked by circles. Surface structure changes are visible even by the naked eye and they were fully confirmed by SEM/ EDX analysis. The plasma treated surface is smooth without any remarkable corrosion structures (see SEM image on the bottom right). Corrosion layers consist mainly of various oxides [18]. The EDX analysis showed that some elements coming from corrosion layers (like sulphur or chlorine) were fully removed. Surface concentrations of sodium and aluminum were slightly decreased due to the plasma treatment. On the other hand, concentrations of potassium and calcium increased because these elements originate from the bulk glass. This is in the agreement with the processing technology used in the given locality where mainly these elements were added as a potash and lime during the glass manufacturing.

The second example is the application of the multielectrode device on another ancient glass vessel (see figure 8). Obtained results are similar as in the precedent case. Also in this second example, the successful restoration of the original surface was reached and corrosion layers were successfully removed.

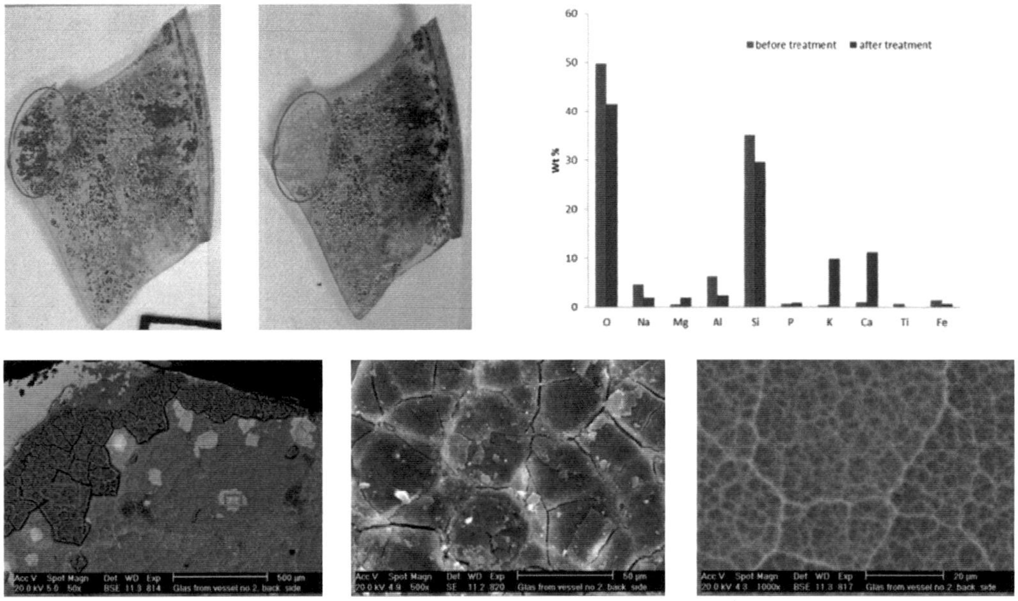

Figure 8. Plasma treatment of another medieval glass vessel. Legend is the same as for figure 7.

Finally, figure 9 shows result of the application on a glass bead. The restoration of the original blue color is well remarkable. Detailed SEM-EDX analyzes showed more or less the same results as in both precedent cases.

Figure 9. Medieval glass bead before (left) and after plasma treatment (right).

4. Conclusion

Two different plasma systems have been used for the corrosion removal. The low-temperature low-pressure hydrogen based plasmas were used to study the influence of processes and discharge conditions on corrosion removal from metallic artificially prepared samples. The capacitive coupled RF discharge in the continuous or pulsed regime was used at operating pressure of 100–200 Pa. To be able to verify the transfer of this laboratory study to practice in the conservation of ancient archaeological objects, the model corroded samples with and without sandy incrustation were prepared. The plasma treatment process was monitored by optical emission spectroscopy of OH radical that is produced during the decomposition of oxides contained in corrosion layers. Plasma treatment conditions were optimized for the selected most frequent materials of original metallic archaeological objects (iron, bronze, copper, and brass). The special focus was kept for the samples temperature because it was necessary to avoid any metallographic changes in material structure. Results show that object maximal temperature is higher if the object has more planar shape and increases if incrustation layers are presented. Incrustation also leads to the shielding effect, and thus the plasma treatment must be applied in more steps where the mechanical removal of incrustation must be employed between particular plasma treatments. Application of the pulsed plasma regime increases the removal energy efficiency and decreases the maximal sample temperature, but there are some limits dependent on the treated material, especially in case of bronze where temperature over 150 °C is undesirable. The removal of oxides is mainly caused by reactions with neutral species while chlorides are decomposed mainly by interactions with hydrogen charged particles.

The low pressure hydrogen plasma is not applicable for objects with the very broken structure or for non-metallic objects due to non-uniform heat stress. Due to this fact, the plasma generated directly in liquid phase was applied on selected original ancient glass objects. The successful corrosion removal was obtained without any damage of the treated objects. The detailed research in this new direction will be a subject of further studies.

Acknowledgements
Authors wishing to acknowledge the Ministry of Culture of the Czech Republic, project No. DF11P01OVV004.

References
[1] Guichen G 1999 *Museum Internat.* **51** 4–6
[2] Gilberg M 1987 *J. Americ. Inst. Conserv.* **26** 105–120
[3] Dillman P, Beranger G, Piccardo P and Matthiesen H 2007 *Corrosion of Metallic Heritage Artefacts – Investigation, Conservation and Prediction of Long Term Behaviour* (Woodhead Publishing, Cambridge)
[4] Moore M 2001 *J. Conserv. Museum Stud.* **7** 1–19
[5] Hamilton D L 1999 *Methods of Conserving Underwater Archaeological Material Culture* (Conservation Files: ANTH 605, Conservation of Cultural Resources I. Nautical Archaeology Program, Texas A&M University, World Wide Web, http://nautarch.tamu.edu/class/ANTH605)
[6] Benešová J 2011 *Conservation and Restoration of Metals: Protection of Cultur Herritage Objects made of Metals and Alloys* (Brno: Technical Museum Brno, Metodic Center of Conservation, 648 p.) – in Czech
[7] Schmidt-Ott K 2008 *Application of Plasma Reduction on Iron and Silver Artefacts*, In *Ligas Metálicas: Investigação e Conservação,* ed. Ferreira da Silva A C and Homem P M (Faculty of Arts of the University of Porto)
[8] Daniels V 1981 *Studies Conservat.* **26** 45–49
[9] Veprek S, Patscheider J and Elmer J 1985 *Plasma Chem. Plasma Proces.* **5** 201–209
[10] Veprek S, Eckmann Ch and Elmer J Th 1988 *Plasma Chem. Plasma Proces.* **8** 445–466
[11] Schmidt-Ott K and Boissonnas V 2002 *Studies Conservat.* **47** 81–87
[12] Klíma M, Zajíčková L and Janča J 1997 *J. Swiss Archeol. Art History* **54** 31–33
[13] Řádková L, Sázavská V and Krčma F 2012 *Proc. European Corrosion Congress 2012*, Istanbul,1265–1270
[14] Rimmer M, Watkinson D and Wang Q 2012 *Studies Conservat.* **57** 29–41
[15] Rašková Z, Krčma F, Klíma M and Kousal J 2002 *Czech. J. Phys.* **52** D927–D932
[16] Řádková L, Fojtíková P, Kozáková Z, Krčma F, Sázavská V and Kujawa A 2014 *Rom. Rep. Phys.* – accepted
[17] Fojtíková P, Řádková L and Krčma F 2014 *Cental. Europ. J. Chem.* – accepted
[18] Melcher M and Schreiner M 2004 *Anal. Bioanal. Chem.* **379** 628–639

Spectroscopic study of plasma during electrolytic oxidation of magnesium-aluminium alloys

J Jovović
University of Belgrade, Faculty of Physics, Studentski trg 12-16, 11001 Belgrade, Serbia

E-mail: jjovica@ff.bg.ac.rs

Abstract. Plasma during Electrolytic Oxidation (PEO) of magnesium-aluminium alloys is studied in this work by means of Optical Emission Spectroscopy (OES). Spectral line shapes of the H_β, Al II 704.21 nm and Mg II 448.11 nm line are analyzed to measure plasma electron number density N_e. From the H_β line profile, two PEO processes characterized by relatively low electron number densities $N_e \approx 10^{15}$ cm^{-3} and $N_e \approx 2 \times 10^{16}$ cm^{-3} were discovered while the shape and shift of Al II and Mg II lines revealed the third process characterized by large electron density N_e = (1-2) × 10^{17} cm^{-3}. Low N_e processes, related with breakdown in gas bubbles and on oxide surface, are not influenced by anode material or electrolyte composition. The ejection of evaporated anode material through oxide layer is designated here as third PEO process.
Using the Boltzmann plot technique, electron temperature of 4000 K and 33000 K is determined from relative intensities of Mg I and O II lines, respectively. Several difficulties in the analysis of spectral line shapes are met during this study and the ways to overcome some of the obstacles are demonstrated.

1. Introduction

Plasma electrolytic oxidation (PEO) is high-voltage anodizing technique which produces a stable oxide coating on the surface of metals and metallic alloys, see e.g. [1] and references therein. PEO is characterized by local formation of plasma, as indicated by the presence of micro-discharges on the metal/oxide surface accompanied by gas evolution. Short-lived micro-discharge events determine thermal and chemical conditions and play an important role in the phase formation, composition and morphology of metal oxide coating [2,3,4].

The study of PEO is often realized using OES, see e.g. [5] and references therein. This approach is however followed by difficulties due to space and time inhomogeneity of micro-discharges appearing randomly across the anode surface, see e.g. [6]. In that case, the recorded spectra represent time integrated low intensity radiation for which long exposure times during spectroscopic observations of PEO are necessary. Hence, the application of OES for PEO diagnostics is mostly limited to spectra identification and observation of temporal evolution of spectral lines in a visible and near UV region.

The PEO spectra recorded from aluminium [5], tantalum [7] and titanium [8] samples in various electrolytes were recently studied. The analysis of the H_α and the H_β line shapes showed that two plasma processes are always present in PEO micro-discharges. During PEO of aluminium, the third discharge process appears followed by evaporation of anode material and formation of aluminium plasma [5]. With tantalum [7] and titanium [8], local evaporation of anode does not occur and consequently spectral lines of anode material were not detected. In this study the focus was on the third process – metal plasma ejection through oxide layer. Therefore, two magnesium-aluminium alloys are used as anode material. The selected alloys have dominant concentration of either Mg or Al, the elements with almost same low melting temperature (Mg: T_m = 650 °C, Al: T_m = 660 °C). This is of importance for metal plasma generation after evaporation of anode material. Also, Mg and Al have much lower melting point than Ti (T_m = 1668 °C) and Ta (T_m = 3017 °C), which do not melt during PEO process [7,8]. The electron number density (N_e) is measured from the shape of the H_β line and from the Stark broadening parameters of Mg II and Al II lines while electron temperature (T_e) required for N_e measurement will be determined from relative intensities of Mg I and O II lines using Boltzmann plot (BP) technique.

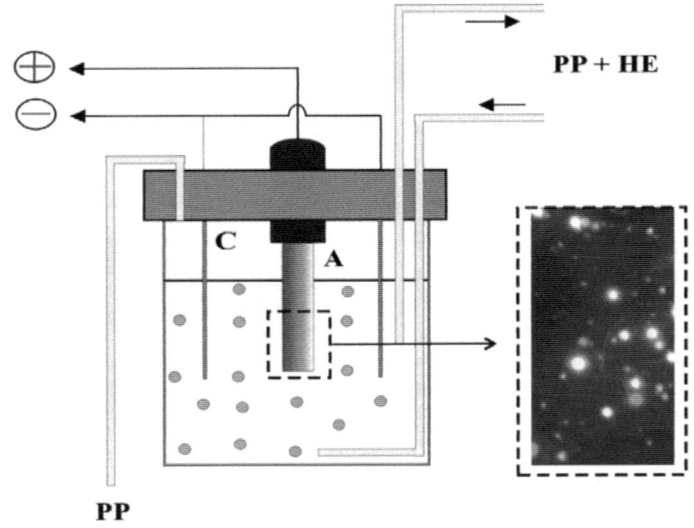

Figure 1. Schematic description of the PEO chamber and photo of micro-discharges recorded after third minute of PEO on Al-alloy anode. C – cathode, A – anode, PP – peristaltic pump, HE – heat exchanger.

2. Experimental

Magnesium-aluminium alloys AZ31 (wt.%, Al 2.5-3.5, Zn 0.6-1.4 and Mg balance) or AA5754 (wt.%, Mg 2.6-3.4, Si 0.4, Fe 0.4, Mn 0.5, Cr 0.3, Zn 0.2, and Al balance) were used as anode material in two separate experiments. Further, these two alloys will be named as Mg-alloy and Al-alloy, respectively. Rectangulary shaped anodes (dimensions 30 mm × 5 mm × 0.25 mm) were placed in electrolytic chamber, see Figure 1, leaving an active surface area of 1.5 cm^2. Two platinum wires (5 cm long and 1 mm in diameter) were used as cathodes. The PEO was carried out at 100 mA/cm^2 current density in voltage range (350-450) V. During PEO, the electrolyte circulated through the chamber-reservoir system (peristaltic pump and heat exchanger) and its temperature (≈ 20 °C) was measured in the close vicinity of the anode. The water solution of 4 g/l $Na_2SiO_3{\cdot}5H_2O$ + 4 g/l KOH (pH ≈ 12.8) was used as an electrolyte.

Spectroscopic measurements were performed with three different spectrometers, see Table 1. The spectrum in a broad wavelength range (380 nm to 850 nm) was obtained by adding subsequent 43 nm spectral intervals recorded with spectrometer No. 1. This spectrometer was equipped with thermoelectrically cooled ICCD (-40 ° C). The H_β and the spectra of O II and Mg I lines used for BP were recorded with spectrometer No. 2, while Mg II and Al II lines were recorded with spectrometer No. 3, see Table 1. The spectrometers Nos. 2 and 3 were equipped with thermoelectrically cooled CCD (-10° C) detector. All spectral lines were recorded after third minute of PEO when the voltage slowly increases with time [2]. The full-width-at-half-maximum (FWHM) of Gaussian-like instrumental profile of spectrometer No. 2 in the first diffraction order was 0.030 nm. The instrumental FWHM of spectrometer No. 3 in the first and in the second diffraction order was 0.028 nm and 0.014 nm, respectively. The image of anode surface was projected with unity magnification to the entrance slit of spectrometer using an achromatic lens (focal length 75.8 mm). A standard coiled-coil tungsten halogen lamp was used for a wavelength sensitivity calibration of all spectrometer-detector systems.

Table 1. Characteristics of spectrometers and CCD detectors.

	Spectrometer 1	Spectrometer 2	Spectrometer 3
Focal length (m)	0.3	0.67	2
Grating (grooves/mm)	1200	1800	651
Linear dispersion in first diffraction order (nm/mm)	2.7	0.83	0.74
Detector	iCCD (-40°C)	CCD (-10°C)	CCD (-10°C)

3. Plasma diagnostics

Before the procedure of plasma diagnostics is explained, the short description of PEO processes is given. Intensive gas generation is observed after first several seconds from the beginning of PEO while after about one minute micro-discharges become visible, see Fig. 2 in [6,9]. With the time, micro-discharges become brighter and larger, while their number is reduced, see Fig. 2 in [9] and [12].

Integral PEO spectra with Mg- and Al-alloy are presented in Figs. 2a and 2b, respectively. Apart from intensity, no other difference in the investigated current density range (50-100) mA/cm^2 is observed. Since there is no change in relative line intensities or line shapes for the same observation time of the PEO, the study of PEO with Mg- and Al-alloy anode was performed only for the electrode current density of 100 mA/cm^2.

3.1 Electron temperature

One of important parameters for PEO characterization and evaluation of N_e is free electrons plasma temperature T_e. It is assumed that under our experimental conditions, electron excitation temperature T_{exc} measured from relative line intensities using BP technique is equal to T_e. This implies that upper energy levels of lines used for the BP are populated in accordance with Boltzmann distribution of free electrons. The discussion on the fulfillment of partial local thermal equilibrium in PEO microdischarges and criterion for the BP application is given in [5]. However, it is justified to apply BP technique only if spectral lines used for T_e diagnostics are optically thin. For that purpose we compared intensity ratio of Mg I lines within a multiplet with corresponding line strengths ratio, see e.g. [10], and a good agreement was found. This is an indication that Mg I lines used for BP are optically thin, see e.g. [11-13]. The spectral lines of O II were so weak that self-absorption test was not necessary to perform.

The BPs of Mg I and O II lines recorded during PEO of Al-alloy anode are given in Figs. 3a and 3b, respectively. The list and data of used Mg I and O II lines are given in Ref [6]. From the slope of linear fit through data points in Fig. 3a, T_e = 3970 K ± 4 % from Mg I lines is determined. This temperature is very low in comparison with T_e measured from O II lines, T_e = 33070 K ± 12 %, see Fig. 3b. The estimated T_e uncertainties are the result of intensity measurements and transition probabilities. An absolute error is impossible to determine without knowledge of line intensity evolution and decay during time of single breakdown. This statement is relevant for all other reported data. For Mg-alloy anode, the Mg I lines BP has the same slope and thus same temperature, as in the case of Al-alloy anode. With the same Mg-alloy anode we were unable to measure T_e from O II lines since their intensity was very weak.

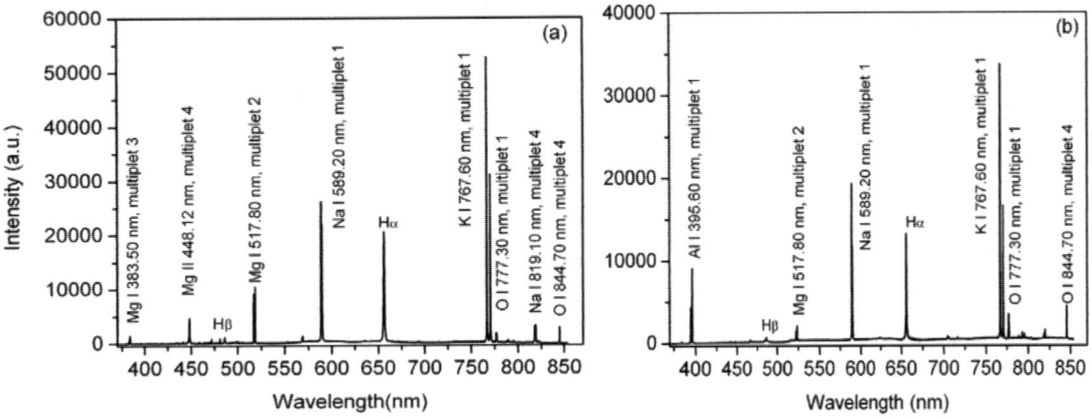

Figure 2. Typical PEO emission spectra in the spectral range 380 nm - 850 nm with: (a) Mg-alloy anode, (b) Al-alloy anode.

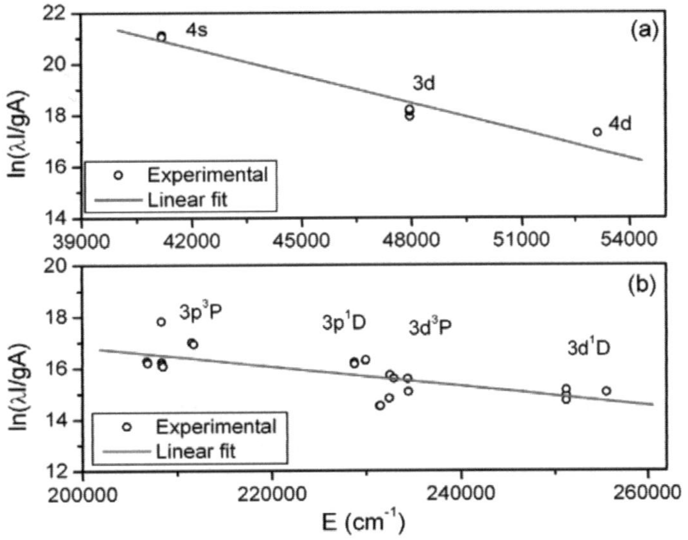

Figure 3. Boltzmann plot of: (a) seven Mg I lines and (b) twenty five O II lines, recorded during PEO with Al-alloy anode. The list of Mg I and O II lines with corresponding atomic data are given in [6].

Low T_e measured from Mg I lines is in a good agreement with recent results: T_e = 3300-7000 K reported in [14] and T_e = 3300 K measured from the BP of W I lines [10]. T_e determined from Mg I and O II lines is further used for N_e diagnostics from the width and/or shift of Al II 704.21 nm and Mg II 448.12 nm line and the H_β profile. High T_e from O II lines is in agreement with T_e results reported for PEO with pure Al anode [5]. The main difficulty in the interpretation of measured T_e data is how to determine the time of plasma evolution and/or decay when the Mg I lines used for BP are emitted. This difficulty will be better understood after reporting procedure and results of N_e measurement, see below.

3.2 Electron number density

It is important that the spectral line selected for N_e measurement is not self-absorbed, that it has good signal to noise ratio and does not interfere strongly with neighboring lines. For the selected line, reliable theoretical Stark broadening data must be available. As in previous PEO studies [5,7,8], first two hydrogen lines of Balmer series are selected as potential candidates for N_e plasma diagnostics. It was shown that the H_α and the H_β can't be fitted with single profile but with two Lorentz profiles. It is well known that self-absorbed line profile may also be, to the certain extent, approximated with Lorentz profile [13]. Therefore, the self-absorption test described in [7] is applied here as well. In addition to the H_α, the procedure also involves the two-profile fit of the H_β. The test revealed the H_α profiles distorted by self-absorption and, if used for diagnostics, 4-5 times larger N_e would be detected in plasma. This is the reason why the H_β line profile was used for plasma N_e diagnostics in this study.

The H_β profile recorded from Mg-alloy anode and fitted with two Stark profiles obtained using the procedure described in [15] is presented in Fig. 4. The theoretical data used for fitting the H_β profile are taken from [16] and two plasma N_e values, $(1.2 \pm 0.3) \times 10^{15}$ cm^{-3} and $(2.3 \pm 0.6) \times 10^{16}$ cm^{-3}, were determined. With the Al-alloy, strong AlO molecular band is interfering with the H_β profile, see Fig. 2b in [5]. This interference is the obstacle to use the H_β profile recorded during PEO of Al-alloy.

Non-hydrogenic ion line profiles were recorded and analyzed as well during the course of this study. The experimental procedure for N_e diagnostics from non-hydrogenic line profiles is described in detail elsewhere [12]. During PEO with Al-alloy, the shape of Al II line belonging to $4s^3S - 4p^3P^0$ multiplet 3, see the inset in Fig 5a, is used for diagnostic purposes. As for Mg I lines, see Section 3.1, the line self-absorption check is carried out using relative line intensities within multiplet. Within experimental uncertainty, strongest and medium intensity line show theoretical line intensity ratio 5:3, see line strength values in [10]. Thus, the strongest Al II line at 704.21 nm is selected for N_e diagnostics. The appearance of a small bulge on the blue wing of Al II lines, which makes lines asymmetric, was detected. Since the profiles of ion lines are not expected to be asymmetric, see e.g. [17], we carried out search for the cause of asymmetry. After the comparison of Al II 704.21 nm line profile with other spectral lines, the spectrometer was excluded as potential source of Al II lines asymmetry. The fitting of experimental profile with two Lorentz profiles having different width and shift was performed as the next step, see Fig. 5b. The two profiles fit describes much better Al II 704.21 nm line profile than single Lorentz profile, compare residue of Figs. 5a and 5b. The line shift of Al II 704.2 nm line is measured as well. This was done in respect to Ne I 705.92 nm line. As predicted by the theory [17], the Al II line shift is towards shorter wavelengths.

The best candidate for N_e diagnostics during PEO of Mg-alloy was $3d^2D - 4f^2F^0$ multiplet 4 with an average wavelength 448.12 nm, see profile recording in Fig. 6. The presented line shape consists of two close line profiles at 448.11 nm and 448.13 nm. This wavelength difference however does not cause difficulty in an analysis of the overall profile since Stark broadened lines within multiplet have same profile shape, width and shift [17-19]. The influence of small wavelength difference of 0.02 nm to the line width measurement is within the precision of this experiment. As for Al II 704.21 nm line, Mg II 448.11 nm line profile was fitted first with one Lorentz profile, see Fig. 6a. The experimental profile shows blue asymmetry at the line pedestal with both wings exceeding best fit Lorentz profile, see also residue in Fig. 6a. There is no logical explanation for isolated ion line profile (energy level separation between upper energy level and nearest perturbing level is larger than the line width) to exhibit such behavior. Therefore, we performed the fitting of experimental profile with two mutually shifted profiles, see Fig. 6b. Two Lorentz profiles with different width and shift significantly improve the fitting procedure, compare residues in Fig. 6a and 6b.

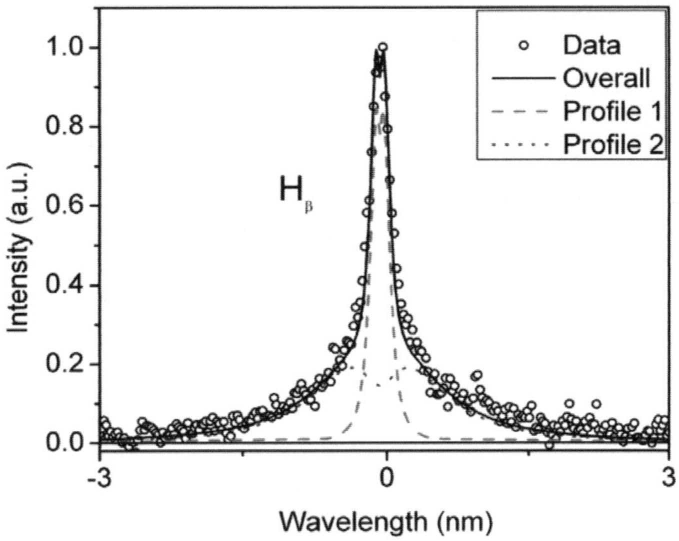

Figure 4. The H_β profile recorded during PEO with Mg-alloy anode and best fit with two Stark profiles. The instrumental profile halfwidth is 0.03 nm.

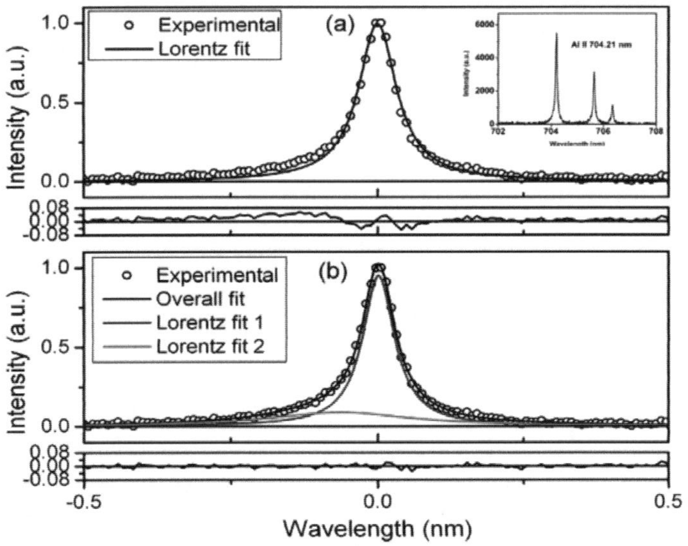

Figure 5. The experimental profile of Al II 704.21 nm line: (a) single best Lorentz fit and (b) two mutually shifted Lorentz fit. The line was recorded during PEO with Al-alloy anode. The instrumental profile halfwidth is 0.028 nm.

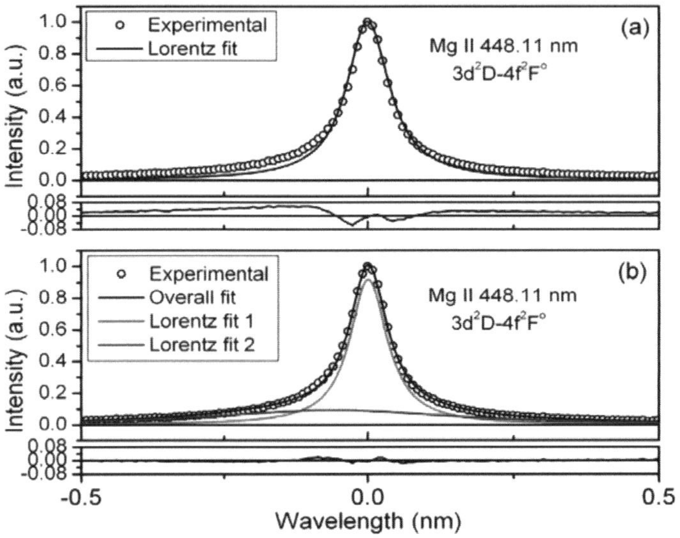

Figure 6. Mg II $3d^2D - 4f^2F^0$ line at 448.11 nm: (a) single best Lorentz fit and (b) two mutually shifted Lorentz fit. The line was recorded during PEO with Mg-alloy anode. The instrumental profile halfwidth is 0.028 nm.

4. Results and Discussion

The results of N_e measurement from halfwidths and shift of Mg II 448.11 nm and Al II 704.21 nm lines are shown in Table 2. Since N_e is determined from time and space averaged line profiles, it is difficult to estimate uncertainty of reported data. An estimated error in Table 2 is the result of experimental scatter, fitting procedure and uncertainty of theoretical data. An explanation of the procedure for N_e measurement from Mg II and Al II lines shall be given. First, the experimental line halfwidths are corrected for the contribution of Van der Waals broadening [20]:

$$w_{VW}[nm] = 5.925\times10^{14}\times K_1\times K_P\times p[mbar]\times T_g[K]^{-0.7} \quad (1)$$

$$K_1 = \lambda^2\times\langle R^2\rangle^{2/5} \quad (2)$$

$$K_P = \alpha^{2/5} / \mu^{3/10} \quad (3)$$

where p is the pressure, T_g gas temperature, λ is the wavelength of spectral line, $\langle R^2\rangle$ is the difference between the squares of the expectation values (in units of a_0) of the coordinate vector of the radiating electron in the upper and lower state of the transition, α is the mean atomic polarizability of the neutral perturber and μ is the atom-perturber reduced mass in a.m.u. Then, N_e is estimated from the comparison of experimental halfwidth and shift with Stark halfwidth or shift evaluated using formulas [11]:

$$w_t(N_e,T_e) \cong 2w_e(T_e)\times10^{-17}\times N_e \quad (4)$$

$$d_t(N_e,T_e) \cong d_e(T_e)\times10^{-17}\times N_e \quad (5)$$

In the above equations, $w_e(T_e)$ and $d_e(T_e)$ are electron impact half halfwidth and shift, respectively, calculated for $N_e = 10^{17}$ cm^{-3} and $T_e = 4000$ K [17]. Van der Waals halfwidth and shift are calculated under assumption: $p = 1000$ mbar and $T_e = T_g$ (T_g gas temperature). The perturber was Al or Mg atom depending on alloy. It was mentioned earlier, in relation to T_e measurement from the BP of Mg I lines, that time interval during the ejection of metal plasma when Mg I lines are emitted is unknown. This time period may differ from the period when Mg II and Al II lines appear. In our case one has no other choice, since only several Mg I lines are available for T_e measurements while appropriate set of Mg II lines was not detected. In relation to the complex shape of Mg II and Al II lines used for N_e diagnostics, see Figs. 5 and 6, one can expect following steps in metal plasma evolution and decay. After breakdown and anode material evaporation, hot and dense metal plasma is ejected through micro-channel formed in oxide layer. In this time period observed Mg II and Al II lines are broad and shifted due to high plasma N_e. After ejection from micro-channel, plasma expands and cools down. This causes decrease of N_e and consequently narrowing of line profile and decrease of line shift. This narrower profile forms upper part of the overall profiles in Figures 5 and 6.

Finally, the atomic lines appear and quench at the end of this process. In relation to metal plasma ejection, the comment on low $T_e = 3300$ K measured from BP of W I lines [5] is given. During PEO of pure aluminium samples, tungsten was present in electrolyte solution [5] and the excitation of W I lines is a result of hot aluminium metal plasma interaction with electrolyte. This explains relatively low temperature measured from W I lines.

Assuming Saha equilibrium, an attempt is made to calculate T_e from N_e values presented in Table 2 using Saha equation for plasma at atmospheric pressure (electrolytic chamber is at atmospheric pressure). From N_e values for Mg- and Al-alloy of 1.6×10^{17} cm^{-3} and 1.2×10^{17} cm^{-3}, T_e of 7400 K and 6700 K is evaluated, respectively. Since the pressure at the opening of micro-channel, where the plasma of highest N_e is most likely ejected may be larger than one atmosphere, lower T_e for the same N_e would be determined from Saha equation. For example, T_e for Mg-alloy would decrease from 7400 K for one to 6000 K for ten atmospheres. In spite of pressure uncertainty, T_e in metal plasma evaluated from Saha equation is comparable with temperature derived using BP technique what is expected for Saha equilibrium.

Table 2. The results for N_e in 10^{16} cm^{-3} obtained from measured halfwidths and shift of Mg II 448.11 nm and Al II 704.21 nm lines.

Line	Anode	
	Mg-alloy	Al-alloy
Mg II 448.11 nm		
Profile 1	2.2 ± 1.0	*
Profile 2	16.0 ± 7.2	*
Al II 704.21 nm		
Profile 1	*	2.8 ± 1.3
Profile 2	*	12.0 ± 5.4
Shift	*	9.0 ± 4.0

Finally, let us summarize the main features of PEO microdischarges. The surface of metal-oxide layer has irregular micro structure, see e.g. Fig. 7 in [7], with sharp points where large electric fields appear and induce breakdown in gas bubbles always present in an electrolyte during PEO. This is the first process with lowest N_e determined from the H_β profile, see Section 3.2. The second process with $N_e \approx 2.2 \times 10^{16}$ cm^{-3}, determined also from the H_β profile, is related to the breakdown through dielectric oxide layer from the surface towards anode. The third process, deduced from Mg II 448.11 nm and Al II 704.21 nm lines, appears after breakdown of dielectric and is followed by ejection of metal plasma of anode material. With Ti and Ta anode, only breakdown through dielectric layer occurs, while melting and ejection of anode material does not occur. In such case, plasma has lower viscosity and lower weight than metal plasma and consists of molecules, atoms and ions belonging to water vapor and ablated dielectric material [6,7].

To conclude, let us point out that the determination of time evolution and time of appearance of each process numbered here from 1 to 3 is a key task for further advances in understanding PEO processes. This would require time resolved spectroscopic study of a single micro-discharge what is not an easy task.

Acknowledgements
The author gratefully acknowledges the discussions with N.Konjević and S. Stojadinović. This work is supported by the Ministry of Education and Science of the Republic of Serbia under Project No. 171014.

References

[1] Yerokhin A L, Nie X, Leyland A, Matthews A, Dowey S J 1999 *Surf. Coat. Technol.* **122** 73.
[2] Sundararajan G, Rama Krishna L 2003 *Surf. Coat. Technol.* **167** 269.
[3] Moon S, Jeong Y 2009 *Corros. Sci.* **51** 1506.
[4] Narayanan T S N S, Park I S, Lee M H 2014 *Progress in Materials Science* **60** 1.
[5] Jovović J, Stojadinović S, Šišović N M, Konjević N 2011 *Surf. Coat. Technol.* **206** 24
[6] Stojadinović S, Vasilić R, Petković M, Belča I, Kasalica B, Perić M, Zeković Lj 2012 *Electrochim. Acta* **59** 354
[7] Stojadinović S, Jovović J, Petković M, Vasilić R, Konjević N 2011 *Surf. Coat. Technol.* **205** 5406
[8] Stojadinović S, Vasilić R, Petković M, Zeković Lj. 2011 *Surf. Coat. Technol.* **206** 575
[9] Jovović J, Stojadinović S, Šišović N M, Konjević N 2012 *J. Quant. Spectrosc. Radiat. Transfer,* **113** 1928
[10] http://physics.nist.gov/PhysRefData/ASD/lines_form.html
[11] Konjević N, Wiese W L 1976 *J. Phys. Chem. Ref. Data* **5** 259
[12] Konjević N 1999 *Phys. Reports* **316** 339
[13] Konjević N, Roberts J R 1976 *J. Phys. Chem. Ref. Data* **5** 209
[14] Hussein R O, Zhang P, Nie X, Xia Y, Northwood D O 2011 *Surf. Coat. Technol.* **206** 1990
[15] Žikić R, Gigosos M A, Ivković M, Gonzales M A, Konjević N 2002 *Spectrochim. Acta Part B* **57** 987
[16] Stehle C, Hutcheon R 1999 *Astron. Astrophys. Suppl. Ser.* **140** 93
[17] Griem H R 1974 Spectral Line Broadening by Plasmas, New York: Academic Press
[18] Wiese W L, Konjević N 1982 *J. Quant. Spectrosc. Radiat. Transfer* **28** 185
[19] Wiese W L, Konjević N 1992 *J. Quant. Spectrosc. Radiat. Transfer* **47** 185
[20] Konjević N, Ivković M and Sakan N 2012 *Spectrochim. Acta B* **76** 16

Journal of Physics: Conference Series **565** (2014) 012014 doi:10.1088/1742-6596/565/1/012014

Diagnostics of laser-induced plasma by optical emission spectroscopy

M Cvejić

Institute of Physics, University of Belgrade, P.O. Box 68, 11080 Belgrade, Serbia

E- mail: marko.cvejic@ipb.ac.rs

Abstract. The procedure for diagnostics of laser induced plasma (LIP) by optical emission spectroscopy technique is described. LIP was generated by focusing Nd:YAG laser radiation (1.064 nm, 50 mJ, 15 ns pulse duration) on the surface of pellet containing among other elements lithium. Details of the experimental setup and experimental data processing are presented. High speed plasma photography was used to study plasma evolution and decay. From those images optimum time for plasma diagnostics is located. The electron number density, N_e, is determined by fitting profiles of Li I lines while electron temperature, T_e, was determined from relative intensities of Li I lines using Boltzmann plot (BP) technique. All spectral line recordings were tested for the presence of self-absorption and then if optically thin, Abel inverted and used for plasma diagnostic purposes.

1. Introduction

Laser-induced breakdown spectroscopy (LIBS) is an emerging analytical technique for elemental analysis of various materials in all aggregate states. Driven by rapid technological development of pulsed lasers, compact and high-performance spectrometers and detectors, LIBS has found promising applications in many areas, such as elemental analysis in hazardous environments, material recycling, quality control in the industry, biomedical applications, environmental survey, interplanetary exploration and art conservation [1-4].

Recently, wider use of Laser Induced Plasma (LIP) as a plasma source for line shape studies has been noticed. There are several reasons for this interest. The main advantage over other plasma sources is simplicity and reproducibility of spectral line shapes for variety of elements very difficult to introduce or excite in traditionally used plasma sources. Some of these new spectroscopic data are of importance not only for theory testing but for diagnostics and development of LIBS analytical techniques, which require plasma diagnostics for further improvements and applications, see e.g. [3, 4, 5]. The determination of plasma parameters, in the first place electron temperature, T_e, and electron density, N_e, are of importance for the application of LIBS. The main difficulty in characterization of laser-induced plasma arises from its spatial and temporal inhomogeneity, which has to be carefully inspected before analyzing line shape and application of relevant plasma diagnostics. This can't be achieved without high speed photography, line self-absorption test and taking into account plasma inhomogeneity along the observation direction.

In this work, the procedure for LIP diagnostics by optical emission spectroscopy is presented. LIP is generated by focusing Nd:YAG laser to the solid target. Spatially resolved plasma optical emission is recorded at different delay times from the laser pulse. The back mirror technique for self-absorption test is described and applied. The experimental setup and processing of spectral recordings are described in detail. As a result, spatially and temporally resolved electron density and electron temperature distributions are obtained.

2. Experimental setup and data processing

The details of the experimental setup and procedure were described in detail recently [5] and therefore only a short description will be given for completeness. The schematic diagram of used experimental setup is given in figure 1. The plasma was induced by Nd:YAG laser radiation at 1064 nm having 15 ns pulse width and 50 mJ pulse energy. The laser beam was perpendicularly directed toward target surface and focused by means of a 100 mm focal length lens. Laser induced plasma is generated in front of a solid state surface. The sample target was laboratory made from the homogenous powder mixture of Al_2O_3 900 mg : Li_2CO_3 400 mg : $MgCO_3$ 100 mg. The target was formed by compressing powder mixture into a cylindrical pellet (8 mm diameter and 2 mm thick). The ratio of target constituents was determined empirically by aiming to avoid strong self-absorption of studied lines. Spectroscopic measurements were performed in air at atmospheric pressure. The image of plasma plume was projected with 1:1 magnification, using lens with 170 mm focal length, onto the entrance slit of the spectrometer (Shamrock sr-303i, Czerny–Turner type, focal length 303 mm) supplied with three gratings 300, 1200, and 2400 grooves/mm. Quartz lens, 50 mm diameter with a focal length of 100 mm, flat mirror and mechanical shutter were mounted behind LIP at the opposite side from spectrometer to verify whether line self-absorption is present. The plasma radiation was recorded with the ICCD detector (Andor Technology, model DH720-18F-63, with 1024 × 256 pixels, 26 × 26 μm^2 pixel size, 18 mm intensifier diameter), mounted at the exit slit plane of spectrometer. The ICCD was operated by a pulse generator (DG-535, Stanford Research Systems). Spectral lines were recorded with 2400 grooves/mm grating. The instrumental profile is close to Gaussian with FWHM of 0.093 nm. Instrumental profiles are measured using low pressure Hg and Ne spectral calibration lamps. The wavelength sensitivity of spectrometer, ICCD system and all optical elements is calibrated against standard tungsten coiled-coil filament quartz halogen lamp (EG&G, model 597-1). The focal plane of laser radiation was set 1 mm below the target surface. The diameter of impact spot at the sample surface was 200 μm. The target was rotated or translated after 16 laser shots. Each recording is an average of 10 accumulations, each one obtained with 16 laser shots. The plasma plume was observed perpendicularly to the laser beam and parallel to the target surface at the distance of 0.5, 1 and 1.5 mm from the target surface.

The recorded spectral image represents spectral radiance emitted by the observed LIP layer as a function of the lateral position and wavelength. The radiation detected by each pixel of the ICCD matrix corresponds to an integrated intensity of the plasma emission along the line-of-sight across the radius of the plasma plume. An example of spectrum is given in figure 2a.

Spectral lines were recorded in "imaging" mode. First column in the matrix of image recording with ICCD camera represents wavelengths while in the first row are lateral positions. In this way other columns represent spectra at different lateral positions and other rows lateral intensity distribution at different wavelength. The program procedure used for experimental data processing is described in the next paragraphs.

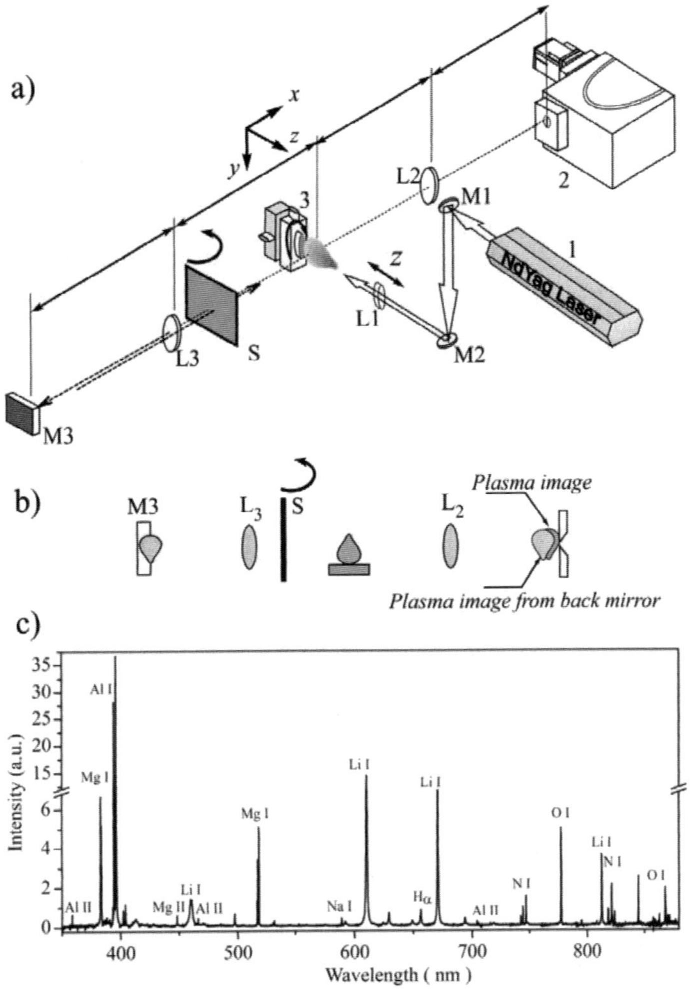

Figure 1. a) Experimental setup for the study of laser induced plasma; b) schematic setup for self-absorption testing; 1 – Nd:YAG laser, 2 – spectrometer, 3 – target holder; L1 , L2, L3 – lenses; M1, M2 –laser beam deflecting mirrors at 45°, M3 – flat mirror, S –mechanical shutter. c) Typical overall spectra

In the first step matrices of experimental data are imported along with correction factor (apparatus function). After intensity and wavelength correction, the image is cropped by specifying cropping limits along wavelength and along lateral axis, see figure 2a. Next step is symmetrisation, which is done by reflecting the upper part of the plasma image about its symmetry axis and averaging with the lower part. This step is important because of the Abel inversion, which can only be applied to plasma with cylindrical symmetry. The image matrix after symmetrisation is shown in figure 2b.

Next step is data smoothing, which is a valuable tool for noise removal and represents an important step for preparation of image matrices for self-absorption test. By specifying number of points and polynomial order of Savitzky – Golay filter, the noise can be removed while preserving spectral line shape and lateral intensity distribution. By testing various combinations of filter parameters the optimal smoothing can be selected. After the selection of filter parameters, the matrices recorded with and without back mirror, see figure 1 and figure 2c and 2d are smoothed along lateral and then, optionally, along wavelength axis.

The most important is the third section, see figure 2e, dealing with self-absorption correction (SAC). The theoretical background is given in [5,6]. The input parameters for SAC procedure are ratios of continuum at different lateral positions d, $R^c(d)$, recorded with and without back mirror and ratios of recorded line profiles, $R_\lambda(d)$, with and without back mirror. The advantage of this procedure: the ratios for each lateral position are calculated and then the correction factor, $K_\lambda(d)$, for each lateral position determined. The ratios of continuum are calculated in the following way. For every data column of matrix recorded without or with mirror far wing intensity values are averaged and set to be continuum intensity. Here "far wing" is the wavelength region 3-5 half widths away from line centre. Next step is the calculation of line intensity ratios along lateral position $R_\lambda(d)$. Using these values with the following formula (1), the correction factor is evaluated.

$$K_\lambda(d) = \frac{\ln \dfrac{R^c(d)-1}{R_\lambda(d)-1}}{1-\dfrac{R_\lambda(d)-1}{R^c(d)-1}} \tag{1}$$

The optical depth $k_\lambda l\ (d)$ is calculated also for different lateral positions. Here k_λ is the absorption coefficient and l is the length of a homogenous plasma layer. If $k_\lambda l\ (d)$ is smaller than unity, spectral profiles were reconstructed to the optically thin case, see e.g. [6]. The spectral lines with $k_\lambda l\ (d)$ approaching unity were not used in further procedure. The example of SAC for the line profile originating from the plasma axis (centre of matrix) is given in figure 2g and examples of SAC for the lateral profiles at line maximum (460.28 nm) are shown in figure 2h.

The final step of experimental data processing is Abel inversion, see figure. 2f. The measured lateral intensity after the correction for self-absorption or optically thin lateral intensity distribution is converted into radial intensity with the help of the Abel inversion performed in two steps [7]. In the first step, the experimental lateral intensity distribution is fitted with Jacobi orthogonal polynomials and in this way, the experimental data distribution is represented by an analytical curve. The advantage of this technique is the possibility of testing and selecting best type and order of Jacobi polynomials. The choice of polynomial and selection of the polynomial degree depends on the overall shape of experimental data and noise contribution. Once the optimal analytical curve is determined, the Abel integral is easily solved in the next step. The described two step procedure of Abel inversion works also well when the experimental lateral profiles have a dip in the middle.

3. Plasma evolution and decay

Serious difficulties for spectroscopic diagnostics of LIP may arise from spatial inhomogeneity and time variation of plasma shape, which is simultaneously followed by the change of plasma parameters. Consequently, the emission of ablated material in plasma plume changes intensity as well. These difficulties are even more pronounced if one intends to perform spatially and temporally resolved measurements. In that case, it is of greatest importance to choose optimum observation time for plasma diagnostics, such as: delay time after laser irradiation of target, gate width and time with good radial plasma symmetry. To achieve this goal plasma images at different delay times after the laser

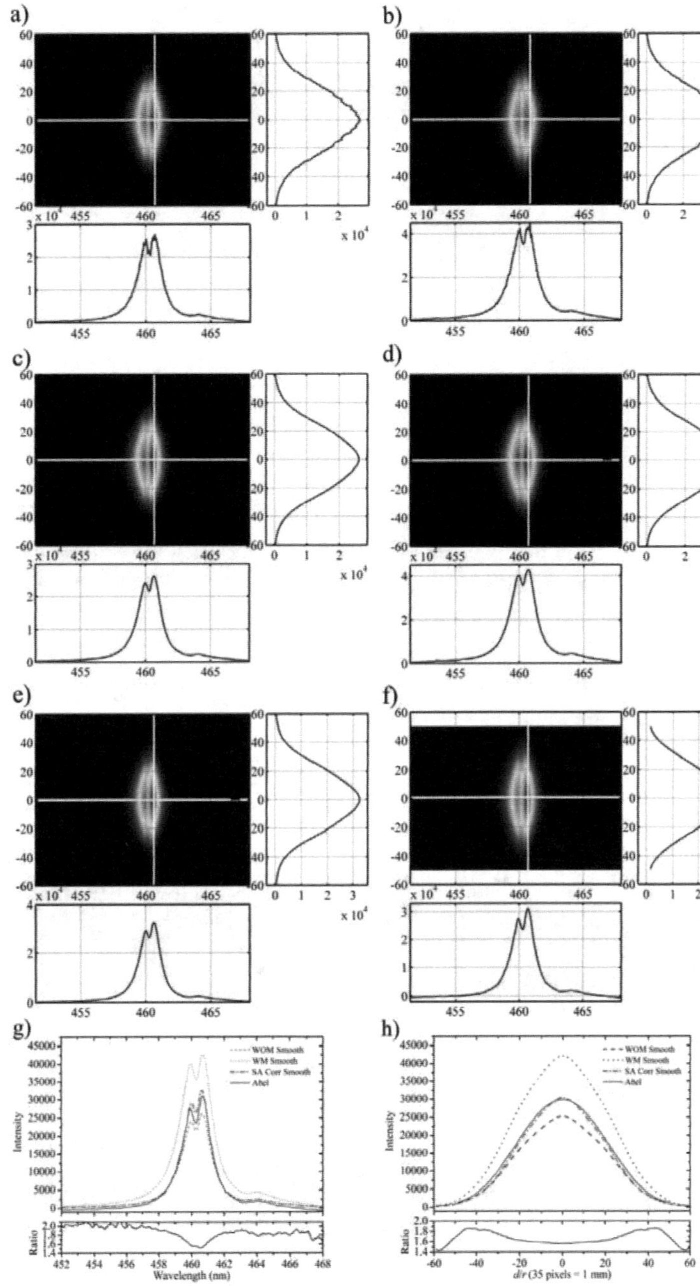

Figure 2. The steps of experimental data processing on the example of Li I 460.28 nm line with forbidden component: a) data matrix recorded without back mirror (WOM) after corrections for apparatus function, wavelength correction and crop; b) WOM after symmetrisation; c) WOM after smoothing; d) data matrix recorded with back mirror (WM) after steps a), b), c); e) data matrix after self-absorption (SA) correction; f) data matrix after Abel inversion. Examples of self-absorption correction and Abel inversion: g) line profiles from plasma axis h) lateral distributions at line maximum.

pulse are recorded, see figure 3, with an objective (focal length 75 mm) mounted with macro bellows on the ICCD camera (1024×1024 pixels). Each pixel covers the area of 0.011 × 0.011 mm. From recorded images it is evident that until 1 μs, plasma dimensions and shape were rapidly expanding. After 1 μs, plasma shape was changing much slower and thus, plasma is more appropriate for diagnostics. To avoid time integration and inhomogeneity, which may affect spectral line profiles, special care was taken to select delay time and gate width. Finally the optimum gate width of 200 ns and delay times for plasma observation of 1, 2, 3, 4, 5, 7 μs after laser pulse were selected. For these delay times and with 200 ns gate width, plasma is assumed to be quasi-stationary. The selected gate width and delay times were appropriate also to avoid strong plasma continuum emission and to enhance signal-to-background ratio.

To obtain spatial distribution of plasma parameters, three different cross sections of plasma plume at z = 0.5; 1; 1.5 mm (starting from target surface) were observed. Because of unity magnification of optical system, the thickness of observed cross sections is equal to the slit width (10 μm).

There are certain constrains for the application of experimental self-absorption test with back mirror, see Fig. 1. Those constrains are related to plasma expansion speed and finite speed of light. At early phase of plasma expansion (up to 700 ns after laser pulse) estimated average plasma speed is $4.3 \cdot 10^3$ m/s so during round trip of plasma light towards and from back mirror (2.66 ns), the observed plasma layer moved along plasma axis for 11 μm, what is more than slit width, 10 μm. One microsecond after laser irradiation of target, the plasma expansion speed is significantly lower (less than 500 m/s) so for the same round trip of light the observed plasma layer moved less than 1μm, so the mismatch between emitting and reflecting volume is only 10%. Thus, by selecting later times for plasma observation and by using reasonably narrow gate width the self-absorption test with back mirror is considered valid.

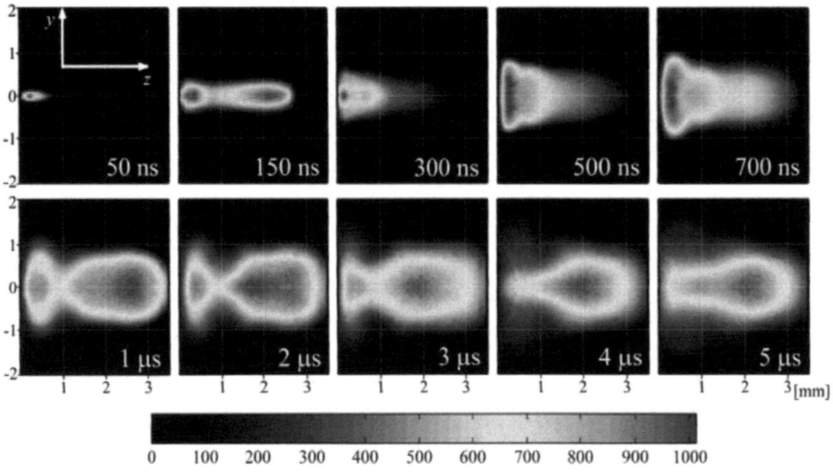

Figure 3. Temporal and spatial evolution of plasma plume. For the first row of images the exposure time (gate width) was 10 ns while for the second row the exposure time was 100ns. Each image represents an average of 5 laser shots. The emission intensity within images is scaled according to the colour map shown below images.

4. Results and discussion

For the electron temperature (T_e) determination Boltzmann plot (BP) technique was employed, using relative intensities of Li I lines given in Table 1. Prior to calculation of T_e, the issue of BP technique applicability under LIP conditions was addressed. The discussion about BP technique and partial local thermal equilibrium (pLTE) existence in LIP plasma is given in recent publications [5,8]. The upper levels of transitions used for T_e determination have to be in pLTE in order to have valid T_e diagnostic by the Boltzmann plot technique. Applying the same procedure as in [5,8] determined critical quantum number at pLTE limit is 2, calculated using eq. (7.77) from [9] for lowest N_e and T_e. This means that all Li I lines with the upper level principal quantum number 3 (ground state + next level) or higher can be used for the T_e determination using the BP technique. From Table 1 one can see that all three Li I lines fulfil pLTE condition. On the other hand, the so called McWhirter criterion, commonly used for estimation of LTE presence in LIP, gives value of $N_e = 6.38\cdot10^{15}$ cm^{-3} (eq. (12) in [10], see also [11]) for critical electron density above which LTE is established. According to this criterion, all levels of Li I could be used under given experimental conditions. These two considerably different results indicate that, generally speaking, criteria for LTE assessment are not well defined in literature, see also [10].

A typical Boltzmann plot with Li I lines is shown in Fig. 4a, while radial distributions of T_e are given in Fig 4b,c,d. The estimated uncertainty of T_e measurements is in the range 8% to 12% depending upon delay time i.e. intensities of Li I lines.

Table 1. Transition array, multiplet, wavelengths, lower and upper energy levels, upper level statistical weight for Li I lines used for the Boltzman plot [12]

	Transition	Multiplet	Wavelength (nm)	E_i (cm^{-1})	E_k (cm^{-1})	g_k	A (10^8 s^{-1})	S
1.	$1s^22p$-$1s^24s$	^{2}P°– ^{2}S	497.166	14 903.66	35 012.06	2	3.460 10^{-2}	0.42
			497.175	14 904.00	35 012.06	2	6.918 10^{-2}	0.84
2.	$1s^22p$-$1s^25s$	^{2}P°– ^{2}S	427.307	14 903.66	38 299.50	2	1.59 10^{-2}	0.122
			427.313	14 904.00	38 299.50	2	3.17 10^{-2}	0.244
3.	$1s^22p$-$1s^25d$	^{2}P°– ^{2}D	413.262	14 904.00	39 094.93	4	1.81 10^{-2}	0.252
			413.262	14 904.00	39 094.94	6	1.09 10^{-1}	2.27
			413.256	14 903.66	39 094.94	4	9.04 10^{-2}	1.26

After all necessary line tests and, if necessary corrections, like for the case of self-absorption, interference with neighbouring lines, the experimental line profile of Li I 497.17 nm line was fitted with an asymmetric profile and from the best fit N_e is determined. The procedure for N_e determination from an experimental line profile is described in [5]. For the fitting of experimental profiles electron impact width w_e and ion broadening parameter A from [13] were used. Typical fitting results are presented in figure 5. Radial distributions of N_e obtained in this way are shown in figure 6. From figure 6 it can be seen that N_e has highest value at early delay time and it is decreasing at later delay time. Also N_e is higher at plasma centre and it is slowly decreasing towards plasma periphery.

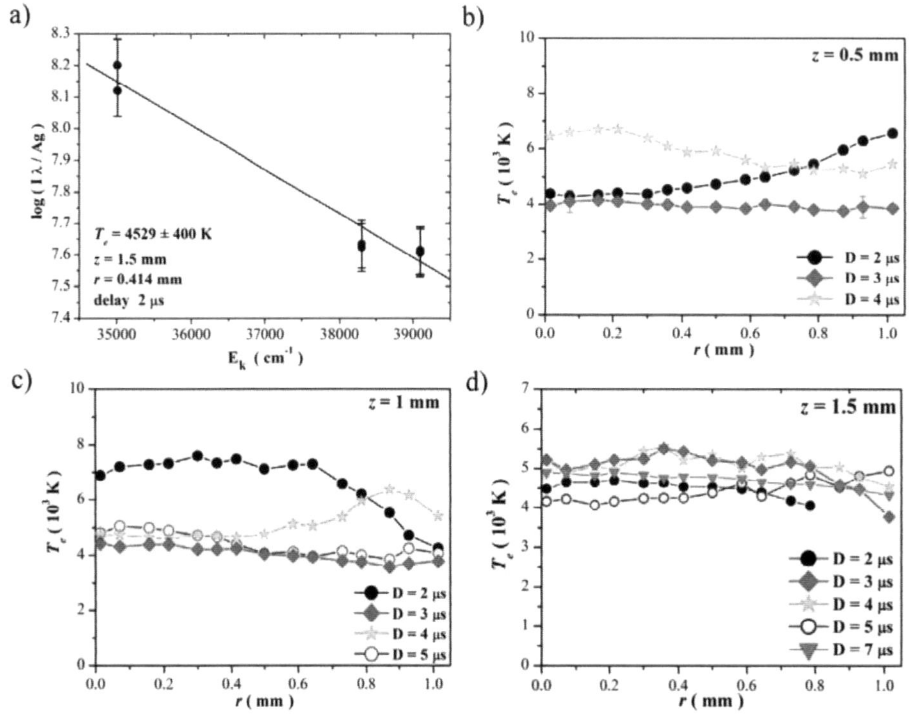

Figure 4. a) Typical Boltzmann plot of Li I lines. Radial distributions of electron temperature at different delay times (D) at the observation distances b) z = 0.5 mm, c) z = 1 mm, d) z = 1.5 mm from the target surface.

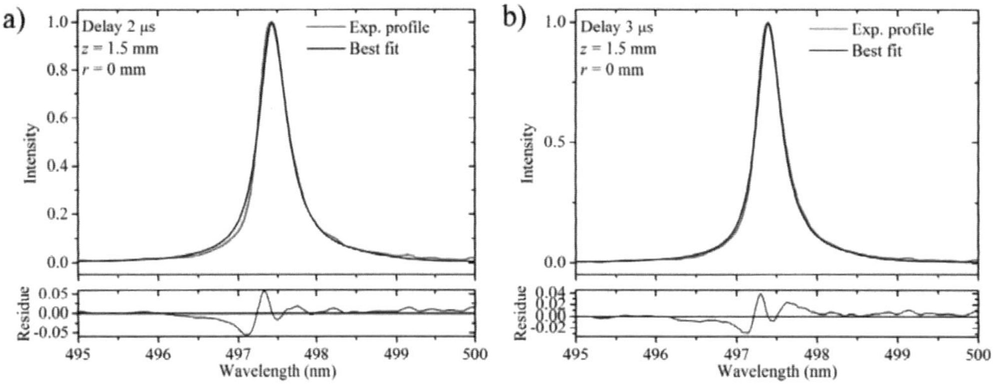

Figure 5. Best fit examples of the Li I 497.17 nm line, for the delay times of a) 2 μs, b) 3 μs. Both experimental profiles originate from the plasma axis at z = 1.5 mm from the target surface.

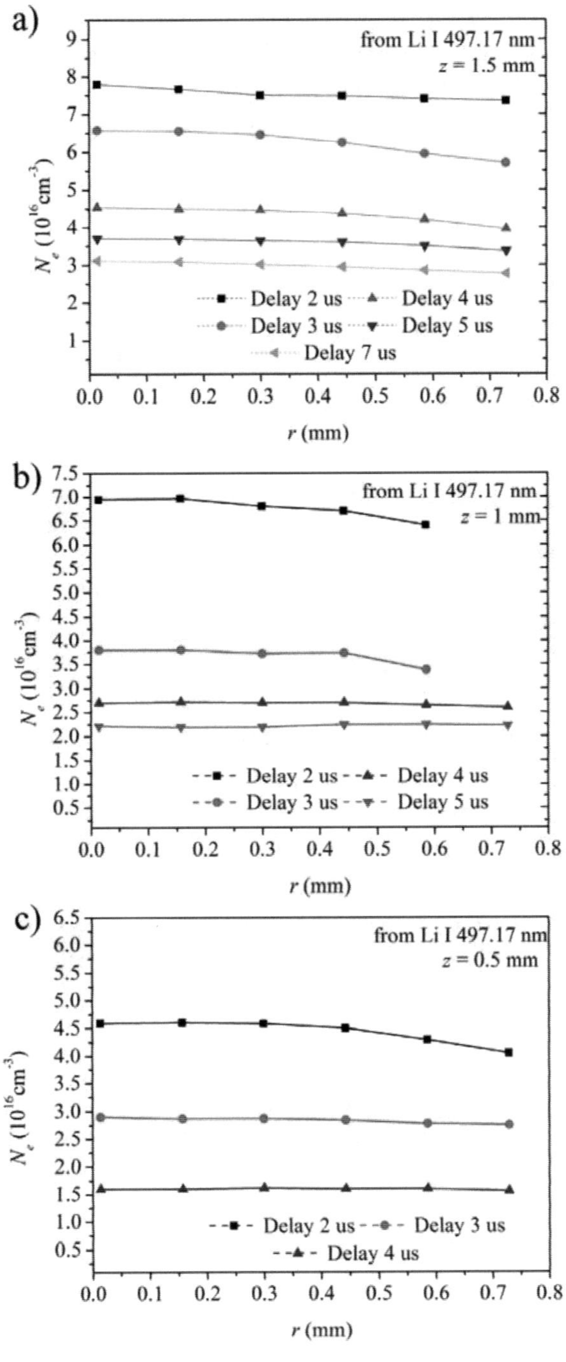

Figure 6. Radial distributions of N_e at different delay times at a) z = 1.5 mm, b) z = 1 mm, c) z = 0.5 mm from the target surface.

5. Summary and conclusions

Spectroscopic diagnostics of LIP is not an easy task. Inherent drawbacks of the source, in the first place inhomogeneity and plasma transient nature are pointed out, followed by the proposed methods to overcome those constraints. The results of spectroscopic diagnostic study of LIP are presented. LIP was generated by focusing Nd:YAG laser radiation (1064 nm, 15 ns pulse duration) on the surface of the pellet containing among other elements lithium. High speed plasma photography was used to determine optimum time for plasma diagnostics. The high speed photography is followed by the tests for the presence of line self-absorption. If spectral lines were optically thin or could be corrected to an optically thin case, the Abel inversion procedure was applied.

All necessary steps towards time and space resolved acquisition and diagnostics are given on the example of Li neutral lines. Some of the results are shown on graphs 4,6. Abel inverted profiles were used to determine radial line intensity distribution for N_e measurement. The line profiles of Li I lines were fitted with asymmetric profile convoluted with Gaussian. The radial distributions of N_e values are shown in Fig. 6. The electron number density, N_e is measured by fitting profiles of Li I 497.17 nm line while radial distribution of electron temperature, T_e, was determined from relative intensities of Li I lines using Boltzmann plot (BP) technique.

References

[1] Michael A P M 2010 *Spectrochim. Acta Part B.* **65** 185
[2] Singh J, Thakur S (ed) 2007 *Laser-Induced Breakdown Spectroscopy* (Elsevier)
[3] Cramers D A, Radziemski L J 2006 *Handbook of Laser-Induced Breakdown Spectroscopy* (London, John Wiley & Sons)
[4] Hahn D W, Omenetto N 2010 *Appl. Spectroscopy* **64** 335A
[5] Cvejić M, Gavrilović M R, Jovićević S, Konjević N 2013 *Spectrochim. Acta Part B* **85** 20
[6] Konjević N 1999 *Phys. Rep.* **316** 339
[7] Djurović S 1999 *Journal of Research in Physics* **28** 153
[8] Cirisan M, Cvejić M, Gavrilović M R, Jovićević S, Konjević N, Hermann J 2014 *J. Quant. Spectrosc. Radiat. Transfer* **133** 652
[9] Griem H R 1997 *Principles of Plamsa Spectroscopy* (Cambridge University Press)
[10] Cristoforetti G, Giacomo A De, Dell'Aglio M, Lenaioli S, Tognoni E, Palleschi V, Omenetto N 2010 *Spectrochim. Acta Part B* **65** 86-95
[11] McWhirter R W P 1965 *Plasma Diagnostic Techniques* Huddlestone R H, Leonard S L (New York, Academic Press)
[12] Kramida A, Ralchenko Yu, Reader J and NIST ASD Team 2012 NIST Atomic Spectra Database, http://physics.nist.gov/asd
[13] Griem H R 1974 *Spectral line broadening by Plasmas* (New York, Academic Press)

Electrical breakdown in air: new experiments and statistical and numerical models

A P Jovanović
Department of Physics, Faculty of Sciences and Mathematics, University of Niš, Višegradska 33, P.O. Box 224, 18001 Niš, Serbia

E-mail: alexandar.bmf@gmail.com

Abstract. The electrical breakdown and discharges in air are of great significance for science and technology due to wide applications. Besides a breakdown itself, the pre-breakdown phenomena and relaxation after the breakdown have a significant role. In this report the analysis of DC glow discharge and its relaxation processes is presented. Statistical analysis of time delay distributions is presented and the statistical model based on the mixture distribution is applied to data measured with different electrodes from which electron yield is calculated. The occurrence of mixture distributions is physically explained based on analysis of the cathode surface. Also, by analyzing the memory curve measured in synthetic air the early and late relaxation of DC discharge are discussed. The processes responsible for the memory effect are identified and corresponding rate coefficients are determined. In order to confirm that the correct particles are identified, the 2D numerical model for relaxation is developed.

1. Introduction

The electrical breakdown and discharges as well as post-discharge in air are of great significance for science and technology due to wide applications for plasma devices operating in the pulsed or periodic regime, for surface treatment and modification, for plasma actuators or in medicine [1-7]. A majority of papers cover the electrical discharge in air at the atmospheric pressure [8-10], and less of them discuss low-pressure cases [11-15].

Electrical breakdown is not instantaneous process and upon application of sufficient voltage some breakdown time delay exists. The time that elapses from the moment of application of voltage greater than the static breakdown voltage U_s to the breakdown itself is the breakdown time delay t_d. It consists of the statistical time delay t_s (time that elapses from the application of sufficient voltage to the appearance of the free electron that initiates breakdown) and the formative time delay t_f (time from the occurrence of initiating electron to the collapse of applied voltage and occurrence of a self-sustained current) [16]. The stochastic nature of the breakdown time delay was experimentally proven by Zuber, while von Laue derived an exponential distribution for the breakdown time delay [17, 18]. A strict mathematical transition from binomial distribution of electron occurrence in the inter-electrode space to exponential distribution for the statistical time delay was derived by Kiselev [19] and in this paper he explicitly stated that using Gaussian distribution is incorrect. Nevertheless, Gaussian and Gauss-exponential distributions of statistical time delay were obtained experimentally [14, 15, 20].

Besides the time delay distributions, the memory curve can be used for the analysis of the processes in relaxation. The memory curve was measured for the first time in the argon diode, when was noticed that the mean value of the electrical breakdown time delay depends on the relaxation time (between the cessation of the previous discharge and the next application of the working voltage) [21, 22]. Since then, this method was used for tracing the active particle decay in afterglow of inert and molecular gases and their mixtures, including air [21-26].

In this paper the statistical analysis of the time delay distributions is presented and the mixture distribution for statistical time delay of electrical breakdown, based on the Gaussian and exponential distributions, are proposed and applied to experimental data. The relation for electron yield for cases of mixture distributions is proposed. The occurrence of mixture distributions in the case of carbon steel electrodes is explained by desorption of surface charges from the oxide on the cathode and enlarged surface area, which is confirmed by scanning electron microscope (SEM), energy dispersive X-ray spectroscopy (EDX) and atomic force microscopy (AFM) measurements. By using the parameters of distribution and proposed relation, electron yield is determined. The afterglow of DC discharge in synthetic air is discussed by analysing results of the time delay measurements. The dominant processes are determined by applying analytical and numerical models and corresponding coefficients are calculated.

2. Experimental details

The time delay measurements were carried out on a gas tubes filled with synthetic air at the pressure of 300 *Pa*. Synthetic air is a mixture of 78 percent of nitrogen with less than 5 *ppm* of impurities and rest of oxygen with less than 3.5 *ppm* of impurities. The discharge tubes are made of borosilicate glass (8245, Schott technical glass) with volume $V \approx 300\ cm^3$ and with characteristic diffusion length $\Lambda = 1.12\ cm$. The measurements of distributions of a statistical time delay were carried out with electrodes made of carbon steel (ASTM/AISI A414E) and stainless steel (AISI 304) with diameters $D = 6\ mm$ and variable inter-electrode distance d. Prior to instalment electrodes were lapped and polished with polishing paste with the grain size 1 μm and after that cleaned in ultrasonic bath in methanol and then in distilled water. The gas tube was pumped down to $10^{-3}\ Pa$, degassed at 650 K and then filled with the synthetic air at $p = 300\ Pa$. Time delay measurements were carried out with 100 data in a single series with glow time $t_g = 1\ s$. For the memory curve measurements only the electrodes made of stainless steel were used. The memory curve was measured by varying relaxation time τ (afterglow period) from 1 *ms* to approx. 15 minutes. The tube was protected from the external light during the measurements. More details about experiment can be found in [15].

3. Analysis of statistical time delay distributions

The exponential distribution for the statistical time delay derived by Kiselev [19] covers the cases of low electron yield. The exponential distribution of statistical time delay is given by following relation:

$$f(t_s) = \frac{1}{\overline{t_{se}}} e^{-\frac{t_s}{\overline{t_{se}}}} \tag{1}$$

where $\overline{t_{se}}$ is the mean of statistical time delay for exponential distribution. From the parameters of this distribution the effective electron yield can be determined [18] from relation:

$$Y_{eff} = \frac{1}{\overline{t_{se}}}. \tag{2}$$

The effective electron yield represents the electron production rate in the gas gap Y multiplied by the breakdown probability P and is given by $Y_{eff} = YP$ [15]. Based on relation for electron yield $Y = \gamma \Gamma S$ it is possible to estimate number density of particles that has lead to secondary electron emission (ions, residual neutral active states, etc.) where γ is the secondary electron yield, Γ is the particle flux and S is the area of the front cathode surface.

Nevertheless, in the cases of high effective electron yield, the occurrence of Gaussian and Gauss like i.e. slightly asymmetric distributions with tail are reported [14, 15, 20]. In order to physically base these distributions, the transition from binomial to Gaussian distribution is given in [14]. By using de Moivre-Laplace theorem for the cases when number of initiating electrons is very large and probabilities for occurrence and non-occurrence of effective electron is not too close to zero (i.e.

effective electron yield *YP* is high), the binomial distribution of electron occurrence is approximated by the Gaussian distribution [14]. In this way the Gaussian distribution used for statistical time delay is physically based and the electron yield can be determined from parameters of the distribution. The Gaussian distribution of statistical time delay has the following form:

$$f(t_s) = \frac{1}{\sqrt{2\pi}\sigma_g} e^{-\frac{(t_s - \overline{t_{sg}})^2}{2\sigma_g^2}} \tag{3}$$

from which electron yield can be determined from parameters of the distribution:

$$Y_{eff} = \frac{\overline{t_{sg}}}{\sigma_g^2} \tag{4}$$

where $\overline{t_{sg}}$ is the mean and σ_g is the standard deviation of the statistical time delay obtained by fitting procedure [14].

In order to describe occurrence of Gauss like distributions the mixture of Gaussian and exponential distribution was proposed [20]. The mixture distribution is the mixture of two or more distributions with certain weights and in the case of statistical time delay with *n* components has the following form [27]:

$$f(t_s) = \sum_{i=1}^{n} a_i f_i(t_s) \tag{5}$$

where $a_i > 0$, $i = 1..n$ $\sum_{i=1}^{n} a_i = 1$, t_s is the statistical time delay and a_i are the mixing weights. In the cases where more than one subpopulation of data exists in the whole population, the mixture distributions are often used [27, 28]. For example, the mixture distributions are applied for description of electric tubes and radio transmitter failures [28, 29], where every component in mixture corresponds to one cause of failure. In the case of statistical time delay the occurrence of mixture distributions implies on existence of more than one subpopulations of initiating electrons. For the case of statistical time delay the mixture of Gaussian and exponential distributions is first proposed in ref. [20]:

$$f_{GE}(t_s) = a_G \frac{1}{\sqrt{2\pi}\sigma_g} e^{-\frac{(t_s - \overline{t_{sg}})^2}{2\sigma_g^2}} + a_E \frac{1}{\overline{t_{se}}} e^{-\frac{t_s}{\overline{t_{se}}}} . \tag{6}$$

Since these distributions are based on binomial distribution of electron occurrence [14, 19] by starting from electron yield for Gaussian distribution, the relation for mixture of Gaussian and exponential distributions is derived [15]:

$$Y_{eff} = \frac{a_G \overline{t_{sg}} + a_E \overline{t_{se}}}{a_G^2 \sigma_g^2 + a_E^2 \overline{t_{se}}^2} . \tag{7}$$

This relation is more general than relations for single components and in limiting cases, when $a_G \to 0$ or $a_E \to 0$, tends to the relation for single component of mixture distribution, i.e. exponential or Gaussian one. By using proper parameters, this equation can be applied to the mixture of different components, i.e. two Gaussian distributions or two exponential distributions. Previous relation (7) can be generalized for the mixture of *n* distributions [15]:

$$Y_{eff} = \frac{\sum_{i=1}^{n} a_i \overline{t_{si}}}{\sum_{i=1}^{n} a_i^2 \sigma_i^2} \tag{8}$$

where a_i represents the weights of components of distribution, $\overline{t_{si}}$ is the mean value and σ_i is the standard deviation of the statistical time delay of a single component [15].

As example for the use of mixture distribution, the results published in [15] are presented in following text. The experimental data measured in synthetic air with carbon steel and stainless steel electrodes were analyzed and compared in order to physically explain occurrence of mixture distribution [15]. The measurements are carried out at working voltage $U_w = 475V$, glow current $I_g = 500\mu A$ and at glow time $t_g = 1s$. The static breakdown voltages were $U_s = 396V$ for the tube with carbon steel electrodes and $U_s = 420V$ for the one with stainless steel electrodes. For the relaxation time $\tau = 17ms$ the distribution of the data measured in the case of the sample with carbon steel electrodes has Gauss like shape (figure 1a) while the data measured in the second sample with stainless steel electrodes follow exponential distribution (figure1b). The mixture of the Gaussian and exponential distribution (6) is applied to fit the experimental data (figure 1a) and the following parameters of distribution are obtained $\overline{t_{sg}} = 4.45\times10^{-6}s$, $\sigma_g = 2.15\times10^{-6}s$, $\overline{t_{se}} = 1.02\times10^{-5}s$, while the weights of distributions are fixed and chosen to be $a_G = 0.78$ and $a_E = 1 - a_G = 0.22$, respectively. From the relation (7) the effective electron yield is calculated to be $Y_{eff} = 8.27\times10^5\ 1/s$. The sample with stainless steel electrodes is modelled by exponential distribution (1), mean statistical time delay is determined $\overline{t_s} = 4.23\times10^{-4}s$ and corresponding effective electron yield is calculated $Y_{eff} = 2365.8\ 1/s$.

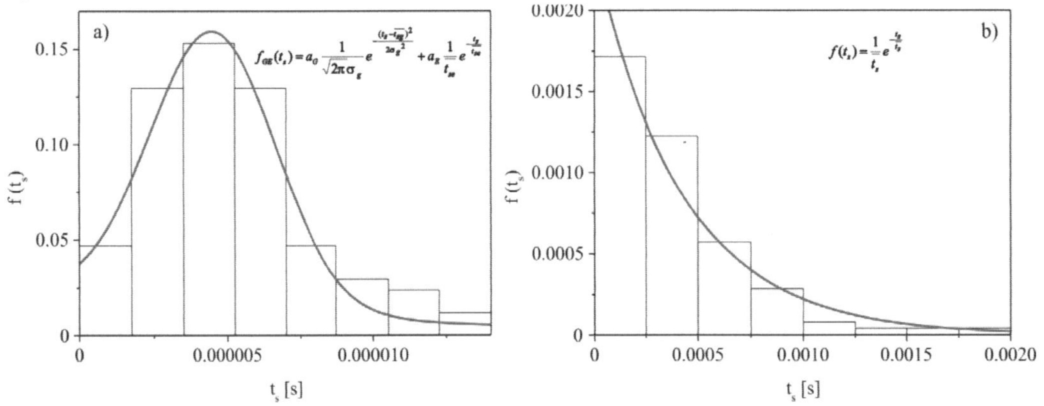

Figure 1. Statistical time delay distributions at relaxation time $\tau = 17ms$ for the sample with a) carbon steel and b) stainless steel cathode [15].

The appearance of mixture distribution and increased effective electron yield indicate existence of additional initiating mechanism. The only difference between two analyzed samples is in cathode material therefore one subpopulation of electrons must originate from the surface initiating mechanism. In order to test this statement the SEM, EDX and AFM measurements are conducted (figure 2). The dark spots on SEM image (figure 2a) indicate the substance with dielectric properties, probably an iron oxide. Also, granular structure of the oxide layer can be noticed, which leads to quite enlarged effective surface area. The bright spots on the tips of granules are caused by the enhanced secondary electron emission. In EDX spectrum (figure 2b), besides iron and manganese which are constituents of the alloy, the oxygen peak is observed. Based on SEM image and EDX spectrum, it can be concluded that inhomogeneous mixture of different iron oxides is formed on the cathode surface. The iron oxide has dielectric properties and can retain surface charges. In order to better analyze the cathode surface AFM is used and from the 3d topography image (figure 2c) the characteristic granular structure similar as in SEM image is observed. The root-mean-squared (RMS) surface roughness

$R_{RMS} = 63.23nm$ is determined for the $30\times 30\,\mu m$ area presented in figure 2c. This confirms that high electron yield is caused by the combined effect of surface charges and enlarged effective surface area [15].

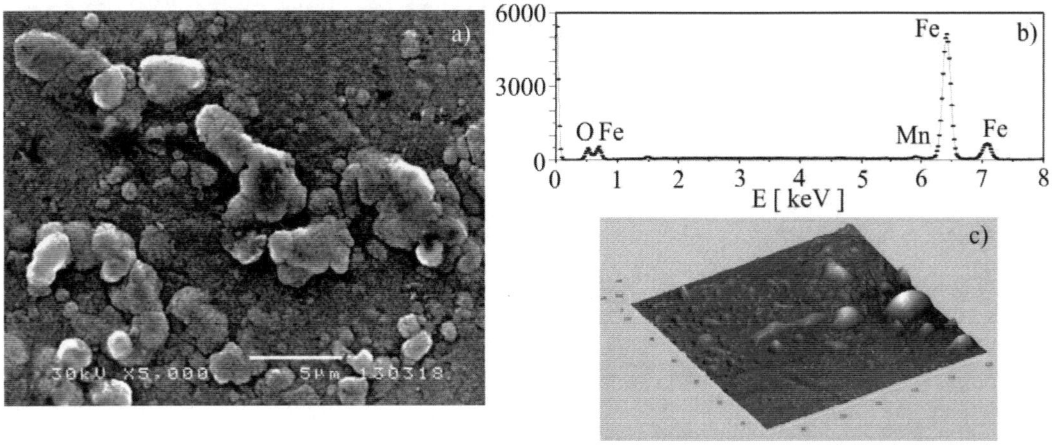

Figure 2. a) SEM image b) EDX spectrum and c) 3d topography of carbon steel cathode [15].

4. Relaxation of DC discharge in synthetic air

The time delay measurements are often used to analyze relaxation in various gases in order to determine the dominant processes and corresponding rate coefficients [21-26]. In this report the relaxation of synthetic air discharge is analyzed by the means of time delay measurements. The memory curve, mean time delay as a function of relaxation time $\overline{t_d}(\tau)$ [21], is measured and the analytical and numerical models are applied. In figure 3 the memory curve measured on the tube filled with synthetic air at the pressure of 300 *Pa*, with electrodes made of stainless-steel, inter-electrode distance *d*=6 *mm*, at working voltage $U_w = 500V$, glow current $I_g = 300\mu A$ and glow time $t_d = 1s$ is presented. The three characteristic regions can be identified on the memory curve.

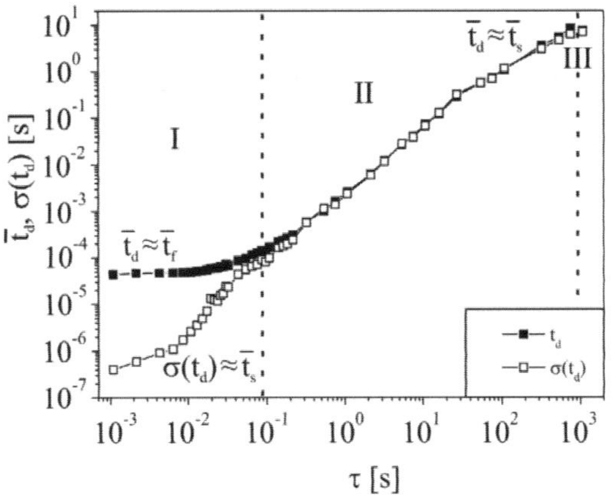

Figure 3. Memory curve measured in synthetic air at the pressure of 300 *Pa*, with stainless-steel electrodes, inter-electrode distance *d*=6 *mm*, at working voltage U_w=500 *V*, glow current I_g=300 *μA* and glow time $t_g = 1s$.

The first region between 1ms and 90 *ms* characterizes with the slow rise of mean time delay corresponding to formative time $(\overline{t_f} \approx \overline{t_d})$ and fast rise of mean of the statistical time delay $(\overline{t_s} \approx \sigma(t_d))$. The second region from 90 *ms* to 1000 *s* is less steep, with the slower rise of the time delay and at the 1000 *s* the saturation of time delay can be observed. It is assumed that in the first region the dominating particles are probably ions remained from the previous discharge. At the times greater than 90 *ms* the ion number density is most likely depleted, therefore influence of the neutral particles can be reasonable explanation of this part of memory curve. The saturation at 1000 *s* indicates some constant production of initiating electrons and can be explained by background radiation caused by natural radioactivity and cosmic radiation. Further, each of these regions is discussed.

The first region of the memory curve, with the fast rise of time delay, is presented in semi-logarithmic plot (figure 4), from which the three different slopes can be observed. The ions remained from the previous discharge are still present in inter-electrode space and can cause emission of initiating electrons. Based on an analytical model from [26] the connection between exponential rise of statistical time delay and diffusion losses of ionic particles is established. Therefore, this part of memory curve can be explained by change of diffusion regimes of ionic particles causing emission of initiating electrons similarly as in [26]. The characteristic decay frequencies ν are determined from the slopes, from which the effective diffusion coefficients $D_{eff} = \nu\Lambda^2$ are determined, where Λ is characteristic diffusion length.

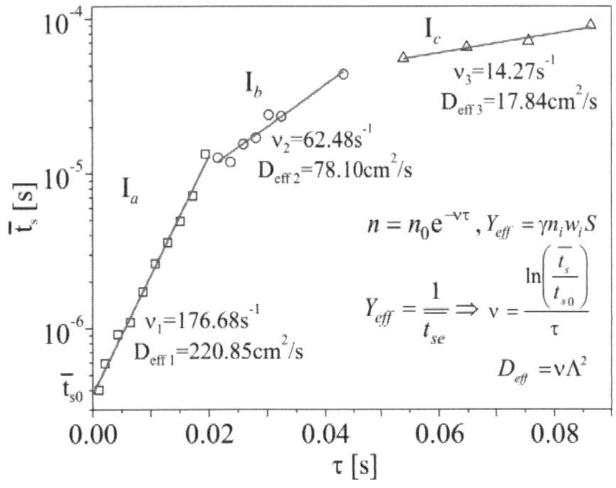

Figure 4. The first region of the memory curve in synthetic air presented in semi-logarithmic plot. By using the relations shown in the figure, the connection between statistical time delay and diffusion of ions is established. The three different slopes imply existence of the change of diffusion regimes. From the relations, the effective diffusion coefficients are determined.

The following step in the analysis is identification of the dominant ions leading to secondary emission of electrons. By taking into account a work function of stainless steel $e\phi_i \approx 4.5\, eV$ [30] and ionization energies of ions present in stable glow in synthetic air it is concluded that only N_2^+ (E_i=15.57 *eV*) and O_2^+ ions (E_i=12.5 *eV*) [31] can lead to secondary electron emission. However, from the rate coefficient for conversion of N_2^+ into O_2^+ ion ($\approx 6\times10^{-11}$ cm^3/s), it can be concluded that N_2^+ density must fall earlier than O_2^+ density. Therefore, O_2^+ ion is dominant in this part of memory curve. The reasonable agreement of determined diffusion coefficients with the ones from other experiments is reached (i.e. 690 $cm^2 Torr/s$ at the electron temperature 4000K measured in time interval from 0.7 to 3 *ms* which gives 52 $cm^2 Torr/s$ at 300 *K* in ref. [12] and 122 $cm^2 Torr/s$ in time interval from 0.02 to 0.5 *ms* in ref. [13]). In order to confirm previous analysis, 2D numerical model is developed and applied.

The two-dimensional model for relaxation, consisting of diffusion equation with productions and losses of corresponding particles, is solved in cylindrical coordinates in order to describe afterglow in synthetic air:

$$\frac{\partial n}{\partial t} - D\left[\frac{1}{r}\frac{\partial}{\partial r}\left(r\frac{\partial n}{\partial r}\right) + \frac{\partial^2 n}{\partial z^2}\right] = S \qquad (9)$$

where n is the particle number density (electron, ion or neutral), D is the diffusion coefficient and S is the term representing productions and losses of corresponding particle in relaxation. The system of differential equations is discretized by using the implicit finite difference method. System of equations is further solved by a successive over relaxation (SOR) method. The used mash has 30×30 nods, while time and spatial steps are set to be $dt=10^{-7}\ s$ and $dr=dz=0.1cm$, respectively. Model includes 5 ionic species N_2^+, O_2^+, N_4^+, O_4^+, O^-, metastable states $N_2(A\ ^3\Sigma_u^+)$, $N_2(a'\ ^1\Sigma_u^-)$ and N and O atoms. The model includes 25 processes of dominant particles with rate coefficients taken from refs. [32-34], where the most dominant process is conversion of nitrogen ions in oxygen ions. The diffusion and mobility data for ions were taken from [35-37] and for neutral particles were calculated from relation given in ref. [38]. The variation of N and O atom number densities in early relaxation is neglected in order to speed up the calculations.

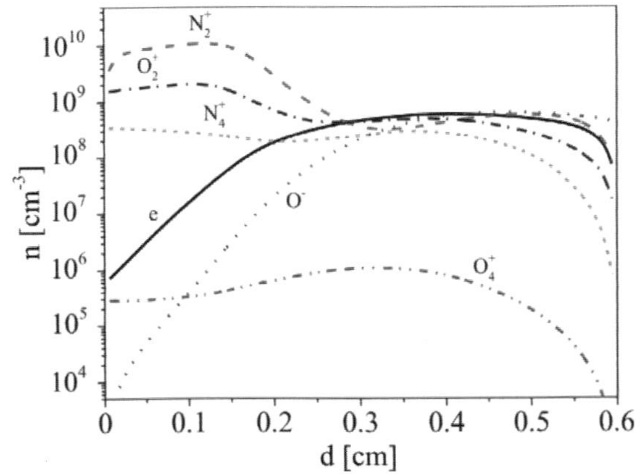

Figure 5. Axial number density profiles of ionic species in stable glow discharge in synthetic air obtained from the one dimensional fluid model used as initial densities for modelling of relaxation.

The initial densities are calculated from a one-dimensional fluid model consisting of the zero momentum (particle continuity equation) of Boltzmann's equation and Poisson equation in which all relevant particles for the afterglow are included as well as ionization terms as described in [39]. The stationary densities of charged particles in glow discharge are used as initial densities for modelling of the relaxation are determined (figure 5). Based on the calculated stationary axial profiles, the initial two-dimensional profiles in the tube are calculated by "filling up" the tube where the densities inside inter-electrode space are taken as a constant and 2D model was running until the stationary densities in whole tube are obtained. Since at this pressures and relaxation times the diffusion losses are dominant, in order to fit the experiment different relations for the change of diffusion regimes were used. For the transition regime of electron diffusion the following relation was used [40]:

$$D_{se}=D_a\left(\frac{D_e+\Lambda^2en_e\mu_e/\varepsilon_0}{D_a+\Lambda^2en_e\mu_e/\varepsilon_0}\right) \tag{10}$$

where $D_a=(\mu_i/\mu_e)D_e$ is ambipolar diffusion coefficient, n_e is electron number density, Λ is characteristic diffusion length of the tube, μ_e is electron mobility, μ_i is mobility of dominant O_2^+ ion, ε_0 is vacuum permittivity and e is elementary charge. When the ratio of characteristic diffusion length and Debay radius λ_D falls to $\Lambda/\lambda_D<100$, the diffusion regime changes into transition regime and for ions is given by the following relation for effective diffusion coefficient [41, 26]:

$$D_{eff}=D_i\left(1+\frac{T_e}{T_i}\right). \tag{11}$$

where D_{eff} is effective diffusion coefficient, D_i is ion free diffusion coefficient, T_e is electron temperature and T_i is ion temperature (equal to gas temperature T_g). In order to correctly describe the electron temperature decay, the two-exponential decay was used, where slowed down decay due to superelastic collision with metastables and vibrational excited states was assumed, similarly as in pure nitrogen [42]. Finally, the last part of memory curve was fitted with the effective diffusion coefficient dependence $D_{eff} \sim \Lambda / \lambda_D$.

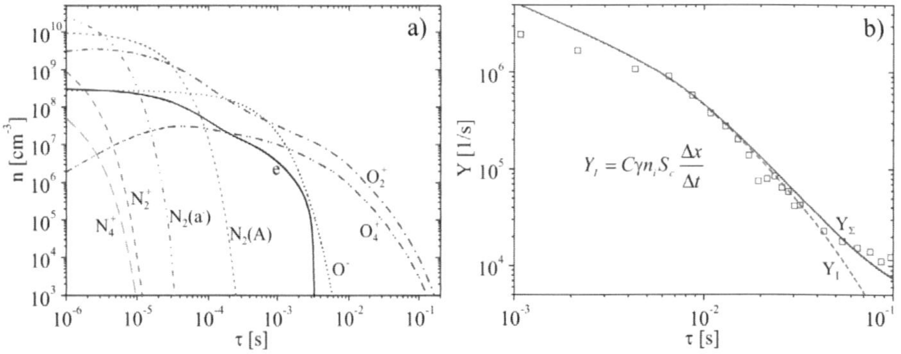

Figure 6. a) The evolution of particle number densities in relaxation and b) the electron yield (symbols) fitted with the numerical model where (dashed line) is contribution of ionic particles and (solid line) sum of contributions.

From the figure 6a it can be seen that due to intensive conversion of nitrogen ions the dominant particles in early relaxation are oxygen ions, i.e. O_2^+ and O_4^+ ion. The experimental data for electron yield are fitted well with diffusion loss of this ion (figure 6b). The discrepancy noticed in the first several milliseconds is most probably caused by the influence of formative time. At around 60 ms the contribution of neutral particles had to be included to describe experimental data because ion density has fallen at these relaxation times. Since the of neutral particles slowly decrease, contribution is almost constant, the fall of ions contribution of electron yield still describes fall of total electron yield.

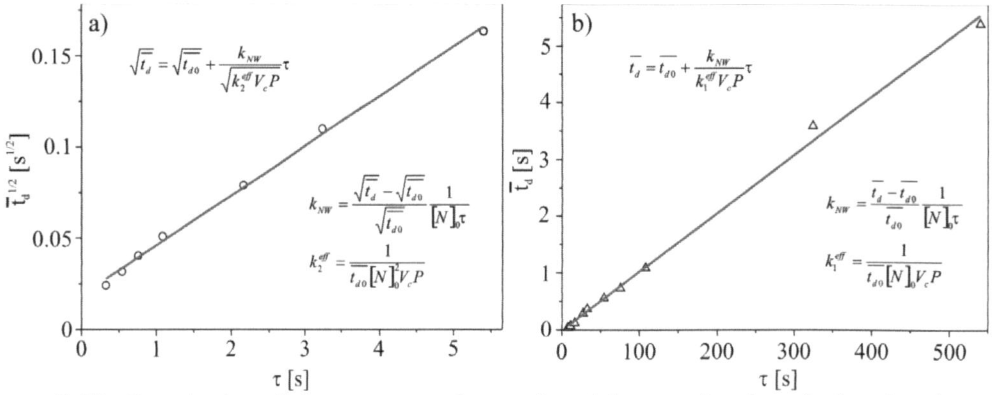

Figure 7. The linearization of a) square root of mean time delay as a function of relaxation time and b) mean time delay as a function of relaxation time.

The second part of memory curve is in interval 90 *ms*-1000 *s* in which number densities of the ionic species are depleted, therefore only neutral particles (most probably nitrogen atoms) can lead to emission of initiating electrons. In the ref. [25] the memory effect in pure nitrogen with the iron cathode is explained by the surface recombination of nitrogen atoms. Based on recombination energy of nitrogen and oxygen atoms it is more probable that nitrogen atom recombination is responsible for the memory effect in synthetic air, like in the case of pure nitrogen [25]. At our conditions the volume recombination of nitrogen atoms is negligible (volume recombination coefficient $\sim 10^{-34} cm^3 / s$), hence only possible loss process is surface recombination. By linearizing the data in corresponding plots [25], the order of reactions is determined (figure 7). The two different slopes indicate change of mechanisms i.e. change of the second order into the first order surface recombination.

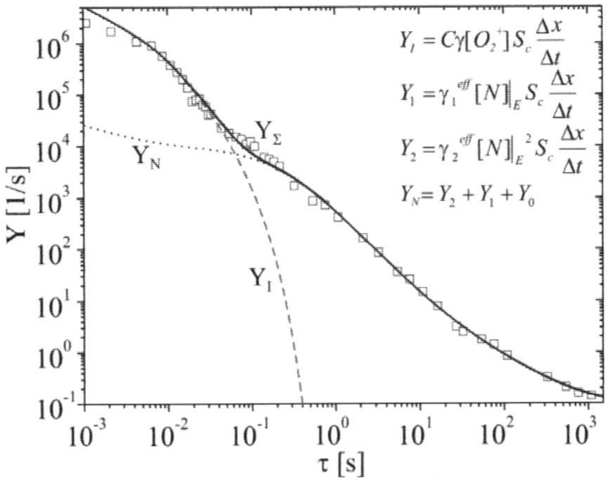

Figure 8. The experimental data for electron yield in relaxation and numerical fit calculated from the 2D numerical model. The relations for electron yield used to fit the experimental data are given in the figure ($[N]|_E$ is number density of nitrogen atoms in front of the cathode surface). The first part of curve in late relaxation is explained by the second order recombination of nitrogen atoms on the cathode and after 6 *s* the first order recombination becomes dominant.

In order to better describe late afterglow in synthetic air, the two-dimensional model is developed. It consists of diffusion equation (9) for nitrogen atom with included surface loss processes. The equation is discretized with explicit finite difference method. By using the recombination coefficient on glass walls $\gamma_{NW} = 4\times10^{-11}\ cm^3/s$, as well as the second $\gamma_2^{eff} = 1.86\times10^{-25}\ cm^3/s$ and first $\gamma_1^{eff} = 1.06\times10^{-14}/s$ order recombination coefficients on electrodes, satisfactory agreement with the experiment is reached (figure 8). The values of surface recombination coefficients on glass are larger, while the surface recombination coefficients on cathode are lower in comparison to values determined in nitrogen [25], which can be explained by the presence oxygen atoms in air and different cathode material (i.e. stainless steel and iron). The change of order is explained with the fall of nitrogen atom density.

5. Conclusion

The analysis of statistical time delay distributions and memory curve measured in synthetic air is presented. The relations for mixture distributions of statistical time delay and for electron yield are generalized starting from a binomial distribution of initiating electron occurrence. The memory effect in synthetic air is explained with diffusion losses of O_2^+ ion as well as surface recombination of nitrogen atoms and corresponding coefficients were determined.

Acknowledgments

The author is grateful to the Ministry of Education, Science and Technological Development of the Republic of Serbia for the financial support (project ON171025), M. Miljković for SEM and EDX and R. Gajić and B. Vasić for AFM measurements.

References

[1] Shao T, Sun G, Yan P, Wang J, Yuan W, Sun Y and Zhang S 2006 *J. Phys. D: Appl. Phys.* **39** 2192
[2] Fang Z, Qiu Y and Kuffel E 2004 *J. Phys. D: Appl. Phys.* **37** 2261
[3] Pandiyaraj K N, Selvarajan V, Deshmukh R R, Yoganand C P, Balasubramanian S, Maruthamuthu S 2013 *Plasma Sci. Technol.* **15** 56
[4] Shin J, Narayanaswamy V, Raja L L and Clemens N T 2007 *AIAA Journal* **45** 1596
[5] Soloshenko I A, Tsiolko V V, Khomich V A, Shchedrin A I, Ryabtsev A V, Bazhenov V Yu and Mikhno I L 2000 *Plasma Phys. Rep.* **26** 792
[6] Kutasi K, Pintassilgo C D, Coelho P J and Loureiro J 2006 *J. Phys. D: Appl. Phys.* **39** 3978

[7] Pointu A, Ricard A, Dodet B, Odic E, Larbre J and Ganciu M 2005 *J. Phys. D: Appl. Phys.* **38** 1905
[8] Yurgelenas Yu V and Wagner H-E, 2006 *J. Phys. D: Appl. Phys* **39** 4031
[9] Sakiyama Y, Graves D B, Chang H, Shimizu T and Morfill G E 2012 *J. Phys. D: Appl. Phys.* **45** 425201
[10] Steinle G, Neundorf D, Hiller W and Pietralla M 1999 *J. Phys. D: Appl. Phys.* **32** 1350
[11] Aleksandrov N L, Anokhin E M, Kindysheva S V, Kirpichnikov A A, Kosarev I N, Nudnova M M, Starikovskaia S M and Starikovskii A Yu 2012 *J. Phys. D: Appl. Phys.* **45** 255202
[12] Spencer M N, Dickinson J S and Eckstrom D J 1987 *J. Phys. D: Appl. Phys.* **20** 923
[13] Dobrov W I and MacDonald A D 1969 *Phys. Fluids* **12** 1618
[14] Marković V Lj, Jovanović A P, Stamenković S N, Popović B Č 2012 *Europhys. Lett.* **100** 45002
[15] Jovanović A P, Popović B Č, Marković V Lj, Stamenković S N and Stankov M N 2014 *Eur. Phys. J. Appl. Phys.* **67** 20801
[16] Morgan C G 1978 *Electrical Breakdown of Gases,* edited by J M Meek and J D Craggs (Chichester: John Wiley & Sons) pp. 656-688
[17] Zuber K 1925 *Ann. Phys.* **381** 231
[18] von Laue M 1925 *Ann. Phys.* **381** 261
[19] Kiselev J V 1965 *Proc. 7th Int. Conf. Phenomena in Ionized Gases*, edited by D Tošić, (Beograd:Gradjevinska knjiga, Belgrade, Yugoslavia) (in Russian) p. 838
[20] Marković V Lj, Gocić S R and Stamenković S N 2006 *J. Phys. D:Appl. Phys.* **39** 3317
[21] Bošan Đ A 1956 Industrial R. and D. Report (Development of surge arresters), TD PO2, Comp Elektronska Industrija, former Zavodi RR (Niš)
[22] Bošan Đ A *Invited Lecture in 16th Summer School and Int. Symp on the Physics of Ionized Gases* (Belgrade, Yugoslavia) 1993 (Nova Science Publishers Inc. 1996, pp. 15-40, eds. Milosavljević M and Petrović Z) (and references therein)
[23] Bošan Đ A and Pejović M M 1979 *J. Phys. D: Appl. Phys.* **12** 1699
[24] Kudrle V, LeDuc E and Fitaire M 1999 *J. Phys. D: Appl. Phys.* **32** 2049
[25] Marković V Lj, Pejović M M and Petrović Z Lj 1996 *Plasma Chem. Plasma Proc.* **16** 195
[26] Marković V Lj, Petrović Z Lj and Pejović M M 1997 *Plasma Sources Sci. Technol.* **6** 240
[27] Titterington D M, Smith A F M and Makov U E 1985 *Statistical analysis of finite mixture distributions*, (Chichester, John Wiley & Sons)
[28] Acheson M A and Mcelwee E M 1951 The Sylvania Technologist **4** 38
[29] Mendenhall W and Hader R J 1958 Biometrika, **45** 504
[30] Michaelson H B, 1977 *J. App. Phys.* **48** 4729
[31] Braithwaite N St J 2000 *Plasma Sources Sci. Technol.* **9** 517
[32] Kossyi I, Kostinsky A Yu, Matveyev A A and Silakov V P 1992 *Plasma Sources Sci. Technol.* **1** 207
[33] Guerra V, Loureiro J 1997 *Plasma Sources Sci. Technol.* **6** 373
[34] Guerra V, Sá P A and Loureiro J 2001 *J. Phys. D: Appl. Phys.* **34** 1745
[35] Bekstein A, Benhenni M, Yousfi M, Ducasse O and Eichwald O 2008 *Eur. Phys. J. Appl. Phys.* **42** 33
[36] Bekstein A, Yousfi M, Benhenni M, Ducasse O and Eichwald O 2010 *J. Appl. Phys.* **107** 103308
[37] Nelson D, Benhenni M, Eichwald O, and Yousfi M 2003 *J. Appl. Phys.* **94** 96
[38] Debal F, Bretagney J, Jumet M, Wautelety M, J Dauchot P and Hecq M 1998 *Plasma Sources Sci. Technol.* **7** 219
[39] Jovanović A P, Stankov M N, Marković V Lj and Stamenković S N 2013 *Europhys. Lett.* **104** 65001
[40] Guerra V, Sa P A and Loureiro J 2001 *Phys. Rev. E* **63** 046404
[41] Phelps A V 1990 *J. Res. Natl. Inst. Stand. Technol.* **95** 407
[42] Guerra V, Sa P A and Loureiro J 2004 *Eur. Phys. J. Appl. Phys.* **28** 125

27th Summer School and International Symposium on the Physics of Ionized Gases (SPIG 2014) IOP Publishing
Journal of Physics: Conference Series **565** (2014) 012016 doi:10.1088/1742-6596/565/1/012016

Measurements and models of transient and stationary regimes of glow discharge in argon

M N Stankov
Department of Physics, Faculty of Science and Mathematics, University of Niš, Višegraska 33, PO box 224, 18001 Niš, Serbia

E-mail: marjansstankov@gmail.com

Abstract. The experimental and theoretical analyses of different regimes of argon DC glow discharge are reported. The experiments were carried out on the argon gas tube with a plane-parallel electrode system made from OFHC (oxygen-free high thermal conductivity) copper. Modelling of the static breakdown voltages was performed by simple fluid model. The applicability of fluid models for modelling of $I-U$ (current-voltage) characteristics at different values of pd (pressure times inter-electrode distance) is tested. The formative time delays are determined from experiment and compared to modeled values obtained by [1D] and [2D] fluid models. The memory curve $\overline{t_d}(\tau)$ (the dependence of the mean value of breakdown time delay on the relaxation time) is presented and the main processes responsible for the memory effect were determined by applying the analytical and numerical models.

1. Introduction

The application of all types of gas discharges in various fields of technology is very common, like for plasma etching, deposition and implantation of thin films, surfaces cleaning from the oxides, metallic and organic contaminants, laser and light sources technology, gas switches development, medical treatment, etc [1-7]. Beside the experimental research, different analytical and numerical models are used very commonly for better understanding of physical processes in various types of gas discharges. Numerical models used for glow discharge modelling can be classified as fluid, particles and hybrid models. Numerical modelling by computer was used for the first time for discharge simulation in 1958 [8] and after that its applicability has rapidly increased [9-12]. The investigation of breakdown time delays is also very useful for the analysis of glow discharges. The information about kinetics of atomic and molecular species remaining from the preceding glow and reactions rate coefficients may be estimated by the analysis of memory curve (the dependence of the mean value of breakdown time delay on the relaxation time $t_d(\tau)$).Memory effect in inert (argon) and molecular gases (nitrogen) was explained by long lived metastable states remaining from the preceding glow [13, 14]. However, the effective lifetimes of metastable states are very short due to quenching and memory effect was explained by the process of surface recombination of nitrogen atoms on the tube walls, while secondary electrons initiating breakdown are produced on the cathode surface [15-17].

The results of measurements and modelling are presented in this report. The theoretical description of different fluid models is presented in section 2, while the details of the experiment are given in section 3. In section 4, Paschen curve modelling is performed by simple fluid model and results are compared to experimental data. The applicability of different fluid models for *I-U* characteristics modelling is analysed in section 5. The different analytical and numerical models are applied for the modelling of formative time delays and results are presented in section 6. The analysis of memory curve is given in section 7 and finally, short conclusion is presented in section 8.

2. Different fluid models

Different fluid models are often used for glow discharge modelling and the most common are simple and extend fluid models [18-20]. In the simple fluid model electron transport and kinetic coefficients are the function of reduced field [19], while in the extended fluid model these coefficients are the function of mean electron energy [20]. Although the fluid model with nonlocal ionization agrees very well with experiments and hybrid models [21], it is used less frequently than the previous mentioned models.

The simple fluid model consists of continuity equation and Poisson's equation:

$$\frac{\partial n^{ele,ion}}{\partial t} + \nabla \mathbf{\Gamma}^{ele,ion} = S, \quad (1)$$

$$\nabla^2 \varphi = -\frac{e}{\varepsilon_0}(n^{ion} - n^{ele}), \quad (2)$$

where n is the number density of electrons and Ar^+ ions, $\mathbf{\Gamma}$ is the flux of particles, S is a source term, φ is a potential and ε_0 is the dielectric constant. The source term S is expressed by relation:

$$S = \alpha \Gamma^{ele} - \beta n^{ele} n^{ion}, \quad (3)$$

where α is Townsend ionization coefficient and β is the coefficient of recombination.

The extend fluid model, beside the continuity and Poisson's equations, has the electron energy balance equation:

$$\frac{\partial (n^{ele} u^{ele})}{\partial t} + \frac{5}{3}\nabla \mathbf{\Gamma}_u^{ele} = -e\mathbf{E}\cdot\mathbf{\Gamma}^{ele} - k_i u^{ion} n^{ele} N, \quad (4)$$

where u^{ele} is mean electron energy, $\mathbf{\Gamma}_u^{ele}$ is the electron energy flux, $\mathbf{E}$ is the electric field, k_i is ionization rate coefficient, $u^{ion} = 15.8\,eV$ is the ionization energy of argon atoms and N is number density of neutral gas.

The equations for fluid model with nonlocal ionization are the same as for the simple fluid model (1, 2). The ionization by fast electrons is included in this model and the source term has the form:

$$S_{fast}(x) = \begin{cases} \Gamma^{ele}(0)\alpha \exp(\alpha x) & \text{for } x < d_c \\ \Gamma^{ele}(0)\alpha \exp(\alpha d_c)\exp(-(x-d_c)/\lambda) & \text{for } x \geq d_c, \end{cases} \quad (5)$$

where $\Gamma^{ele}(0)$ is the electron flux at cathode, x is the distance from cathode, λ is the decay constant and d_c is the length of cathode sheath region where the electric field is strong. The ionization coefficient α used in equation 5 is a function of reduced field over the cathode sheath [21].The decay constant is calculated from the relation $\lambda = \varphi(d_c)/[(pC-d)(\alpha d_c)]$ [21], where $\varphi(d)$ is the cathode sheath voltage and C is the constant. The values of the transport and kinetic coefficients of electron and argon atom ion using for the modelling are taken from [19, 20, 22].

3. Experimental details

The measurements of static breakdown voltages, current-voltage characteristics and time delays are performed on gas tube presented in fig.1. The tube was evacuated down to 10^{-7} *Torr*, baked at 600*K*, and then filled with argon with an impurity level below 1 *ppm* (Matheson Co.) at pressure p =1.5*Torr*. The gas tube is made of borosilicate glass, with volume of $V \approx 60cm3$ and the plane-parallel cylindrical OFHC copper electrodes. The radius of electrodes is $R = 11mm$, while the inter-electrode distance d is variable.

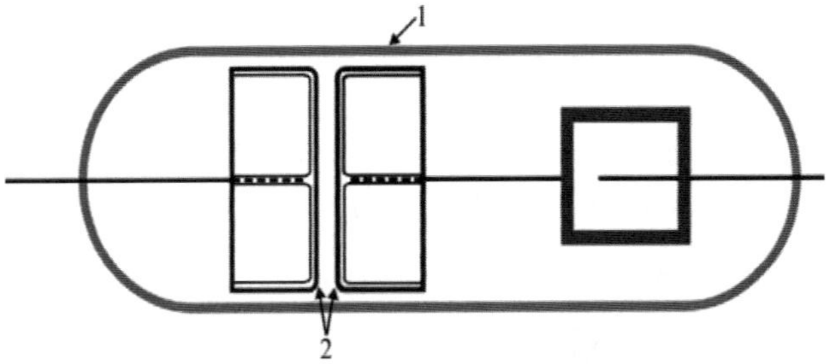

Figure 1. Schematic cross section of the discharge tube:
1- borosilicate glass and 2 – copper electrodes

4. Paschen curve modelling

The minimum voltage required for breakdown is the static breakdown voltage U_s. For the breakdown occurrence, the condition $\gamma(exp(\alpha d) -1) \geq 1$ must be fulfilled , where γ is the secondary electron yield. The simple fluid model is applied for the calculation of static breakdown voltages and the results are compared to the experiment. The equations 1 and 2 are partial differential equation and for solving them finite difference method is used. After discretization, these equations in one and two dimension cases have the following form:

1D:

$$\frac{n_i^{k+1} - n_i^k}{\Delta t} + \frac{\Gamma_{i+1/2}^{k+1} - \Gamma_{i-1/2}^{k+1}}{\Delta x} = S_i^k , \tag{6}$$

$$\frac{\varphi_{i+1}^{k+1} - 2\varphi_i^{k+1} + \varphi_{i-1}^{k+1}}{\Delta x^2} = -\frac{\rho_i^k}{\varepsilon_0}, \tag{7}$$

2D:

$$\frac{n_{i\,j}^{k+1} - n_{i\,j}^k}{\Delta t} + \frac{1}{r_j}\frac{\Gamma_{i\,j+1/2}^{k+1} - \Gamma_{i\,j-1/2}^{k+1}}{\Delta r} + \frac{\Gamma_{i+1/2\,j}^{k+1} - \Gamma_{i-1/2\,j}^{k+1}}{\Delta z} = S_{i\,j}^k , \tag{8}$$

$$\frac{1}{r_j}\frac{\varphi_{i\,j+1}^{k+1} - 2\varphi_{i\,j}^{k+1}}{\Delta r} + \frac{\varphi_{i+1\,j}^{k+1} - 2\varphi_{i\,j}^{k+1} + \varphi_{i-1\,j}^{k+1}}{\Delta z^2} + \frac{\varphi_{i\,j+1}^{k+1} - 2\varphi_{i\,j}^{k+1} + \varphi_{i\,j-1}^{k+1}}{\Delta r^2} = -\frac{\rho_{ij}^k}{\varepsilon_0}, \tag{9}$$

where i and j are the spatial indices, k is the time index, ρ is the charge density , Δt is the time step and Δz and Δr are the spatial steps. Flux terms in equations 6 and 8 are expressed in exponential form introduce by Sharffeter and Gummel [23].The equations (6-9) present the system of linear equations and for solving them successive over relaxation method is used.

The values of electron transport coefficients and argon atom ion mobility are function of reduced field [19, 22], while the diffusion coefficient of argon atom ion is calculated from relation $D^{ion} = 200 / p[cm^2 s^{-1}]$ [19]. Based on the temporal evolution of Ar atom ion number density, the static breakdown voltage is determined. When the applied voltage U_w is equal to or greater than the static breakdown voltage U_s the number density of argon ion increases with increasing time (fig. 2).

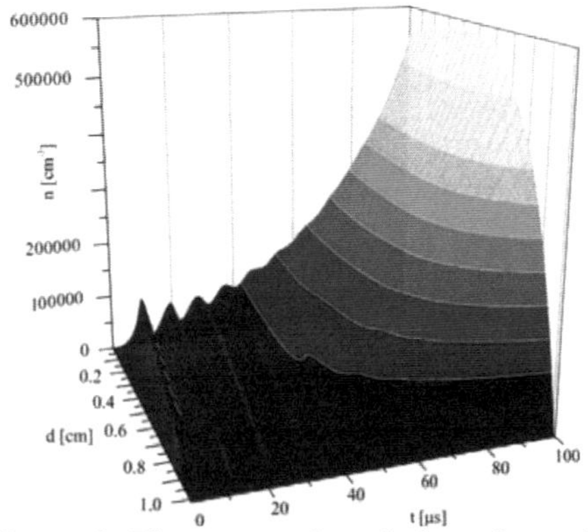

Figure 2. The temporal evolution of atomic argon ion when applied voltage is equal to or greater than the static breakdown voltage $U_w \geq U_s$ (U_s=214V)

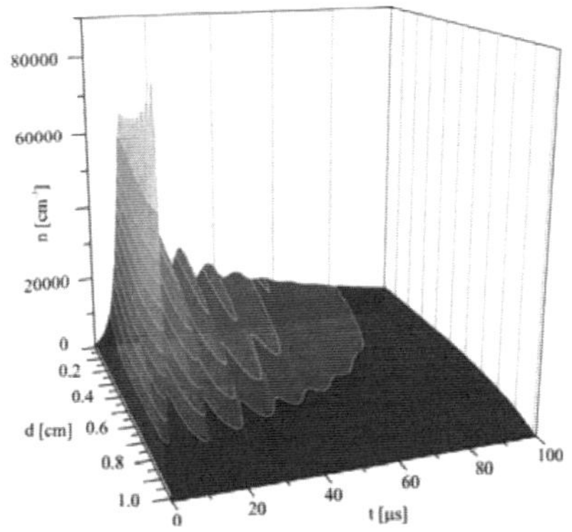

Figure 3. The temporal evolution of atomic argon ion when applied voltage is lower than the static breakdown voltage $U_w < U_s$ (U_s=214V)

In the cases when the applied voltage is lower than static breakdown voltage, the breakdown condition is not fulfilled and the number density of argon ion decreases with increasing time (fig.3). The modelling of Paschen curve is carried out for different values of secondary electron yield and the results are compared to the experimental data. The best agreement is achieved by using the variable secondary electron yield obtained from the breakdown condition for different values of *pd* (fig. 4a). The values for first Townsend ionization coefficient used for calculation of secondary electron yield is obtained by fitting the experimental data [24]. The agreement of secondary electron yield with data from literature [25] is given in fig. 4b.

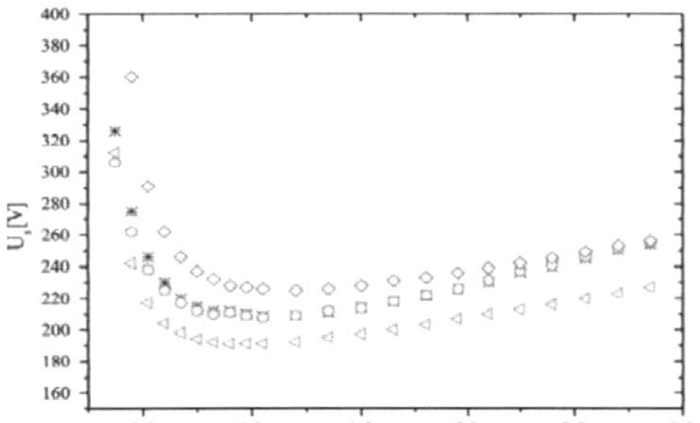

Figure 4a. Paschen curves obtained from experiment and models: - experiment, - modelled with variable γ, -modelled with γ= 0.015, -modelled with γ= 0.007

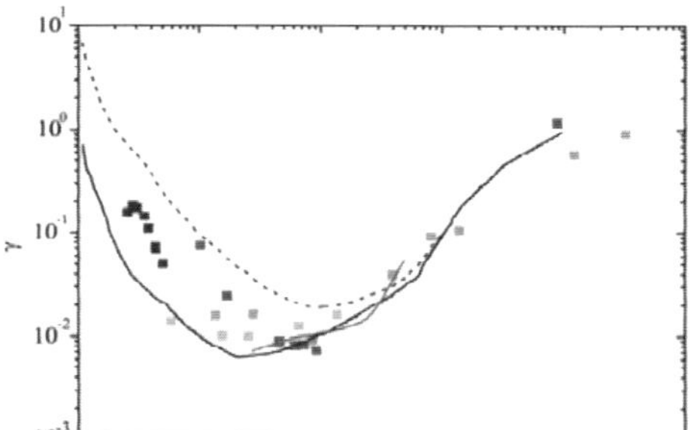

Figure 4b. Secondary electron yield in the function of reduced field *E/N*: red line is the secondary electron yield obtained from the breakdown condition and other symbols are from the paper [25]

5. The applicability of fluid models for modelling of current-voltage characteristics

The current-voltage (*I-U*) characteristics of the gas tube are measured for inter-electrode distances *d*=15*mm*, 12*mm* and 6*mm* (fig 5). In this paper the modelling of *I-U* characteristics is performed by using the one dimensional models. One dimensional models are applicable if all cathode area is occupied by glow, which is the case in the abnormal region of the glow discharge (100-600 *μA*), fig 5.

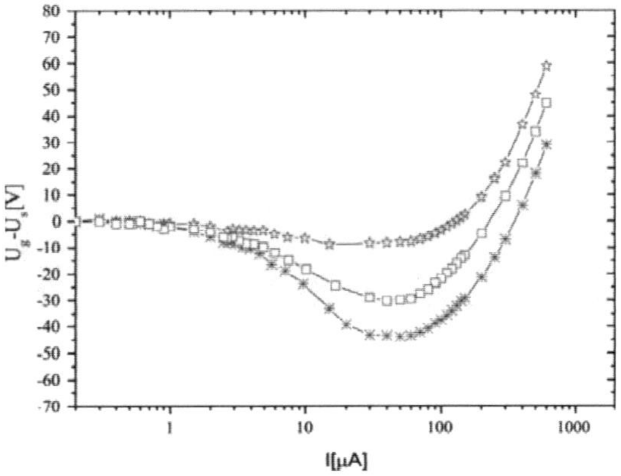

Figure 5. The (I-U) characteristics of glow discharge for different inter-electrode distances: ✳ - *d*=15*mm*, □ - *d*=12 *mm* and ☆- *d*=6 *mm*

The simple, extend and fluid models with nonlocal ionization are applied for *I-U* characteristics modelling and comparison of modelled and experimental data for inter-electrode distances *d*=15 *mm* and 12 *mm* are presented in fig.6. The values of secondary electron yield for these inter-electrode distances are γ=0.00809 and 0.00922, respectively. These values are obtained from the breakdown condition and used for the modelling as a simple approximation, but it should be noted that many authors have shown that secondary electron yield is not constant along I-U characteristics [26-28]. The best agreement with experiment is achieved by applying the fluid model with nonlocal ionization, while the results from other models deviate from experimental data.

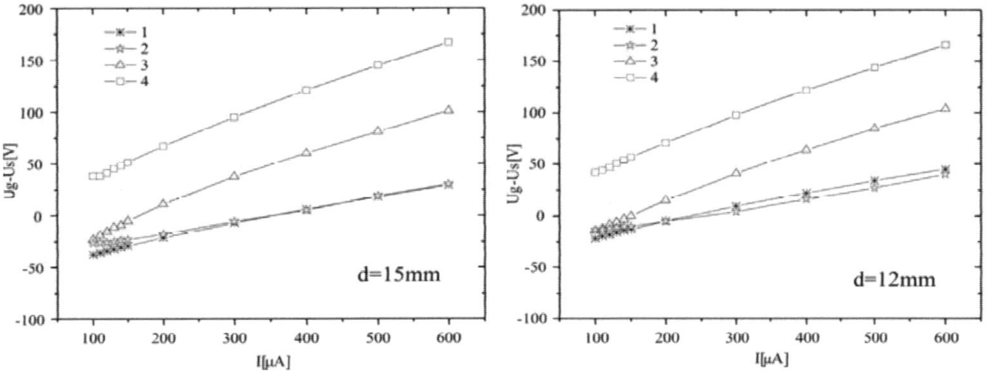

Figure 6. The current-voltage characteristics of abnormal glow discharge (100-600 *μA*) for the inter-electrode distances *d*=15*mm* and 12*mm* obtained from experiment and different models:
1- experiment, 2 - fluid model with nonlocal ionization, 3 - extend fluid model and 4 - simple fluid model

In the case when d =0.6cm, the region beyond the cathode space is very thick and the ionization by fast electrons in this region can be neglected, so the extend fluid model shows good agreement with the experiment (fig.7). However, the results from the simple fluid model still deviate from the experimental data. The value of secondary electron yield in this case is γ=0.00994.

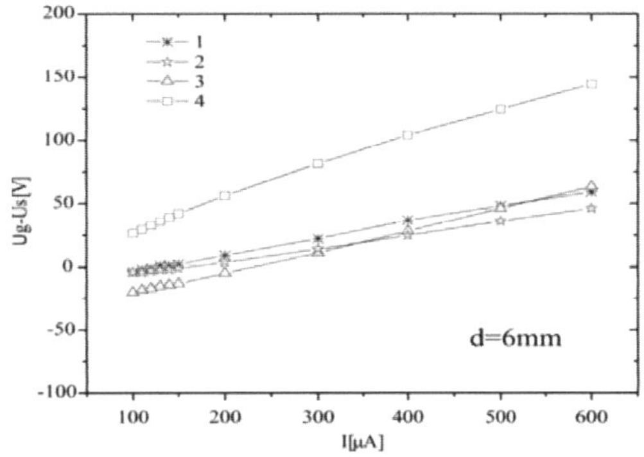

Figure 7. The current-voltage characteristics of abnormal glow discharge (100-600 μA) for the inter-electrode distance d=0.6 cm obtained experiment and different models: 1 - experiment, 2 - fluid model with nonlocal ionization, 3 - extend fluid model and 4 - simple fluid model

6. The comparison of measured and modelled formative time delays

The time that elapses from the moment of applying of voltage greater than the static breakdown voltage U_s to the breakdown occurrence, is denoted as the breakdown time delay t_d. It consists of statistical time delay t_s (from the application of voltage to the appearance of a free electron that initiates the breakdown) and the formative time delay t_f (from this moment to the collapse of the applied voltage and occurrence of a self-sustained current) [13, 14]. The time delay measurements were carried out by applying the step pulses (fig. 8) at different overvoltages $\Delta U = U_w - U_s$ and relaxation times (afterglow period) τ.

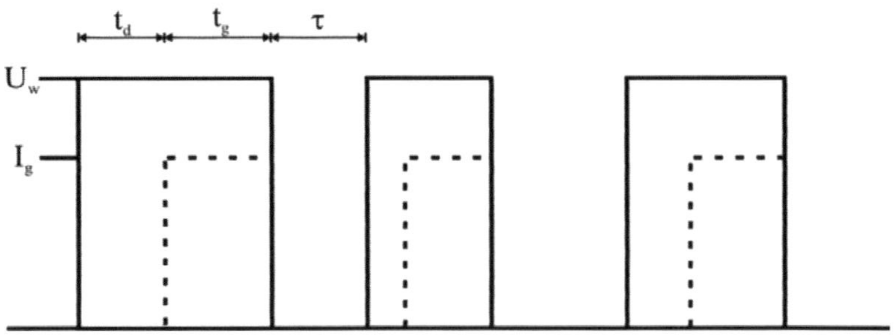

Figure 8. The sequence of pulses applied to the discharge tube: U_w - working voltage, I_g - glow current, t_d - breakdown time delay, t_g - glow time and τ- afterglow period.

The electron current density from the cathode is expressed by the relation [29, 30]:

$$i(t) \approx i_0 + qi(t - t_i) \approx i_0 + q[i(t) - t_i(di/dt)], \tag{10}$$

where i_0 is the current density of the seed electrons, $t_i = d/w_i$ is the ion transit time from anode to cathode and w_i is the ion drift velocity. After integration, the current density is obtained:

$$i(t) = i_0\left[\frac{q}{q-1}\exp(\frac{q-1}{q}\frac{t}{t_i}) - \frac{1}{q-1}\right], \tag{11}$$

and relation for formative time delay has the following [29, 30]:

$$t_f = \frac{q t_i}{q-1} \ln \frac{1+(q-1)(i/i_0)}{q}, \tag{12}$$

where q is the multiplication factor of electron avalanche. The multiplication factor is expressed by the relation:

$$q = \widetilde{\gamma} \frac{\alpha}{\widetilde{\alpha}} [\exp(\widetilde{\alpha} d) - 1], \tag{13}$$

where $\widetilde{\alpha}$ is the electron ionization coefficient with included radial diffusion loss of electrons [31]:

$$\widetilde{\alpha} = \alpha - (D_e / w_e)(2.4/R)^2 . \tag{14}$$

The formative time dependence on overvoltage is presented in fig. 9.

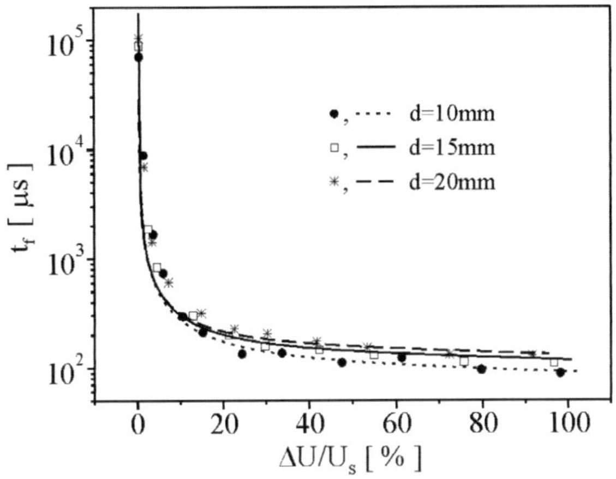

Figure 9. Formative time delays (•, □, ∗) and the corresponding line fits.

For the calculation of formative time delays 1D and 2D fluid models are also applied. In fig. 10 the comparison between modelled and measured formative time delay is presented. The formative time delay calculated from one dimensional fluid model is much shorter than measured ones because the radial diffusion of particles is not included in the model, while in case of two dimensional modelling, a very good agreement with experiment is achieved.

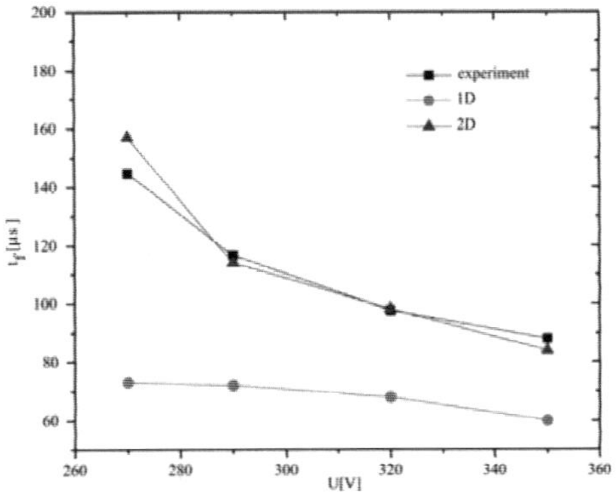

Figure 10. Comparison of formative time delay obtained from experiment and from fluid models.

7. The analysis of memory effect

The afterglow kinetics in argon is studied by the breakdown time delay measurements as a function of relaxation time $\overline{t_d}(\tau)$ (memory curve), fig. 11. The memory effect is the long time variation of the electrical breakdown time delay on the relaxation time (τ). The memory curve is divided in three regimes: in the first regime (early relaxation) the main process of breakdown initiation is secondary electron emission caused by ions, in the second regime (late relaxation) electrons are generated by surface recombination of nitrogen atoms present as impurities [15-17] and the third regime presents the saturation of memory curve caused by the cosmic rays and natural radioactivity.

The mean value of the statistical time delay can be expressed as:

$$\overline{t_S} = \frac{1}{YP}, \qquad Y = \gamma \cdot \Gamma \cdot S, \tag{15}$$

where Y represents a number of generated electrons in the inter-electrode space per second (electron yield), P is the probability of one electron to cause the breakdown [32, 15], γ is the coefficient for secondary electron yield, Γ is the flux of particles and S is the front area of cathode surface. More details about calculation of electron yield can be found in [16].

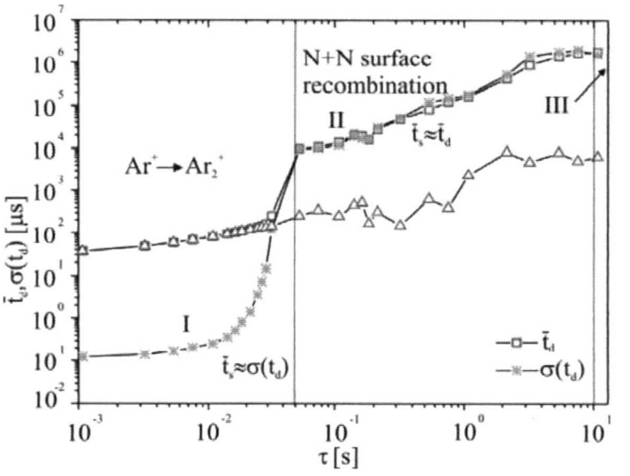

Figure 11. Argon memory curve t_d (τ) and time delay standard deviation $\sigma(t_d)$ for glow current *I=130mA*, working voltage *U=250V* and inter-electrode distance *d=10mm*

The number densities of Ar^+, Ar_2^+ ions and Ar^* ($^3P_2$) metastable state in afterglow is determined by solving the equation:

$$\frac{\partial n}{\partial \tau} - D \cdot \nabla^2 n = S, \tag{16}$$

where D is the diffusion coefficient of particles and S is the term in which the processes of production and loss of particles are included [15]. In the term S different reactions are included for which the rate coefficients are used from [16, 33, 34]. The temporal evolution of number density of nitrogen atoms, present as impurities, is obtained by the equation:

$$\frac{d[N]}{d\tau} = -\gamma_{nw}[N]^2, \tag{17}$$

where $[N]$ is the number density of nitrogen atom and γ_{nw} is the effective second order surface recombination coefficient on glass walls.

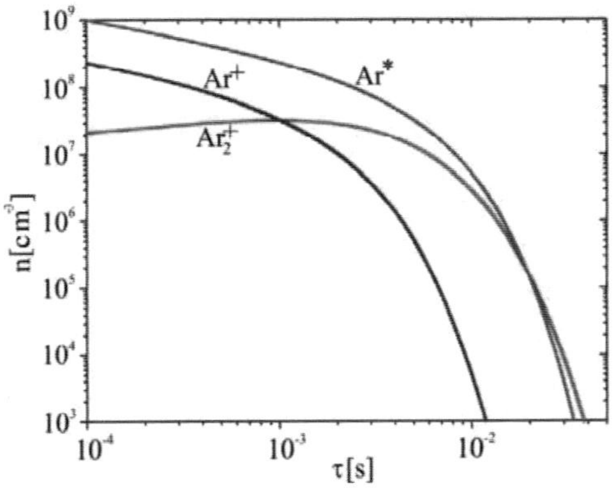

Figure 12. Number densities evolution of Ar^+, Ar_2^+ ions and Ar^* ($^3P_2$) metastable state in afterglow

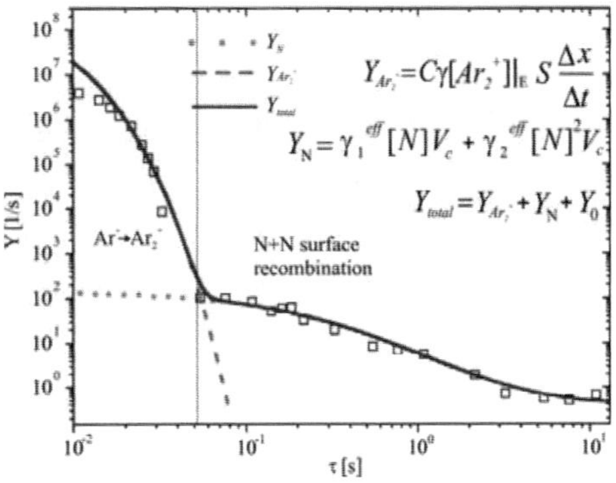

Figure 13. Fitting of the electron yield

In fig.12, the temporal evolution of Ar^+, Ar_2^+ ions and Ar^* ($^3P_2$) metastable state number densities are presented. It can be noticed that the number density of Ar^+ ion decreases due to the conversion of Ar^+ into Ar_2^+ (fig.12). In the fig. 13 the relations for electron yield cause by Ar_2^+ and nitrogen atoms are written [16], where C is the proportionality constant, γ is secondary electron yield cause by Ar_2^+ ion, $[Ar_2^+]|_E$ is the molecular argon ion mean number density in the last grid points next to the front cathode surface, S is a front area of cathode surface, Δx is the grid spacing, Δt is the numerical time step, $[N]$ is the nitrogen atom number density, V_C is the inter-electrode space volume, Y_0 is the electron yield cause by the cosmic and background radiation and γ_1^{eff}, γ_2^{eff} are the effective gas phase rate coefficients for the surface recombination on the cathode and are first and second order, respectively. It is shown that only the decay of molecular argon ions Ar_2^+ can fit the experimental values for the electron yield and explain the early memory effect in argon (fig.13). For longer afterglow times, a very good fit of the electron yield is obtained by the surface recombination of nitrogen atoms present as impurities.

8. Conclusion

For the analysis of glow discharge many different numerical and analytical models were often applied. The Paschen curve is modelled and the best agreement with the experiment is achieved by using the fluid model with variable secondary electron yield. The current-voltage characteristics are measured for different inter-electrode distances and the fluid model with nonlocal ionization gives the best results for all inter-electrode distances. For formative time delays, the results from analytical model with included radial diffusion of electron and 2D fluid model are in a good agreement with the experiment. The memory effect is analyzed and it was found that the early afterglow kinetics is dominated by the decay of molecular argon ions and after that (for longer afterglow times) nitrogen atoms present as impurities are recombined on the cathode surface providing secondary electrons for breakdown initiation.

Acknowledgments
The author is grateful to the Ministry of Education, Science and Technological development of the Republic of Serbia (projects 171025).

References

[1] Lieberman M A and Lichtenberg A J 1994 *Principles of Plasma Discharges and Materials Processing* (New York : Jon Wiley & Sons)
[2] Papas D 2011 *J. Vac. Sci. Technol.* A **29** 020801
[3] Slusher R E 1999 *Rev. Mod. Phys.* **71** S471
[4] Zissis G and Kitsinelis S 2009 *J. Phys. D: Appl. Phys.* **42** 173001
[5] Chang J S, Wang H, Zhang Q G and Qui A 2011 *Plasma Sci. Technol.* **13** 719
[6] Neyts E C, Yusupov M, Verlackt C C and Bogaerts A 2014 *J. Phys. D: Appl. Phys.* **47** 293001
[7] Kong M G, Kroesen G, Morfill G, Nosenko T, Shimizu T, J. van Dijk and Zimmermann J L 2009 *New J.Phys.* **11** 115012
[8] Ward A L 1958 *Phys. Rev.* **112** 1852
[9] Davies A J, Davies C S and Evans C J 1971 *Proc. IEE* **118** 816
[10] Winkler R and Wilhelm J 1988 *Phys. Scr.* **T23** 264
[11] Stankov M N, Petković M D, Marković V LJ, Stamenković S N and Jovanović A P 2014 *Rom. J. Phys.* **59** 328
[12] Kulikovsky A A 1998 *Phys. Lett. A* **245** 445
[13] Bošan Dj A 1978 in *Proceedings of 5th International Conferenc on Gas Discharges* (Liverpool University, Liverpool UK) pp. 273–276
[14] Bošan Dj A and Pejović M M 1979 J. Phys. D **12**, 1699.
[15] Marković V Lj, Petrović Z Lj and Pejović M M 1994 *J. Chem. Phys.* **100** 8514
[16] Marković V Lj, Gocić S R, Stamenković S N and Petrović Z Lj 2005 *Phys. Plasmas* **12** 073502
[17] Huo W G, Jian S J, Yao J and Ding Z F 2014 *Phys. Plasmas* **21** 053505
[18] Fiala A, Pitchford L C and Boeuf J P 1994 *Phys. Rev. E* **49** 5607
[19] Rafatov I R, Akbar D and Bilikmen S 2007 *Phys. Lett. A* **367** 114
[20] Becker M M, Loffhagen D and Schmidt W 2009 *Comput. Phys. Commun.* **180** 1230
[21] Rafatov I, Bogdanov E A and Kudryavtsev A A 2012 *Phys. Plasmas* **19** 093503
[22] Hagelaar G J M and Pitchford L C 2005 *Plasma Sources Sci. Technol.* **14** 722, BOLSIG+ 2005 CPAT: http://www.bolsig.laplace.univ-tlse.fr/download.php
[23] Scharfetter D and Gumell H 1969 *IEEE Trans. electron devices* **16** 64
[24] Kruithof A A 1940 *Physica* **7** 519
[25] Phelps AV and Petrović Z Lj 1999 *Plasma Sources Sci. Technol.* **8** R21.
[26] Donko Z 2001 *Phys. Rev. E* **64** 026401
[27] Phelps A V 2001 *Plasma Source Sci. Tech.* **10** 329
[28] Marić D et al 2002 *Eur. Phys. J. D* **21** 73
[29] Raizer Y P 1991 *Gas Discharge Physics,* (Berlin: Springer)
[30] Marković V Lj, Stamenković S N and Gocić S R 2007 *Contrib. Plasma Phys.* **47** 413
[31] Marković V Lj, Stamenković S N, Gocić S R, Jovanović A P and Stankov M N 2012 in *Proceedings of the 26th Summer School and Int. Symp.on the Physics of Ionized Gases*, editors M. Kuraica and Z. Mijatović (Zrenjanin, Serbia) p. 253
[32] Morgan C G 1978 in *Electrical Breakdown of Gases*, edited by Meek J M and Craggs J D John (Chichester: Wiley & Sons)
[33] Bogaerts A and Gijbels R 1999 *J. Appl. Phys.* **86** 4124
[34] Li Z, Zhao Z and Li X 2012 *Phys. Plasmas* **19** 033510

Journal of Physics: Conference Series **565** (2014) 012017 doi:10.1088/1742-6596/565/1/012017

Study of high-performance plasmas in the Large Helical Device

A Komori and T Morisaki
National Institute for Fusion Science
322-6 Oroshi, Toki, Gifu 509-5292 Japan

E-mail: komori@nifs.ac.jp

Abstract. Recent results from LHD experiments are presented together with the progress of the related fusion technology supporting the experiment. An ultra-long pulse discharge with central electron and ion temperatures of 2 keV and electron density of 1.2 x $10^{19}m^{-3}$ was successfully maintained for 48 minutes. Such a high performance plasma was heated and sustained by electron cyclotron resonance heating (ECH) with newly developed gyrotrons and ion cyclotron range of frequency (ICRF). Due to the continuous heating with the relatively high power of 1.2 MW, the total heating energy during the long pulse discharge reached 3.36 GJ, which is the world's record. The highest central ion temperature of 8.1 keV was also recorded in the last experimental campaign. In these experiments, it is essential to reduce the edge electron density, i.e., edge neutrals. Dedicated wall conditioning was performed to evacuate neutrals in the vessel wall before the main discharge with the high heating power. This process realizes the quite low recycling for particles, which results in the reduction of the edge plasma density.

1. Introduction

For more than 16 years, the Large Helical Device (LHD) has led fusion science and plasma physics as the largest heliotron device in the world [1]. With great efforts and contributions not only from domestic collaborators but also from collaborators all over the world, the LHD has produced many significant outcomes [2-10]. In particular, achievement of the steady-state operation [11], high ion temperature plasma, and three-dimensional edge/divertor physics [12-15] are highly evaluated, and which have been brought about by the advantages and features of the heliotron system with currentless plasma.

In order to realize the fusion reactor, it is necessary to maintain the high performance plasma for more than one year, thus establishment of the steady-state operational technique and understanding of the related physics are mandatory [16-18]. Challenges to achieve long pulse discharges have been made since the early phase of the LHD project [7,11]. Although a heliotron system such as the LHD is intrinsically equipped with confinement property in the magnetic configuration formed only by the external coils, the plasma cannot easily be maintained for a long time. It is often seen that the discharge terminates suddenly due to the increase of neutrals and/or impurities released from the vessel wall [11]. Owing to the adequate discharge cleaning, the ultra-long pulse plasma could successfully be maintained for 48 minutes. Such a dedicated wall conditioning also contributes to the improvement of the plasma performance. As a featured result, the highest central ion temperature of 8.1 keV was obtained in the last experimental campaign with a quite low recycling environment for particles.

In this paper, the research highlights obtained in recent experiments are presented, focusing on the effect of the wall conditioning on plasma performance. After describing the LHD and peripheral apparatus in section 2, experimental results are shown in section 3. Finally, the summary and prospects for the LHD project are given in section 4.

2. Experimental setup

The LHD is the world's largest superconducting heliotron device with poloidal/toroidal period numbers of 2/10, major and averaged plasma minor radius of 3.6 – 4.0 m and 0.6 m, respectively. Three negative-ion-based 180 keV neutral beams (NBs) with total heating power of ~ 16 MW are injected tangentially to generate and heat the plasma. Two positive-ion-based 40 keV NBs with total heating power of ~ 12 MW are also injected perpendicular to the plasma. For additional heating, ion cyclotron range of frequency (ICRF) oscillators and electron cyclotron resonance heating (ECH) devices are installed, which are sometimes utilized for the wall conditioning. For fuelling, the LHD is equipped with four gas puff valves and two frozen H_2 pellet injectors.

Principal diagnostics in the LHD are routinely utilized in various experiments. The Thomson scattering system provides radial profiles of the electron temperature T_e and the electron density n_e. The line averaged density is measured with the far infrared interferometer (FIR). Radiated power from impurities is measured with the bolometer array, and impurity densities with the charge exchange recombination spectroscopy system (CXRS). The visible and VUV spectroscopy systems are also employed to diagnose impurity ions and recycling particles. The plasma stored energy is measured with diamagnetic coils.

3. Experimental results

In this section, three experimental results necessary to realize the helical fusion reactor are mainly presented. Achievements of the ultra-long pulse discharge, high ion temperature discharge, and the neutral particle control with closed helical divertor (CHD), which is closely related to the former two experiments, are included.

3.1. Long pulse discharge

The long pulse discharge experiment has been attempted as one of the main research projects in the LHD [7,11]. The plasma performance and the record of duration time have been improved year by year. A newly installed ICRF antenna, together with conventional poloidal array (PA) antennas, and an improved ECH system have greatly contributed to the progress of the steady-state operation. The new phase controllable ICRF launcher called the "handshake shape-type antenna" (HAS) can control the parallel wave number of the ICRF wave which heats the core plasma efficiently in the dipole mode operation [19]. The HAS antenna is designed to reduce the power loss at the plasma edge or space between the plasma and the chamber wall. A photograph of the HAS antenna is shown in figure 1.

Figure 1. Phase controllable handshake shape-type antenna (HAS) installed in the LHD.

Experiments were carried out at the magnetic axis position R_{ax} = 3.65 m, with toroidal magnetic field B_T = 2.71 T. The helium plasma with minority hydrogen ions was maintained with the HAS antenna and new 77 GHz gyrotrons. The heating power of ICRF and ECH were 0.9 MW and 0.3 MW, respectively. It has been confirmed that, from the power modulation experiment, the HAS antenna in the dipole mode operation has a heating efficiency of more than 90%.

Figure 2 shows the time evolution of line averaged electron density, n_e, central ion and electron temperatures, T_i, T_e, and total input power of IRCR and ECH, $P_{heating}$, of the long pulse discharge. It can be seen that the high temperature plasma of $T_i \sim T_e \sim 2$ keV is maintained for 48 minutes. It should also be noted that relatively high n_e of $\sim 1 \times 10^{19}$ m^{-3} is achieved. High plasma density contributes to mitigate the high energetic ions produced by ICRF waves, which make local hot spots on the plasma facing components (PFCs). Such a hot spot or region on the PFCs sometimes causes the release of large amounts of neutrals or material flakes from the PFCs.

In order to avoid the abrupt discharge termination by such events, thorough cleaning of PFC surface and the wall conditioning were performed. For more effective neutral particle control, i.e., recycling control, however, the closed helical divertor (CHD) with pumping system should be necessary. Details of the CHD experiment are presented in section 3.3.

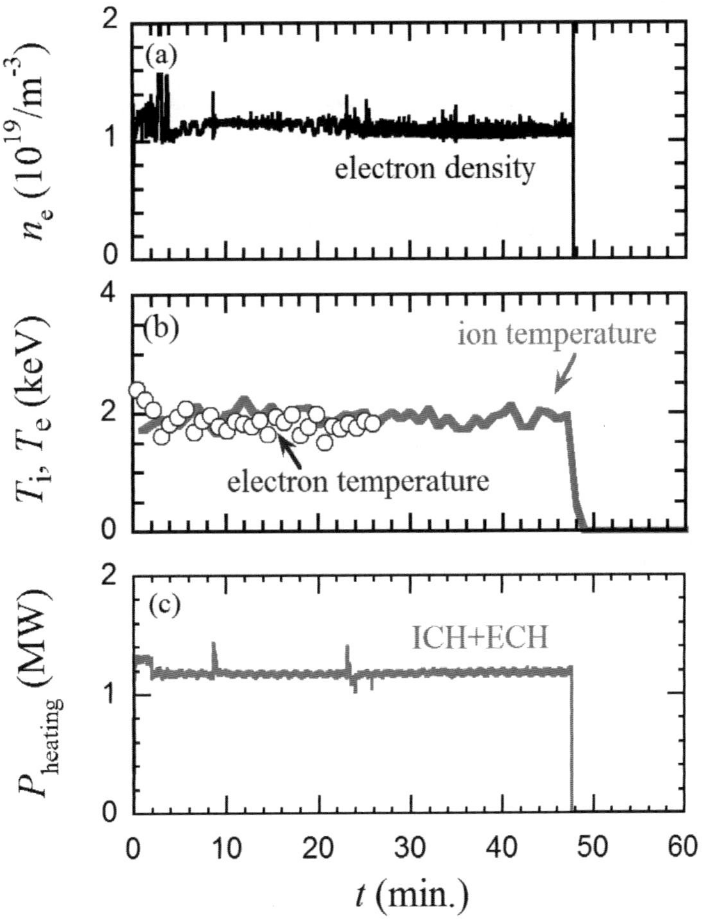

Figure 2. Temporal evolution of (a) line averaged electron density, (b) ion and electron temperatures, and (c) total power of ion heating and electron heating during long pulse discharge for 48 minutes.

3.2. High ion temperature discharge

The high T_i discharge is achieved by a carbon pellet injection into the high power NB heated plasma, after the dedicated wall conditioning [20]. Typical T_i, T_e and n_e profiles after carbon pellet injection are shown in figure 3. In this discharge, the total NB heating power was 23 MW. It can be seen that high T_i more than 7 keV is achieved with the formation of the ion internal transport barrier (ITB) where T_i gradient changes, i.e., gentle outside and steep inside. This is the transition to the improved mode with the ion ITB. Although T_i increases with the formation of ITB, T_e, does not change significantly.

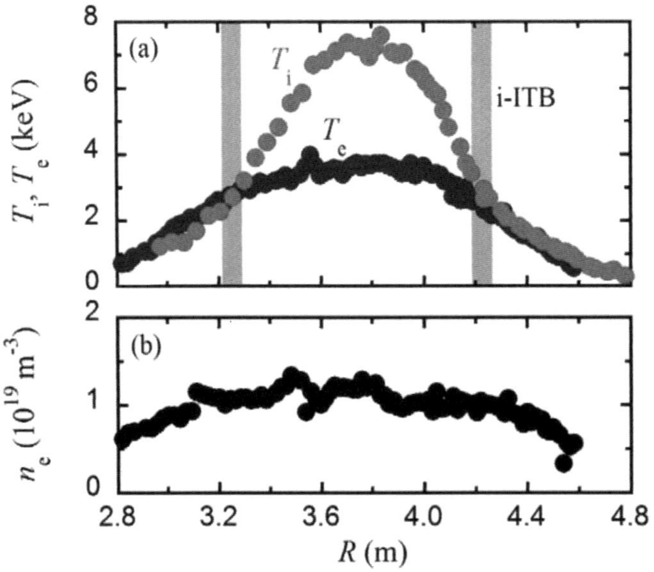

Figure 3. Typical radial profiles of (a) T_i, T_e and (b) n_e, during high T_i discharge.

To obtain the high central T_i, the carbon pellet injection is essential. However, the highest T_i is achieved more than 0.15 sec after the carbon pellet injection. At that time, most of the carbon is almost exhausted from the core region. The carbon density n_C in the core region increases just after the pellet injection, then decreases with the increase of T_i gradient, which is the ion ITB formation. n_C in the core region continues to decrease until the back transition without ITB, then an extremely hollow n_C profile is formed, which is called "impurity hole" [21]. Radial profiles of carbon density 0.18 s and 0.58 s after the pellet injection are shown in figure 4.

According to the heat transport analysis with the TASK-3D code, the ion thermal diffusivity χ_i inside the ion ITB decreases to the neoclassical level, which may be attributed to the significant reduction of anomalous transport in the core region.

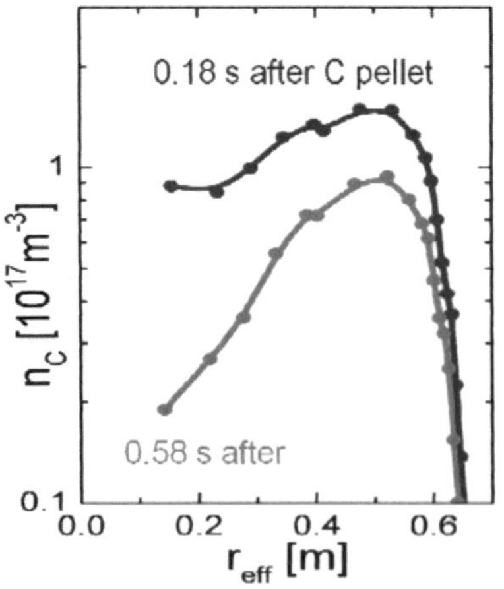

Figure 4. Carbon density profiles 0.18 s and 0.58 s after the pellet injection.

Another effort to obtain the high ion temperature plasma, various edge particle controls, i.e., wall conditionings such as glow discharge cleaning, and titanium getter pumping [22] have been performed so far in LHD. Recently, a new effective wall conditioning technique is proposed [23]. With existence of the magnetic field for the plasma confinement, a series of discharge cleaning with 38 MHz ICRF heating for helium plasma was carried out before the high power NB heated discharges aiming at the high T_i plasma.

Figure 5 shows the time evolution of (a) the NB heating power, P_{NBI}, (b) line averaged electron density, n_{e_fir}, (c) H_α emission intensity, $I_{H\alpha}$, (d) central electron, T_{e0}, and (e) ion, T_{i0}, temperatures at the plasma center, for the high T_i discharge just before (#111165) and after (#111199) the ICRF discharge cleaning. Before the discharge #111199, 30 ICRF discharge cleanings were made. In two discharges shown in figure 5, a carbon pellet was injected at $t = 4.52$ s, followed by a sharp increase of n_{e_fir}. T_{i0} increased until ~ 4.75 s and then gradually decreased. It can clearly be seen, after the repetitive ICRF discharge cleanings, that n_{e_fir} is reduced, together with H_α emission intensity, indicating the decreased hydrogen recycling.

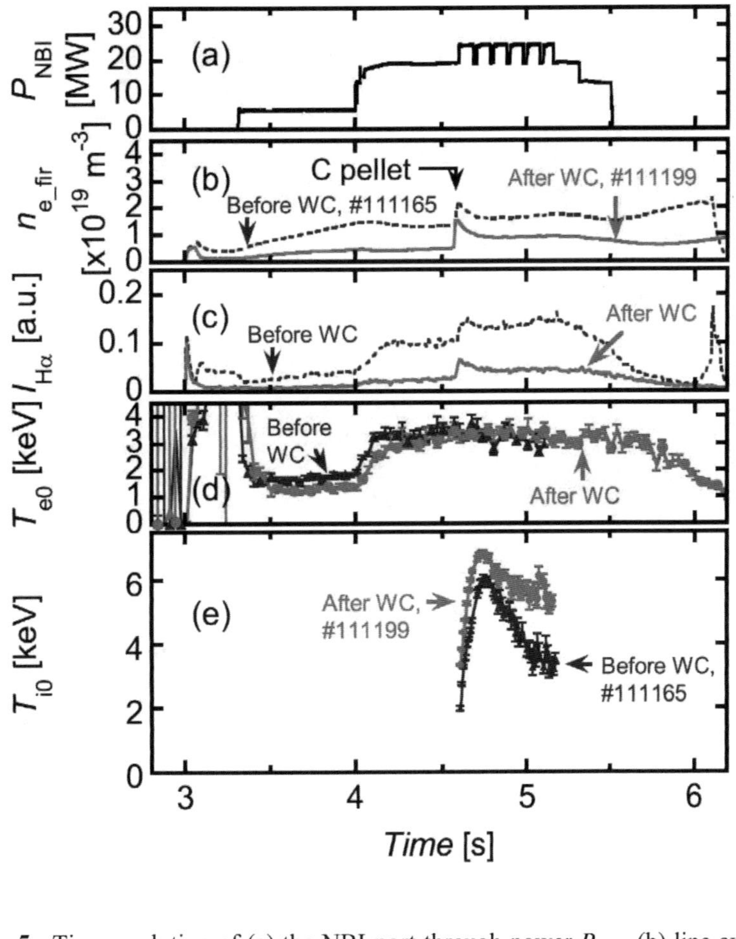

Figure 5. Time evolution of (a) the NBI port through power P_{NBI}, (b) line averaged electron density, $n_{e\ fir}$, (c) H_α emission intensity $I_{H\alpha}$, (d) central electron T_{e0}, and (e) ion T_{i0} temperature, for before (blue) and after (red) ICRF discharge cleaning. From *Plasma and Fusion Research*, 9, 1402050 (2014).

Investigating the effect of the wall conditioning on high T_i discharges, radial profiles of electron density n_e, and ion temperature T_i are compared between before and after the ICRF discharge cleaning, as shown in figure 6. It is found that the n_e profile becomes peaked after the discharge cleaning, although it was quite hollow before the cleaning. The n_e at the center (core region) does not change significantly. This result is closely related to the reduced recycling environment after the ICRF discharge cleaning, which can be seen in the H_α emission signal in figure 5 (c). Low recycling and consequent low edge electron density lead to the increased power deposition of NB on ions in the core region, which results in the achievement of the high central T_i discharge. Actually, T_i after the ICRF discharge cleaning is higher than that before the discharge cleaning, and especially in the central region T_i increases about 1 keV, as shown in figure 6 (bottom).

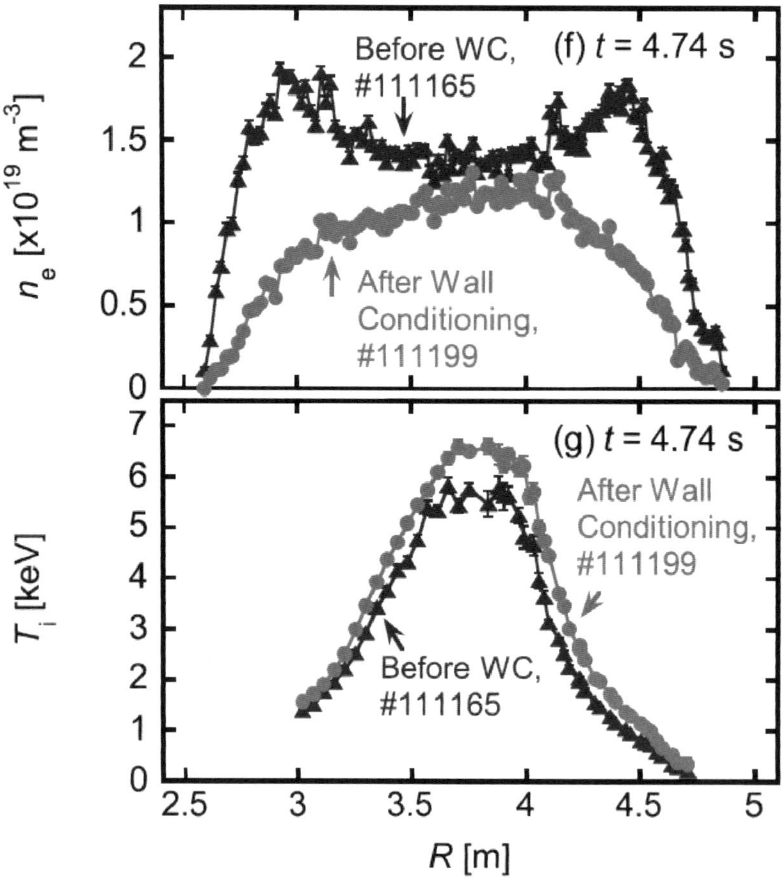

Figure 6. Radial profiles of electron density n_e and ion temperature T_i. Blue and red correspond to before and after ICRF discharge cleaning, respectively. From *Plasma and Fusion Research*, 9, 1402050 (2014).

3.3. Closed helical divertor

As mentioned in the previous sections 3.1 and 3.2, the wall conditioning which reduces the hydrogen recycling and impurity release from PFCs is quite important to achieve the high performance plasma. The new effective wall conditioning technique using ICRF discharge cleaning, presented in section 3.2, is introduced to LHD, in addition to the conventional wall conditioning methods, e.g., baking for PFCs, glow discharge cleaning, titanium getter pumping, boronization (boron coating), etc. From the experimental results, these techniques are known to be quite effective in improving the plasma performance. However, their effects cannot last for a long period of time, thus it is necessary to perform such treatment for PFCs repeatedly. Although this kind of procedure is acceptable for the pulse operation in the research experiment, it is inherently impossible to apply in reactors or reactor-relevant devices such as ITER.

To obtain a long-lasting wall conditioning method, installation of the divertor with active pumping system is one of the solutions, although it is only valid for the hydrogen recycling control. In the LHD, the closed helical divertor (CHD) with cryogenic pumps is being installed [15]. Figure 7 (a) shows a bird's-eye view of a CHD module installed in the inboard side of the LHD vacuum vessel. The CHD consists of ten discrete modules. Following the numerical expectations, it is partially installed on the inboard side of the torus, into which ~ 88% of diverted particles flow. Each module consists of a pair of vertical divertor plates and a "dome" structure, under which the cryogenic pump will be installed. The R&D activity and installation of the cryopump are the most important ongoing tasks in the LHD. The schematic view of its cross section is also presented in figure 7 (b). Geometrical optimization of the divertor plates and the dome was carried out with the 3-D Monte Carlo code "EIRENE" to maximize the neutral compression in the CHD region [24]. It was demonstrated in the simulation that the pressure in the CHD can be increased ~ 13 times higher than that in the open helical divertor (HD). In experiments, three ASDEX-style fast ion gauges are installed under the dome to measure the actual neutral pressures in the CHD. For comparison, another fast ion gauge is also installed in the open HD region which is 180° apart from the CHD in the toroidal direction.

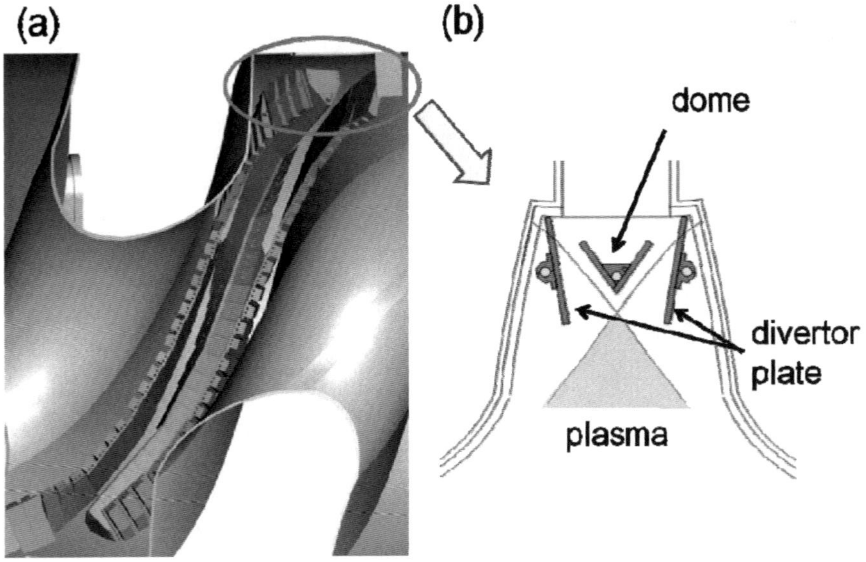

Figure 7. (a) Bird's-eye view of a closed helical divertor and (b) its poloidal cross section.

Preliminary experiments were carried out with neutral beam (NB) heated hydrogen discharges. Due to the continuous gas puffing from 0.1 s after the NB injection, density continued increasing during the discharges for more than 2 s. Figure 8 shows the neutral pressure P_0 in the CHD and the open HD regions as a function of the line averaged electron density n_e for three discharges. In this experiment, the gas inlet was 108° and 144° apart from the open and the CHDs, respectively, in the toroidal direction. It is found that P_0 in both divertors increases monotonically with increase in n_e. Scattering in the low density region before gas puffing was caused by different initial wall conditions for each discharge. It is clearly seen that P_0 in CHD is more than 10 times higher than that in the open HD. This result agrees quantitatively well with that obtained with the numerical simulation, as mentioned above.

Based on these experimental results presented above, we are confident that CHD is invaluable to control neutrals in the peripheral region. To improve the efficiency of the cryopump, we are now energetically promoting the R&D activities, focusing especially on the geometrical design for better conductance and the optimization of the charcoal.

However, we do not think that is the best solution for the edge neutral control and wall conditioning which can be applied for the reactor or reactor relevant devices such as ITER, DEMO, or Helical-DEMO [25]. The pumping system employed in the present CHD in the LHD is a cryopump, which is essentially run by intermittent operation, i.e., it is necessary for the cryogenic pump to regenerate itself after the experiment. The regeneration cycle, of course, depends on the pumping capacity and the experimental condition. It is expected, from the experiment in a test chamber, that the charcoal does not become saturated with hydrogen/deuterium molecules during a series of experiments even over one week. Thus it is planned that the regeneration will be carried out over the weekend, when we start the D-D (deuterium) experiment to be planned in the LHD.

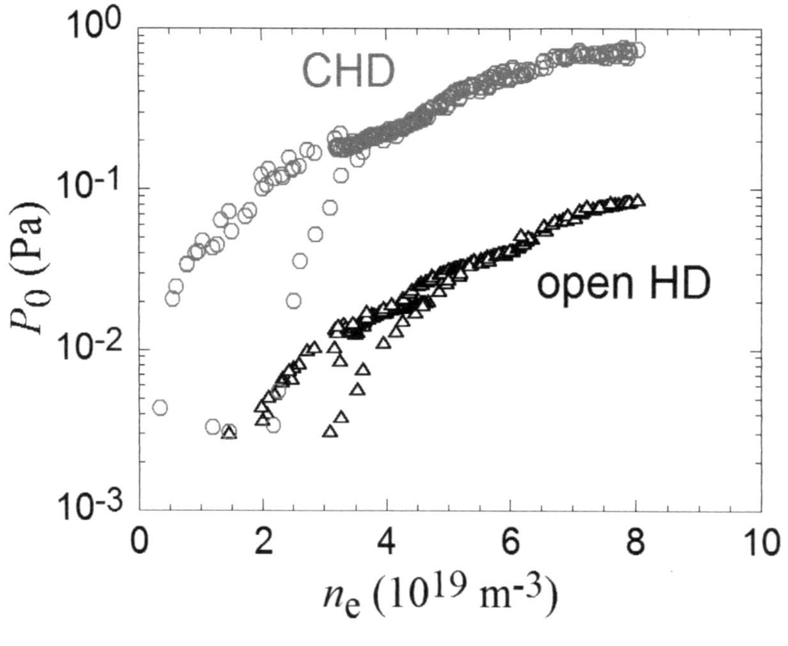

Figure 8. Neutral pressure in open (black) and closed (red) HDs as a function of density.

4. Summary

An ultra-long pulse discharge of high performance plasma was successfully maintained for 48 minutes in the LHD. This is attributed to the progress of the plasma heating technology. The newly installed ICRF antenna can control the parallel wave number of the ICRF wave to heat the core plasma efficiently, and can reduce the power loss at the plasma edge or in the space between the plasma and the chamber wall. Improvement of the ECH system, i.e., gyrotrons and the transmission system including the launcher, has also contributed to the success.

The high ion temperature plasma in the LHD of $T_{i0} \sim 8.1$ keV was recently obtained under the condition of the quite low particle recycling. To achieve such a favorable condition for the core plasma heating, repetitive ICRF discharge cleanings with the existence of the confinement magnetic field were carried out. After the ICRF discharge cleaning, clear reduction of the Hα emission and the edge electron density was observed, and which is the obvious indication of the low recycling. The effect of the ICRF discharge cleaning on plasma performance was experimentally confirmed, i.e., T_{i0} increases from 5.6 to 6.6 keV.

The baffle-structured CHD is being constructed in the LHD, aiming at the active edge plasma control, through the particle recycling control. Comparing with numerical expectations, the neutral compression capability of CHD was experimentally investigated. During the continuous gas puffing discharge, it was observed that the neutral pressure in CHD was more than 10 times higher than that in the open HD. It is expected that a quite low recycling environment can be achieved, if CHD is equipped with the efficient pumping system. However, we should note the quite important fact that the present pumping system of CHD in the LHD, i.e., the in-vessel cryopump, cannot be applied to the reactor, because the cryopump is inherently an intermittent operational equipment. Therefore, for future devices, we must utilize a new pumping system which can be run continuously, or wait for the advent of an innovative pumping concept.

Finally, we summarize the present status of the LHD in figure 9, presenting the fusion triple product as a function of the plasma duration time obtained in various experiments over 16 years. Based on the accumulated knowledge and experiences embedded in the figure, it is expected that the LHD can explore the future aspects of helical reactors.

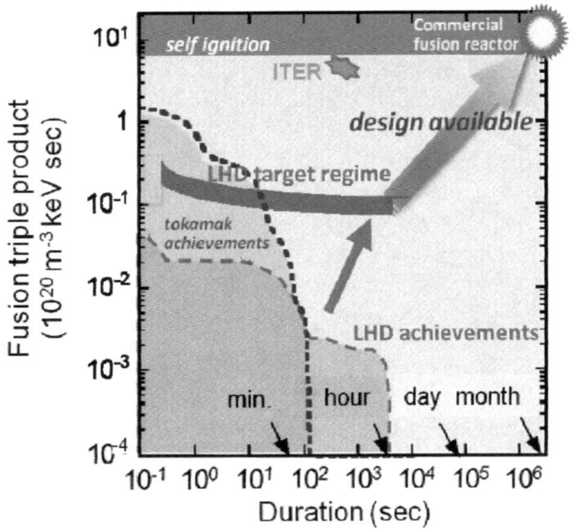

Figure 9. Present status and future targets of LHD.

Acknowledgements

The authors wish to thank Prof. Y. Takeiri, Prof. T. Mutoh, Dr. K. Nagaoka, Dr. H. Kasahara, and Dr. H. Takahashi for their special contributions to this work, and all members of the LHD Experiment Group for their support and fruitful discussions. They also would like to thank all members of the LHD Engineering Group for their support and operation of the machine.

This work is funded by NIFS14ULPP701, NIFS14ULPP801 and partially supported by the Grant-Aid for Scientific Research from MEXT of the Japanese Government.

References

[1] Motojima O. *et al* 1999 *Phys. Plasmas* **6** 1843.
[2] Iiyoshi A. *et al* 1999 *Nucl. Fusion* **39** 1245.
[3] Fujiwara M. *et al* 1999 *Nucl. Fusion* **39** 1659.
[4] Motojima O. *et al* 2003 *Nucl. Fusion* **43** 1674.
[5] Komori A. *et al* 2005 *Nucl. Fusion* **45** 837.
[6] Motojima O. *et al* 2005 *Nucl. Fusion* **45** S255.
[7] Motojima O. *et al* 2007 *Nucl. Fusion* **47** S668.
[8] Komori A. *et al* 2009 *Nucl. Fusion* **49** 104015.
[9] Yamada H. for the LHD Experiment Group 2011 *Nucl. Fusion* **51** 094021.
[10] Kaneko O. *et al* 2013 *Nucl. Fusion* **53** 104015.
[11] Mutoh T. *et al* 2013 *Nucl. Fusion* **53** 063017.
[12] Morisaki T. *et al* 2003 *J. Nucl. Mater.* **313-316** 548.
[13] Morisaki T. *et al* 2010 *Fusion Sci. Technol.* **58** 232.
[14] Kobayashi M. *et al* 2013 *Nucl. Fusion* **53** 093032.
[15] Morisaki T. *et al* 2013 *Nucl. Fusion* **53** 063014.
[16] Sakamoto M. *et al* 2004 *Nucl. Fusion* **44** 693–8.
[17] Li J. G. *et al* 2010 *Proc. 23rd Int. Conf. on Fusion Energy 2010 (Daejeon, Korea, 2010)* (Vienna: IAEA) CD-ROM file OV/1-2 and http://www-naweb.iaea.org/napc/physics/FEC/FEC2010/html/index.htm
[18] Dumont R. *et al* 2012 *Proc. 39th EPS Conf. on Plasma Physics (Stockholm, Sweden, 2012)* P4,005.
[19] Kasahara H. *et al* 2010 *Plasma Fusion Res.* **5** S2090.
[20] Takeiri Y. *et al* submitted to *Plasma Fusion Res.*
[21] Ida K. *et al* 2009 *Phys. Plasmas* **16** 056111.
[22] Masuzaki S. *et al* 2010 *Fusion Sci. Technol.* **58** 297.
[23] Takahashi H. *et al* 2014 *Plasma Fusion Res.* **9** 1402050.
[24] Shoji M., Kobayashi M., Masuzaki S., Watanabe T., Yamada H., Komori A., LHD experiment groups 2009 *J. Nucl. Mater.* **390-391** 490.
[25] Sagara A., *et al* 2012 *Fusion Eng. Design* **87** 594.

27th Summer School and International Symposium on the Physics of Ionized Gases (SPIG 2014) IOP Publishing
Journal of Physics: Conference Series **565** (2014) 012018 doi:10.1088/1742-6596/565/1/012018

Fifty Years of Quasars: Physical Insights and Potential for Cosmology

J W Sulentic[1], P Marziani[2], D Dultzin[3], M D'Onofrio[4], A del Olmo[1]
[1]IAA (CSIC), Spain
[2]INAF Oss. Astronomico Padova, Italy
[3]IA-UNAM, Mexico
[4]Dipartimento di Fisica & Astronomia, Università di Padova, Italy

E-mail: `sulentic@iaa.es`

Abstract. Last year (2013) was more or less the 50th anniversary of the discovery of quasars. It is an interesting time to review what we know (and don't know) about them both empirically and theoretically. These compact sources involving line emitting plasma show extraordinary luminosities extending to one thousand times that of our Milky Way in emitting volumes of a few solar system diameters (log $L_{\rm bol}$= 44.0 - 48.0 erg s^{-1}: D=1-3 light months $\sim 10^3 - 10^4$ gravitational radii). The advent of 8-10 meter class telescopes enables us to study them spectroscopically in ever greater detail.

In 2000 we introduced a 4D Eigenvector 1 parameters space involving optical, UV and X-ray measures designed to serve as a 4D equivalent of the 2D H-R diagram so important for depicting the diversity of stellar types and evolutionary states. This diagram has revealed a principal sequence of quasars distinguished by Eddington ratio (proportional to the accretion rate per unit mass). Thus while stellar differences are primarily driven by the mass of a star, quasar differences are apparently driven by the ratio of luminosity-to-mass.

Out of this work has emerged the concept of two quasars populations A and B separated at Eddington ratio around 0.2 which maximizes quasar multispectral differences. The mysterious 8% of quasars that are radio-loud belong to population B which are the lowest accretors with the largest black hole masses. Finally we consider the most extreme population A quasars which are the highest accretors and in some cases are among the youngest quasars. We describe how these sources might be exploited as standard candles for cosmology.

1. Introduction

Today the existence of a large quasar population is taken for granted. They are simply regarded as the hyperactive nuclei of galaxies observed across cosmic time. In 1963 the reaction was quite different. In the years immediately before the discovery, some blue stars had apparently been detected with radio telescopes. This was a surprise because stars are such weak radio emitters (thermal radio emission from the Sun was only detected around 1946 [1]. These mysterious radio stars were eventually observed spectroscopically and they showed most unusual spectra. Instead of typical stellar absorption line spectra they showed broad emission lines [2, 3]. At first the lines could not be identified but after some time it was realized that they were redshifted Balmer lines (in the first two cases z= $c\Delta\lambda/\lambda_0$= 0.16 (3C273) and 0.42 (3C47)). In the latter case MgII2798 was detected – it had only been detected in the SUN previously. In those early years even the cosmological nature of the quasar redshift was an open question although a gravitational origin for the redshift had been ruled out [4]. If the redshift reflected velocity of recession in an expanding Universe then the quasars were extremely luminous–and small– based on emitting region sizes (a few thousand Schwarzschild radii) inferred from timescale of intensity fluctuations (weeks/months). Figure 1 shows an optical image and spectrum for 3C273.

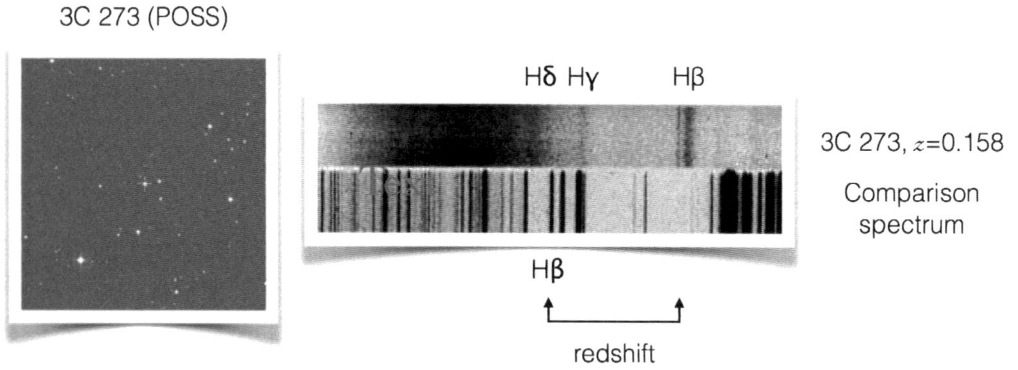

Figure 1. Left: An optical image of 3C273 the first (or one of the first) quasars discovered. It was radio-loud and also showed an optical jet. Right: optical spectrum of 3C273 showing broad Permitted and narrow forbidden emission lines. Optical FeII emission is unusually strong in this relatively nearby quasar–a harbinger of the FeII emission now seen in almost all low z quasars.

Ideas about this new constituent of the Universe were in great flux during these early years. It is important to remember that quasars were discovered 20 years before the advent of CCD detectors and space astronomy was in its infancy. Computers and sophisticated electronic instrumentation were yet to come. The first quasar analog spectra were recorded on glass plates (a nonlinear detector with 1-2% quantum efficiency). It is difficult from the perspective of 2014 to imagine that any progress could have been made. Already in 1967 a quasar with z=2.2 was found (the record is now z=7.08; [5]) showing broad and narrow emission lines as well as absorption lines with different redshifts [6]. This is perhaps one of the reasons why the nature of the redshift was an open question from the beginning. By about 1970 however a consensus had been reached about the nature of such sources – they were powered by gravitational accretion onto a supermassive black hole ([7, 8, 9]; the model: a line emitting accretion disk surrounding a supermassive black hole: AD+BH, [10]). By 1985 an obscuring torus had been added to the standard model ([11] and references therein). A unification model involving a BH+AD+torus structure viewed from different directions can indeed unite much of the AGN diversity (the unifying name for all the manifestations of quasar activity "active galactic nucleus" (AGN) came into use in the early '80s). Fig. 2 sketches the main elements of the standard accretion model. Not all elements are in the cartoon need to be present in the same AGN: the innermost thick structure (the slim disk) is expected to develop only for relatively high accretion rates, above Eddington ratio $\approx 0.2-0.3$. Some type-2 AGN may be sources accreting at very low rate, missing broad line emitting gas and not obscured by a thick torus of clumpy molecular gas [12].

2. DEFINITION OF A QUASAR

Physicists sometimes speak about the need for establishing an operational definition of a phenomenon in order to facilitate development of models employing the known laws of physics. In astronomy, where sources are remote and observations beset with large uncertainties, theoretical speculations are often far ahead of the evidence. Our 1963-64 definition of quasar would involve something like "radio-loud blue variable stellar (i.e. point) source showing broad redshifted emission lines" . The first quasars showed a UV excess reflecting a continuum rising toward shorter wavelengths which is well visible as the dark band short-ward of Hδ on the spectrum of Fig. 1. This kind of spectral energy distribution is similar to the one observed from "hot" white dwarf stars. Searches were rapidly devised to find more very blue stars. Followup spectroscopy could then separate the white dwarf stars from the high redshift quasi-stellar sources. This kind of optical selection quickly demonstrated that radio-quiet quasars outnumbered radio-loud ones. Today only a mysterious 8% of quasars with z <1.0 are found to be radio-loud. "Radio" cannot be part of an operational definition for quasars; neither "blue" because quasars are now found with a wide range of colors. Note that in astronomy what you see is often affected by the intervening medium–all or part of the dispersion in a property might not be intrinsic to a source (i.e. the Sun appears redder when it is observed at low altitude).

The word "stellar" also becomes a problem for our operational definition if we realize that evidence for broad emission lines in the nuclei of some local galaxies had been known since early in the 20th century (hence the label AGN). Most notable are local spiral galaxies with ultra-luminous nuclei [13]. A conference was held at Steward Observatory in 1968 to debate the idea that Seyfert galaxy nuclei might simply be lower luminosity manifestations of the quasar phenomenon. Quasars at high redshift appear stellar because their host galaxies are too faint to be detected. Hubble Space Telescope has revealed evidence of host galaxies around many high redshift quasars. A few claims of naked quasars remain intriguing (e.g. [14]). It turns out that many quasars and Seyfert galaxies require us to remove "broad emission" from the operational definition as well because many of these sources show only narrow (forbidden) lines. Later the discovery of lineless quasars (i.e. BL Lacs) made our original definition even more off the mark [15, 16]. All or much of our problems with definition stem from the likelihood that quasars, unlike stars, are not isotropic sources. They presents different aspects when viewed at different orientations to the line of sight. Today all of the manifestations found before, during and after 1963 are united under the label "active galactic nuclei" (AGN) a term which came into usage in the early '80s. Much detailed history, and a full discussion of nomenclature, can be found in "Fifty Years of Quasars" [17].

X-ray astronomy was born with the launch of UHURU just seven years after quasars were discovered (3C273 had already been X-ray detected: [18]). The best empirical commonality (operational definition?) for AGN may involve the presence of a hard X-ray power-law component (e.g. [19]). It is important to point out that the central AD+BH+torus structure cannot be spatially resolved in any AGN even if interferometric mid-infrared observations are likely to achieve this goal [20]. Even if a few of the lowest redshift examples are eventually resolved it is not clear how much this will advance our understanding of central source structure. All is not lost of course–we can use spectroscopy to resolve the structure in many quasars–most are relatively unobscured essentially down to the BH event horizon (emission lines are seen at radii as small as a few hundred R_G in many sources–and possibly to 2 – 6 R_G via the 6.4keV Fe Kα line in a few cases).

3. CONTEXTUALIZING QUASARS – TOWARDS A QUASAR H-R DIAGRAM?

Given the multiwavelength diversity of AGN properties now observed it should come as no surprise that different groups now focus on specific subtypes (e.g. blazars, radio galaxies, LINERS, narrow line Seyfert 1s) and even on observations in a particular wavelength domain. It can be educational to use the NASA Extragalactic Database (NED) search engine to identify all papers discussing one of the hundred best studied sources. In many cases one finds a strong bias towards radio studies (followed by IR and X-ray) perhaps because after 50 years we are still

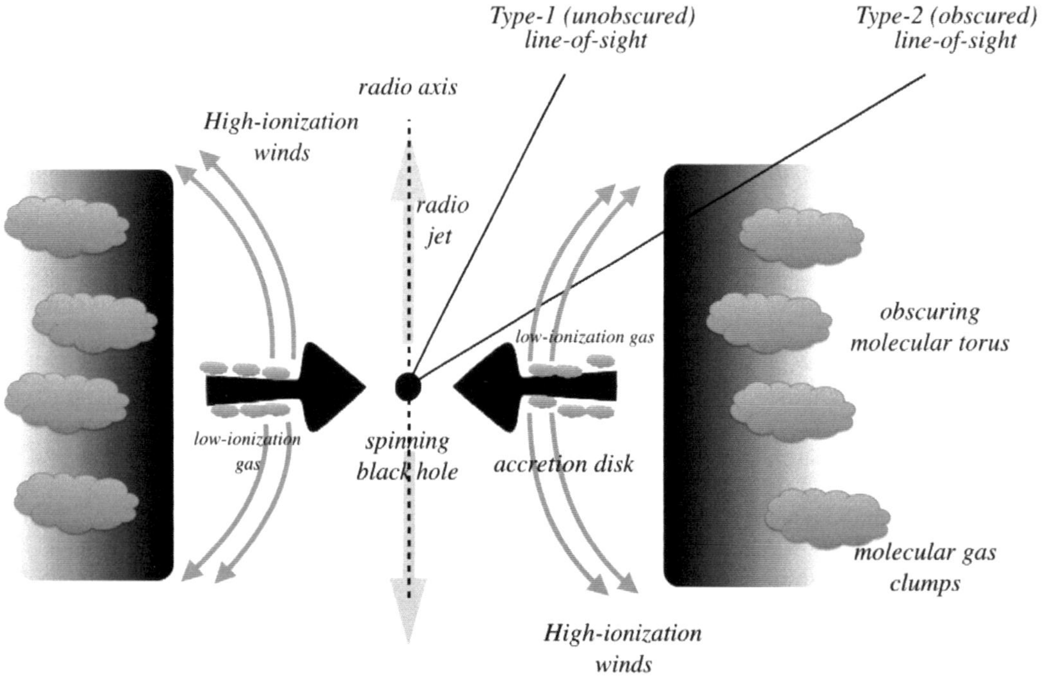

Figure 2. A cartoon illustrating AGN accretion and unification ideas. Broad lines (i.e. quasars) are seen in sources oriented at intermediate viewing angles. Narrow emission lines can be seen in all or most sources oriented near edge-on because the lower density gas extends high enough to be seen above or below the torus. The torus blocks the broad line emission from the accretion disk. The disk is shown with an "arrow" shape indicating that there are many models involving inner radiative and outer gravitational instabilities for this structure. We show a disk with an inner geometrically and optically thick structure (also known as a slim disk [21]) expected to form only for dimensionless accretion rates $\dot{m} > 0.2 - 0.3$. High ionization outflows emerge from the disk, while lower ionization gas may be confined to a flattened configuration that is coplanar with the disk (see, e.g., [22, 23] for reviews and references).

unclear if all quasars pass through a RL phase or if the 8% of RL quasars have fundamentally different source central structure/kinematics diversity

Another important component missing from quasar studies until the '90s (or we would say the 2000s) was an empirical formalism allowing one to contextualize quasar (especially spectroscopic) diversity. When making a NED search one often finds only 1-2 out of 100-500 references for a specific source presenting/discussing optical spectra. One reason for this bias is probably the widespread impression that all quasar spectra are self similar. Nothing could be further from the truth. It is our contention that this misimpression has seriously retarded progress on quasar studies.

A good and simple example of source contextualization involves the H-R diagram for stars (Fig. 3). If all stellar spectra were self similar such a contextualization would not be useful or even possible. But we recognize at least seven – OBAFGKM – principal stellar spectral types organized along a Main Sequence as well as specific regions of the diagram occupied by stars in particular states of evolution. The principal driver is recognized to be stellar mass with different fusion processes and metallicity playing important roles as well. Can the much more energetic quasar phenomenon contain so little spectroscopic diversity that contextualization not useful? How has such an impression come about? In part it reflects the scarcity of good spectroscopic data before 1990. In 1989 only about 60 quasars could be found in the literature with spectra suitable for serious classification but even these revealed interesting diversity in line profile shape [24]. The lack of good spectra fed the impression that all quasars are spectroscopically similar (perhaps in the face of the striking broad redshifted emission lines all the rest seemed minor details). We think that the lack of clarity about quasar spectral diversity is responsible for our failure, up to the present, to develop a standard physical model for the broad line emitting region of quasars (e.g. [25]). The right panel of Figure 3 shows the principal occupation sequence for quasars in a simple plot involving two fundamental optical spectroscopic measures.

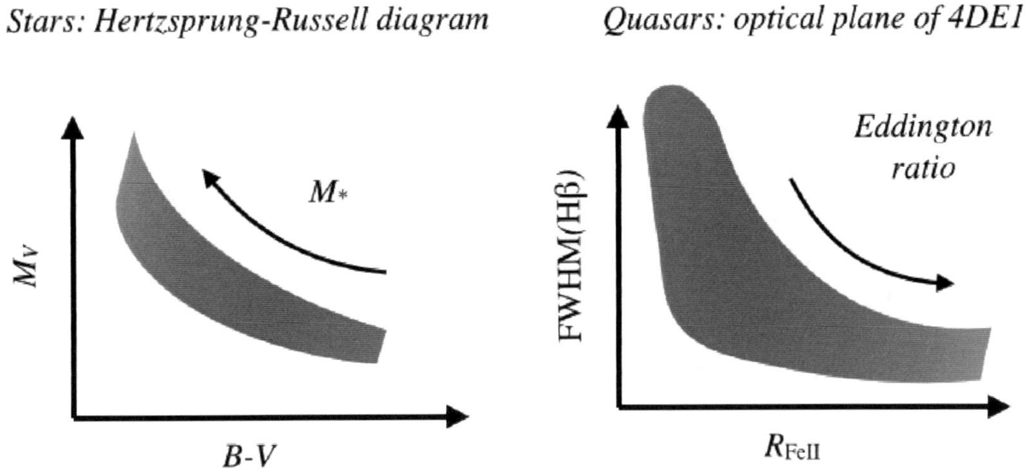

Figure 3. Cartoon representations of the 2D stellar H-R diagram (left) and 2 of the 4 dimensions (i.e. optical plane) of the 4D Eigenvector 1 (4DE1) diagram for quasars. In both cases the schematic main sequence of source occupation is shown with the stellar sequence drive by mass (M) and the quasar sequence driven by Eddington ratio. More dimensions are needed for quasars because, unlike stars, they are not isotropic radiators

Serious progress on the spectroscopic front begins after 1992 with an increase in the quantity and quality (S/N$\sim$20; R$\sim$1000) of quasar spectra thanks mostly to the CCD detector which entered astronomy in 1982. Lick Observatory was ahead of the game in the sense that their IDS (image dissector scanner) system gave them the competitive edge in the 1970s-80s. Most of this early work focussed on the brighter Seyfert 1 nuclei. Quasar spectroscopy had to wait for the CCD era (except for the IPCS system used at Palomar in this period).

From this point onwards let us focus on type 1 AGN (quasars and Seyfert 1 galaxies) which show both high and low ionization broad emission lines. They also almost always show strong blends of permitted FeII emission–often enough that one might want to keep separate the few sources that do not show it. In this way one part of an operational definition for (low redshift) Type 1 quasars can be "broad redshifted low ionization emission lines including optical FeII". The FeII requirement cannot be confirmed in sources beyond z=0.7 unless we: 1) accept UV FeII emission as a surrogate for optical FeII emission and/or 2) follow the increasingly redshifted region of the strongest optical FeII blends near Hβ into the infrared windows. This has been accomplished for about 70 quasars in the range z=1.0-3.7 using the VLT-ISAAC infrared spectrometer [26, 27, 28]. FeII emission is seen in all of them [28] while a study of almost 500 of the brightest low redshift quasars in the northern sky found all but 14 with detectable FeII emission [29].

Our focus on Type 1 sources reflects a working strategy that they are the high accreting **parent population** of the AGN phenomenon. Inclusion of Seyfert 1 sources with quasars requires an addition to the operational definition "stellar sources with and without evidence of a surrounding host galaxy". We assume that the broad emission lines arise in a flattened distribution of gas associated with an accretion disk (AD). The AD provides a dense medium suitable for FeII production. Permitted FeII emission is strong in most type 1 sources and its strength requires a low ionization, high density, and high column density emitting medium. The best candidate for such an emitting region is arguably the AD. There is also evidence that part of the Balmer lines arise in the clouds emitting Fe II. Gravitational accretion onto a supermassive BH is the only know mechanism that can account for the extreme luminosities of the Type 1 sources. Hence the BH+AD paradigm.

Restriction to Type 1 sources is required for this type of approach. Following everything that we know they represent the most unambiguous class of high accreting sources offering the most spectroscopic clues into their nature. Once they are fully characterized and understood one can try to unify them with other more ambiguous classes of AGN with and without broad lines. A further restriction to sources with broad emission lines AND optical FeII emission removes additional ambiguities associated with obscured sources (e.g. Seyfert 1.5) and sources with the narrowest broad lines (e.g. narrow line Seyfert 1 (NLSy1) sources). Even with these restrictions we find impressive spectroscopic diversity. Understanding this diversity –we argue–is a key to development of successful physical models for the BLR of Type 1 sources.

Progress in understanding Type 1 quasars came with the advent of larger samples of quasars with high s/n spectra. A major advance involved analysis of spectra for 87 sources in the Palomar Bright Quasar survey [30]. The spectra allowed accurate measurements of broad and narrow emission line properties. Principal Component Analysis techniques were applied to the correlation matrix representing the measured parameters. This study identified the principal (Eigenvector 1) correlations that exist in the dataset and also showed that source luminosity belonged to Eigenvector 2 (the second orthogonal solution). This work (and also work on X-ray-optical correlations [31]t motivated us to develop the empirical context in which to interpret the spectroscopic diversity of type 1 sources [32, 33]. Fortunately by the mid 90s the Hubble Space Telescope archive was supplying moderate quality UV spectra for more than 130 low z sources.

4. OUR WORK WITH 4D EIGENVECTOR 1 – MAJOR RESULTS

We focussed our efforts on a four dimensional parameter space (4D Eigenvector 1 = 4DE1) [34]. 4DE1 has roots in the PCA analysis of the Bright Quasar Sample ([87 sources, [30]) as well as in correlations that emerged from ROSAT(e.g.,[31]). 4DE1 as we define it involves BG92 measures: (1) full width half maximum of broad Hβ (FWHM Hβ) and (2) equivalent width ratio of optical FeII and broad Hβ: RFe = W(FeIIλ4570)/W(Hβ). We added a [31]-defined measure involving (3) the soft X-ray photon index ($\Gamma_{\rm soft}$) and a measure of (4) CIVλ1549 broad line profile velocity displacement at half maximum ($c(\frac{1}{2})$) to arrive at our 4DE1 space. Other points of departure from BG92 involve our comparison of RQ and RL sources as well as subordination of BG92 [OIII] measures (although see [35, 36]).

The principal diagnostic measures defined above can be interpreted as physical measures involving: 1) velocity dispersion of the low ionization broad line (HIL) region. It is standard to assume (and there are consistency arguments to support it) that these lines arise from a rotating Keplerian disk making FWHM (usually Hβ or MgII2798) a virial estimator of BH mass. 2) RFE involves the relative strength of LILs that are thought to arise in the same AD structure. This ratio serves as an estimator of electron density (n_e) in the LIL gas and the high density requirement for its production supports the AD as the source of the line emission. 3) Γ_{soft} measures the strength of a (thermal) soft X-ray photon index. This is thought to be correlated with the accretion rate. 4) Th CIV shift parameter measures the amplitude of systematic radial motions in the high ionization line (HIL) gas. It is sensitive to the amplitude and geometry of an AD wind.

Many other correlates exist especially for quasars above z=1.0. We focus this summary on low redshift results and the relevant diagnostic measures. Our 4DE1 parameter space immediately provided several clues about the kinematics and physical conditions in the broad line region. Note that the added dimensionality of 4DE1 compared to the stellar H-R diagram is required if for no other reason than the fact that the multiwavelength properties of a quasar are orientation dependent. This is most obvious for RL sources but ample evidence exists that it is also true for the RQ majority. Major clues coming out of 4DE1 are summarized here.

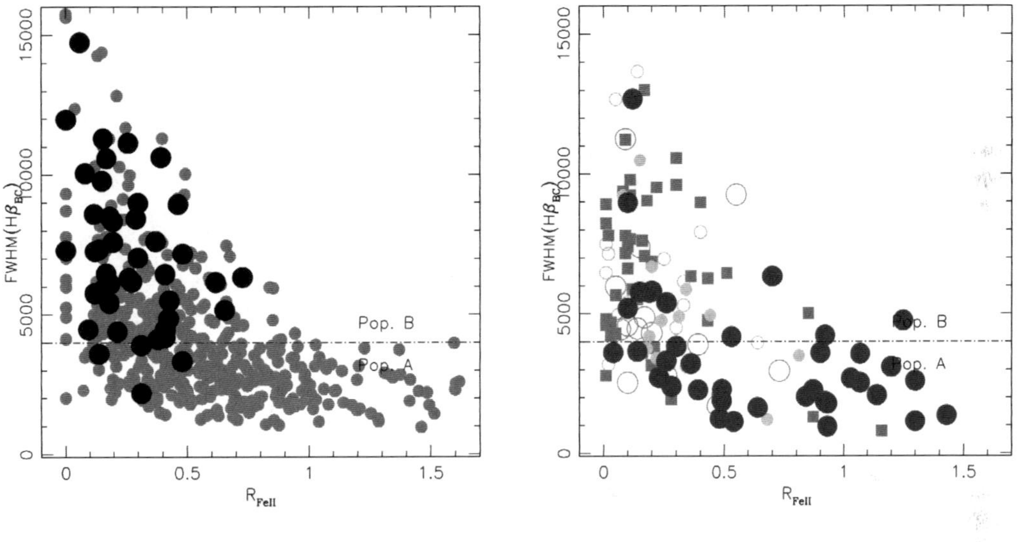

Figure 4. The 4DE1 optical plane involving the width of Hβ and optical FeII strength. The horizontal line marks the boundary between population A (lower) and B (upper) sources. Left: source occupation for 470 SDSS-DR5 quasars (z<0.75) with highest s/n SDSS spectra [37]. Small grey circles represent RQ sources; large black circles lobe-dominated RL sources which show a strong preference for Population B. Right: optical plane for all low z quasars with measurable HST/FOS UV spectra [38]. Different symbols represent the amplitude of the CIV 1549 blueshift at half maximum $c(\frac{1}{2})$. Large blue filled circles involve sources with the largest CIV blueshifts which strongly favor Population A. Large open circles represent sources with a large CIV redshift and grey squares those with no significant line shift [38, 23].

- Figure 4 is the clearest representation of the diversity of optical and UV spectroscopic measures in low redshift sources [33, 34]. Subsequent work suggests that the trend (i.e. source occupation) in Figure 4 (left) is driven by source Eddington ratio (proportional to the dimensionless accretion rate for constant radiative efficiency) convolved with line-of-sight orientation. M_* (stellar mass) drives the H-R diagram main sequence occupation but black hole mass plays a secondary role in 4DE1 [39, 40] in low-z samples.
  Figure 4 (right panel) again shows the 4DE1 optical plane and source occupation for the 470 SDSS-DR5 quasars with highest s/n SDSS spectra. RL sources are distinguished from the

RQ majority (grey dots) with symbols that reflect their radio morphology. Lobe dominated (LD–largely FRII) sources are indicated by open red circles while core-dominated (CD RL) are shown as filled blue squares. The latter are interpreted in orientation-unification scenarios as preferentially aligned LD where the radio jets are pointed along our line of sight. Our default assumption is that the radio jets are aligned perpendicular to the broad line emitting disk. The first major result is that the majority of LD sources occupy a restricted domain relative to the RQ majority of quasars. The domain is also restricted for the CIV and Γ_{soft} measures. This may be telling us that RL quasars are a distinct quasar population with different BLR structure and kinematics. A second result involves a displacement of CD RL sources towards smaller FWHM Hβ and slightly larger RFE. The green arrow in figure 1 indicates the change in median optical 4DE1 measures from LD to CD. This is consistent with the orientation-unification ideas mentioned above and was previously inferred from a correlation between FWHM Hβ and radio core/lobe flux ratio [41, 37].

- The multiwavelength differences between sources at the two ends of the sequence seen in Figure 4 motivated the concept of two quasar populations (horizontal line in Figure 4 left/right marks the Population A-B boundary [37] at FWHM Hβ=4000 km s^{-1}. This value (and RFE=0.5) lie very close to the values identified in a 2D K-S test maximizing the parameter space separation between RL and RQ sources. Many other differences exist between largely RQ Population A sources (below 4000 km s^{-1}) and the mixed RQ/RL population B above this FWHM boundary. Physically the FWHM Hβ boundary (4000 km s^{-1}) corresponds to $L/L_{\rm Edd} = 0.2\pm0.1$ for a source with black hole mass log $M_{\rm BH} \sim$ 8.0. This boundary may represent a critical Eddington ratio where the BLR undergoes a significant change in structure/kinematics. Even if it has no profound physical significance it represents a way to separate quasars into high and low accreting samples. The lack of this kind of discrimination underlies previously mentioned difficulties with attempts at developing a model for BLR physics.
- The 4DE1 context reveals that the much discussed narrow line Seyfert 1 (NLSy1) sources are not a distinct class of quasars with unusual properties. Rather they are simply the most extreme (i.e. narrowest FWHM Hβ) Population A sources. All evidence points toward 4000 km s^{-1} (and not 2000 km s^{-1}) as the more significant boundary for sources with $\log L \leq 46$ [erg s^{-1}]. The limit is expected to increase for more luminous sources because of the weak, but significant, effect of increasing BH mass [42, 43].
  'item Other evidence that Population A and B quasars differ includes a fundamental change in broad line profile shape near the Population A-B boundary [33, 44, 38]. The broad Hβ (characteristic low ionization line at low z) profile is well fit by a symmetric Lorentzian function in most Pop. A sources. Population B sources require a double Gaussian fit including broad (FWHM$\sim$ 5000 km s^{-1}) relatively unshifted component plus a very-broad (FWHM$\sim$10000 km s^{-1}) and redshifted (a few thousand km s^{-1}) component. CIV1549 (characteristic high ionization line at most redshifts) shows strong profile blueshifts and blue asymmetries in Population A sources but is usually relatively symmetric and unshifted in Population B sources. This description is confirmed in median composites involving 50-250 low z sources despite the fact that individual sources show an impressive diversity in profiles shapes/shifts/asymmetries.

We think that the above results offer the key to a physical model of quasars that can unify the intriguing 4DE1 sequence and the Population A-B differences. A major advance in our understanding of quasar physics is coming soon.

5. A PROMISING FUTURE

The Eddington ratio of a quasar is proportional to the ratio of source luminosity to BH mass. If Eddington ratio and BH mass can be derived from some distance-independent measure it would be possible to derive distance-independent quasar luminosities. Quasars are now easily detected out to $z = 4$ and those in the range $1 < z < 3 - 4$ are of greatest interest because the effect of the cosmic matter density is believed to dominate over the repulsive effect of the cosmological constant in this range. Quasars radiating close to the Eddington limit show distinct optical and UV spectral properties that can be recognized in major survey spectra – if the data is contextualized within the 4DE1 formalism. Measures of the Hβ spectral range and the 1900 blend yield selection criteria involving two related ratios: (i) Al III λ1860/Si III] $\lambda 1892 \geq 0.5$ and (ii) Si III] λ1892/C III]$\lambda 1909 \geq$1.0. A little less than 100 sources satisfying these criteria cluster around the Eddington limit with a relatively small dispersion ≈ 0.13 dex [45]. Other 4DE1 correlated parameters related to the X-ray continuum shape have been proposed [46, 47] and may become useful with new data provided by planned space missions such as Athena or ongoing ones like nuSTAR. More than 300,000 spectra have been collected in the SDSS IV - BOSS survey [48]. The optical and UV selection criteria offer the non-negligible advantage of facilitating selection of large samples from presently available data since the frequency of quasars radiating close to $L/L_{Edd}=1$ is estimated to be at least a few percent of the unobscured population. We have shown that such quasar samples can yield independent measures of Ω_M with tight limits even if samples of just a few hundred sources are considered [49, 45].

Current issues go beyond the existence of the dark energy and focus more on its properties. The simplest model for dark energy is a cosmological constant with a fixed equation of state ($p = w\rho$, with fixed $w = -1$). However, the dark energy density may depend weakly on time, according to many proposed models of its nature [50]: a general scalar field predicts w to be negative and evolving with redshift. The strength of a quasar sample also includes the possibility to construct a Hubble diagram uniformly sampling a broad range of redshift. Extreme Eddington quasars are, at least in principle, sources suitable for testing whether the dark energy equation of state is constant or is evolving as function of redshift following selected parametric forms for $w(z)$). After 50 years, and a few failed attempts, are we on the threshold of using quasars for cosmology?

References

[1] Hey J S 1946 *Nature* **157** 47–48
[2] Schmidt M 1963 *Nature* **197** 1040
[3] Schmidt M and Matthews T A 1964 *ApJ* **139** 781
[4] Greenstein J L and Schmidt M 1964 *ApJ* **140** 1–+
[5] Mortlock D J, Warren S J, Venemans B P, Patel M, Hewett P C, McMahon R G, Simpson C, Theuns T, Gonzáles-Solares E A, Adamson A, Dye S, Hambly N C, Hirst P, Irwin M J, Kuiper E, Lawrence A and Röttgering H J A 2011 *Nature* **474** 616–619 (*Preprint* 1106.6088)
[6] Arp H C, Bolton J G and Kinman T D 1967 *ApJ* **147** 840
[7] Zel'dovich Y B and Novikov I D 1964 *Soviet Physics Doklady* **9** 246
[8] Salpeter E E 1964 *ApJ* **140** 796–800
[9] Lynden-Bell D 1969 *Nature* **223** 690–694
[10] Shakura N I and Sunyaev R A 1973 *AAp* **24** 337–355
[11] Nenkova M, Sirocky M M, Nikutta R, Ivezić Ž and Elitzur M 2008 *ApJ* **685** 160–180 (*Preprint* 0806.0512)
[12] Laor A 2003 *ApJ* **590** 86–94
[13] Seyfert C K 1943 *ApJ* **97** 28
[14] Magain P, Letawe G, Courbin F, Jablonka P, Jahnke K, Meylan G and Wisotzki L 2005 *Nature* **437** 381–384 (*Preprint* arXiv:astro-ph/0509433)
[15] Schmidt M 1968 *ApJ* **151** 393
[16] Strittmatter P A, Serkowski K, Carswell R, Stein W A, Merrill K M and Burbidge E M 1972 *ApJL* **175** L7
[17] D'Onofrio M, Marziani P and Sulentic J W (eds) 2012 *Fifty Years of Quasars From Early Observations and Ideas to Future Research* (*Astrophysics and Space Science Library* vol 386) (Berlin-Heidelberg: Springer

Verlag)
[18] Bowyer C S, Lampton M, Mack J and de Mendonca F 1970 *ApJL* **161** L1
[19] Elvis M, Maccacaro T, Wilson A S, Ward M J, Penston M V, Fosbury R A E and Perola G C 1978 *MNRAS* **183** 129–157
[20] Jaffe W, Meisenheimer K, Röttgering H J A, Leinert C, Richichi A, Chesneau O, Fraix-Burnet D, Glazenborg-Kluttig A, Granato G L, Graser U, Heijligers B, Köhler R, Malbet F, Miley G K, Paresce F, Pel J W, Perrin G, Przygodda F, Schoeller M, Sol H, Waters L B F M, Weigelt G, Woillez J and de Zeeuw P T 2004 *Nature* **429** 47–49
[21] Abramowicz M A, Czerny B, Lasota J P and Szuszkiewicz E 1988 *ApJ* **332** 646–658
[22] Marziani P and Sulentic J W 2012 *The Astronomical Review* **7** 040000–57 (*Preprint* 1210.2059)
[23] Marziani P and Sulentic J W 2012 *NARev* **56** 49–63 (*Preprint* 1108.5102)
[24] Sulentic J W 1989 *ApJ* **343** 54–65
[25] Dumont A M, Collin-Souffrin S and Nazarova L 1998 *AAp* **331** 11–33
[26] Sulentic J W, Stirpe G M, Marziani P, Zamanov R, Calvani M and Braito V 2004 *A&Ap* **423** 121–132 (*Preprint* arXiv:astro-ph/0405279)
[27] Sulentic J W, Dultzin-Hacyan D, Marziani P, Bongardo C, Braito V, Calvani M and Zamanov R 2006 *Revista Mexicana de Astronomia y Astrofisica* **42** 23–39 (*Preprint* arXiv:astro-ph/0511230)
[28] Marziani P, Sulentic J W, Stirpe G M, Zamfir S and Calvani M 2009 *A&Ap* **495** 83–112 (*Preprint* 0812.0251)
[29] Zamfir S, Sulentic J W, Marziani P and Dultzin D 2010 *MNRAS* **403** 1759 (*Preprint* 0912.4306)
[30] Boroson T A and Green R F 1992 *ApJS* **80** 109–135
[31] Wang T, Brinkmann W and Bergeron J 1996 *A&Ap* **309** 81–96
[32] Marziani P, Sulentic J W, Dultzin-Hacyan D, Calvani M and Moles M 1996 *ApJS* **104** 37–+
[33] Sulentic J W, Marziani P and Dultzin-Hacyan D 2000 *ARA&A* **38** 521–571
[34] Sulentic J W, Marziani P, Zwitter T, Dultzin-Hacyan D and Calvani M 2000 *ApJL* **545** L15–L18 (*Preprint* arXiv:astro-ph/0009326)
[35] Zamanov R, Marziani P, Sulentic J W, Calvani M, Dultzin-Hacyan D and Bachev R 2002 *ApJL* **576** L9–L13 (*Preprint* arXiv:astro-ph/0207387)
[36] Marziani P, Sulentic J W, Zamanov R, Calvani M, Dultzin-Hacyan D, Bachev R and Zwitter T 2003 *ApJS* **145** 199–211
[37] Zamfir S, Sulentic J W and Marziani P 2008 *MNRAS* **387** 856–870 (*Preprint* 0804.0788)
[38] Sulentic J W, Bachev R, Marziani P, Negrete C A and Dultzin D 2007 *ApJ* **666** 757–777 (*Preprint* 0705.1895)
[39] Marziani P, Sulentic J W, Zwitter T, Dultzin-Hacyan D and Calvani M 2001 *ApJ* **558** 553–560 (*Preprint* arXiv:astro-ph/0105343)
[40] Marziani P, Zamanov R K, Sulentic J W and Calvani M 2003 *MNRAS* **345** 1133–1144 (*Preprint* arXiv:astro-ph/0307367)
[41] Sulentic J W, Zamfir S, Marziani P, Bachev R, Calvani M and Dultzin-Hacyan D 2003 *ApJL* **597** L17–L20 (*Preprint* arXiv:astro-ph/0309469)
[42] Dultzin D, Martinez M L, Marziani P, Sulentic J W and Negrete A 2011 Narrow-Line Seyfert 1s: a luminosity dependent definition *Proceedings of the conference "Narrow-Line Seyfert 1 Galaxies and their place in the Universe". April 4-6, 2011. Milano, Italy.* Proceedings of Science ed et al (Eds) L F
[43] Negrete A, Dultzin D, Marziani P and Sulentic J 2012 *ApJ* **757** 62 (*Preprint* 1107.3188)
[44] Sulentic J W, Marziani P, Zamanov R, Bachev R, Calvani M and Dultzin-Hacyan D 2002 *ApJL* **566** L71–L75 (*Preprint* arXiv:astro-ph/0201362)
[45] Marziani P and Sulentic J W 2014 *MNRAS* **442** 1211–1229 (*Preprint* 1405.2727)
[46] Wang J M, Du P, Valls-Gabaud D, Hu C and Netzer H 2013 *Physical Review Letters* **110** 081301 (*Preprint* 1301.4225)
[47] La Franca F, Bianchi S, Ponti G, Branchini E and Matt G 2014 *ApJL* **787** L12 (*Preprint* 1404.2607)
[48] Pâris I, Petitjean P, Aubourg E, Bailey S, Ross N P, Myers A D, Strauss M A, Anderson S F, Arnau E, Bautista J, et al. C 2012 *ArXiv e-prints* (*Preprint* 1210.5166)
[49] Marziani P and Sulentic J W 2013 *ArXiv e-prints* (*Preprint* 1310.3143)
[50] Carroll S M 2005 Why is the Universe Accelerating? *Observing Dark Energy* (*Astronomical Society of the Pacific Conference Series* vol 339) ed Wolff S C and Lauer T R p 4

Electron excitation and autoionisation cross sections for elements of chemically peculiar stars: Study of bismuth

B Predojević[1,2*], V Pejčev[1], D Šević[1] and B P Marinković[1†]

[1] Institute of Physics, University of Belgrade, POB 68, 11080, Serbia
[2] Faculty of Natural Sciences, University of Banja Luka, Republic of Srpska, Bosnia and Herzegovina

[*]E-mail: bpredojevic@teol.net
[†]E-mail: bratislav.marinkovic@ipb.ac.rs

Abstract. Electron impact excitation from the ground state of bismuth atoms has been studied. A beam of electrons was scattered from a beam of atoms and the intensity of scattered electrons was measured for scattering angles up to 150° and incident electron energies of 40 and 60 eV. Obtained intensities were used for the calculation of relative differential cross sections (DCS). In addition, we recorded the energy loss-spectra at different incident electron energies and scattering angles. These spectra were analysed in order to identify the energy levels of bismuth atom below and above (autoionisation) the first ionization limit in electron spectroscopy. The presence of bismuth was confirmed in spectra of the chemically peculiar (CP) magnetic Ap 73 Dra and HR 465 and nonmagnetic Hg-Mn HR 7775 and χ Lupi stars. The obtained results for relative DCS and identified autoionised energy levels of bismuth were analysed and compared with previous experimental and theoretical data. The connection between our investigations of bismuth and astrophysical measurements are discussed.

1. Introduction

Bismuth is a heavy atom (Z=83) that is very interesting subject for investigations in many fields of physics including electron-atom scattering processes and investigation of chemical structure of the stellar atmospheres. The absorption and emission spectra of bismuth, including hyperfine structure, have been the subject of many detailed investigations in both laboratory environment [1-6] and in astronomy observation [7-11].

Only a few papers contain theoretical calculations [12-14] and experimental data [14] of the differential cross sections (DCSs) for elastic and inelastic electron scattering on bismuth atoms. Williams and Trajmar [14] measured DCSs for elastic and inelastic (excitation of the $6p^3\ ^2D_{3/2}$, $^2D_{3/2}$, $^3P_{1/2}$, and $2p^27s\ ^4P_{1/2}$ states) collisions at E_0 = 40 eV and scattering angles from 0° to 130°. The cross sections were normalised to an absolute scale using the optical oscillator strength (OOS) value by Lvov [15] for the $6p^3\ ^4S_{1/2} \rightarrow 2p^27s\ ^4P_{1/2}$ transition. To our knowledge, no calculations of DCSs exist for the inelastic electron scattering on bismuth atom.

The presence of bismuth features in stellar spectra has been reported in different type of chemically peculiar (CP) stars. The position of CP stars in the Hertzsprung–Russell (HR) diagram is shown in figure 1. Guthrie [7] was the first who has reported and identified the BiI features in magnetic Ap (CP2) star 73 Dra at 472.2 nm. This line corresponds to the $6p^3\ ^2D_{3/2} \rightarrow 6p^27s\ ^2P_{1/2}$ transition and has long been known to belong to the emission spectra of bismuth [1]. Two lines of the BiII (139.39 nm, 143.683 nm) where identified from the IUE (International Ultraviolet Explorer) spectrum of the magnetic Ap star HR 465 [9]. Several strong absorption lines due to BiII were observed in nonmagnetic (CP3) Hg-Mn star HR 7775 in high-resolution spectra obtained with the IUE [8]. The HR 7775 is a part of a multiple-star system that includes the bright star HR 7776 (at a distance of 205") and a close companion (the later to lie at a distance of 0.68" from HR 7775). Using high-resolution spectra obtained from the Goddard High Resolution Spectrograph (GHRS) on Hubble Space Telescope (HSP) Wahlgren *et al* [10] were able to detect absorption features in spectra of the binary nonmagnetic Ap χ Lupi star that contains contributions from AsII and BiII. The strongest transition of BiII in the observed intervals is the $6p^2\ ^3P_2 \rightarrow 6p7s\ ^3P_1$ transition at 190.23 nm. Its nine hyperfine components were studied by Bouzza and Bauche [16]. Selected lines at 143.683 nm (BiII) and 142.34 nm (BiIII) were studied in the ultraviolet spectrum of the χ Lupi [17], with conclusion that the bismuth abundance is only marginally enhanced in this star. Bismuth abundances in HR 7776 and χ Lupi stars have been the subject of an extensive study in ultraviolet region of wavelengths by Wahlgren *et al* [18]. In table 1 the summary of the results from this study is given.

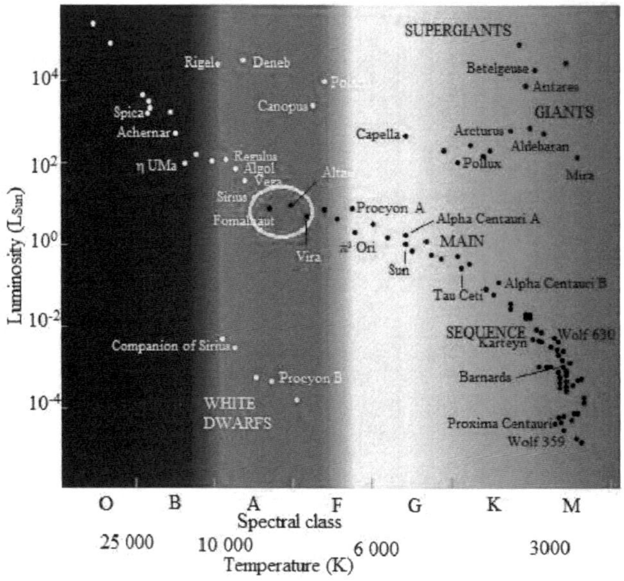

Figure 1. Approximate position of the chemically peculiar (CP) stars on Hertzsprung–Russell diagram

Table 1. Bismuth abundances in HR 7775 and χ Lupi, $\log N_{Bi}$ includes the Solar system value of 0.7 dex

Wavelength (nm)	Ion	HR 7775	χ Lupi
143.6	BiII	< 5.5	< 2.3
190.2	BiII	< 6.0	< 2.7
142.3	BiIII	< 6.0	< 1.7

Very good agreement among derived abundances for different lines for HR 77775 star is evident. In contrast, for χ Lupi a significant disagreement between abundance of BiII and BiIII was obtained and an ionisation anomaly is apparent. Dolk *et al* [19] analysed the spectra of HR 7775 recorder by the Nordic Optical Telescope (NOT) in optical region. Fourteen BiII lines were detected in the range from 407.9 to 680.09 nm. The Bismuth abundance was determined using the strongest lines, 514.4 and 520.9 nm. The obtained value of 5.8 is in good agreement with results calculated by Wahlgren *et al* [18].

Line intensities in autoionising spectra of atoms for both "allowed" and "optically forbidden" transitions are of great importance for determination of physical constants in astrophysical plasma. Ionisation processes via autoionisation could be in many cases more important than the direct one. The contribution of core-excited autoionising states to the total ionisation cross section has been in detail examined theoretically by Kim and Stone, in particular for atoms of the IIIA group (boron, aluminium, gallium and indium) [20] and IVA group (silicon, germanium, tin and lead) [21] of the Periodic Table. They found that the core-excited autoionising states contribute significantly to the total ionisation cross sections for all the numbered atoms except for lead.

In this paper we present relative DCS for electron impact excitation of the $6p^2\ 7s\ ^4P_{1/2}$ state in bismuth at electron impact energies of 40 and 60 eV. In addition, the analysis for the energy loss spectra below and above the first ionisation threshold (autoionisation) is given. The connections between our results and astrophysical measurements are discussed. In section 2, the apparatus is described and the experimental procedure is given. In section 3, results of relative DCS measurements and obtained energy-loss spectra are discussed and presented graphically. Finally, conclusions are formulated in section 4.

2. The experiment and procedure

Experimental arrangement used in our measurements is a conventional crossed-beam electron spectrometer described in our previous works [22,23]. The optics of both the monochromator and the analyser are very similar to the one designed by Chutjian [24]. The analyser can be positioned at angles from -30° up to 150° with respect to the direction of primary electron beam. The electron spectrometer itself can be operated in three different modes: recording of electron energy-loss spectra, scanning incident energy and direct angular distribution measuring of elastically and inelastically scattered electrons. The position of the real zero scattering angle was determined before each angular distribution measurement by checking the symmetry of the scattered electron signal at positive and negative angles, with respect to the un-scattered electron beam. The energy scale was calibrated against the $3s3p\ ^1P_1$ excitation threshold of Mg at 4.345 eV. Using significantly improved energy resolution (50 meV), we did not find the shift of the energy scale due to contact potential difference between the thoriated tungsten filament (work function of Th 3.4 eV) and magnesium plated (3.66 eV) collision chamber. A channel electron multiplier is used for single-electron counting. The operating condition are summarised in table 2.

Table 2. The experimental parameters used in our electron spectrometer

Parameters	Conditions
Impact energy range (eV)	10-100
Energy resolution (meV)	100-150
Uncertainty in energy scale (meV)	300
Angular resolution (^{0})	1.5
Uncertainty in angular scale (^{0})	0.5
Oven temperature (K)	1000
Metal-vapour pressure (Pa)	10
Oven nozzle aspect ratio ($\gamma=d/L$)	0.075
Primary electron current (nA)	10-50
Residual magnetic field in interaction region (double μ-metal shield) (μT)	< 0.1
Background pressure (mPa)	< 5

The metal-vapour beam source consists of a stainless-steel crucible. The crucible is placed in a stainless-steel cylinder co-axially wrapped with two different resistive bifilar heaters (top and bottom, whose temperatures were monitored with two thermocouples), enabling the top of the source to be at approximately 100 K higher temperature than the bottom. This prevents clogging of the nozzle. An additional outer copper cylinder served as a holder for the helical tube of the water-cooler.

In this work, relative DCSs for electron impact excitation of the $6p^2\ 7s\ ^4P_{1/2}$ state from the ground state of bismuth atom are obtained as follows. At a certain incident electron energy and energy-loss (ΔE), the position of the analyser was changed from 2^0 to 150^0 and the angular distribution of scattered electron was measured. The scattering intensity was corrected using the effective path-length correction factors according to the approach of Brinkman and Trajmar [25].

The contributions to the total uncertainties of the relative DCSs arise from statistical uncertainties, uncertainty from effective path-length correction factor (0.06), and estimation of the energy (0.01) and angular (0.10) scale.

3. Result and discussion

We have measured relative differential cross sections for electron-impact excitation of the $6p^27s\ ^4P_{1/2}$ state of bismuth atom from the ground state at the incident impact energies of 40 and 60 eV and the scattering angle from 2° to 150°. The relative DCSs with total uncertainties are presented graphically in figure 2.

In order to compare in shape the presented DCS at 40 eV electron impact energy with DCS obtained by Williams and Trajmar [14] we normalised their DCS to present DCS at 10°. As one can see in figure 2 the agreement in shape is good up to 30°. At higher scattering angles our DCS have three maxima and two minima; while DCS obtained by Williams et al [14] monotonically decreases up to 90° and after that monotonically increases. For the measured electron impact energy of 60 eV our DCSs show two minima, the first at 80° and the second close to 140°.

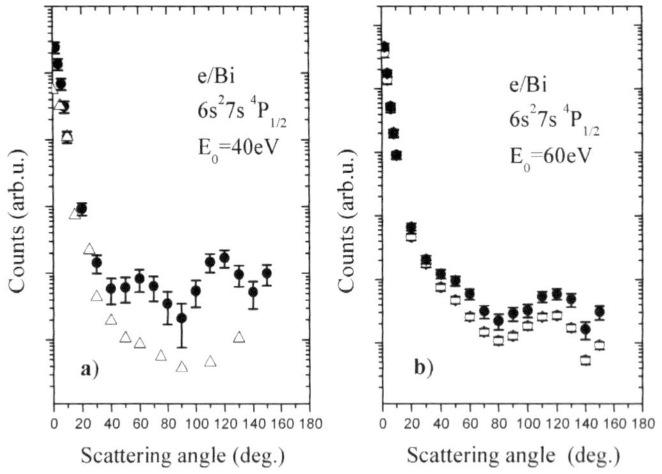

Figure 2. Relative differential cross sections for electron-impact excitation of the $6p^27s\ ^4P_{1/2}$ state of Bi at **a**) 40 and **b**) 60 eV electron-impact energies. •, present; Δ, Williams and Trajmar [14] normalised at present at 10°; ○, absolute DCS for electron-impact excitation of the $6p7s\ ^3P_{0,1}$ state in lead Milisavljević *et al* [26] normalised at present at 10°

Generally, present relative DCSs are very similar in shape to DCSs for the excitation of the 6p7s $^3P_{0,1}$ state in lead atom measured by Milisavljević *et al* [26] (see figure 2b). Also, at 60 eV we observe DCSs similar in shape to DCSs for the excitation of the 6s6p $^1P_1$ state of mercury obtained by Panajotović *et al* [27] (not presented here).

We recorded energy-loss spectra at different scattering angles and electron impact energies. Marinković *et* al [22] presented in their study electron energy-loss spectra at 20 and 100 eV while here we present the energy-loss spectrum at the electron impact energy of 40 eV and scattering angle of 10° in the range of energy losses up to 13 eV (see figure 3). Within the literature on BiI the LS system is mostly employed, so we will retain this scheme in our discussion of level structures. Below the $6p^2\ ^3P_0$ first ionisation limit (7.286 eV) of bismuth, Moore [28] listed 33 levels, which she assigned principally in the series $6p^2ns\ ^{2,4}P$ or $6p^2nd\ ^2D$, with $n \leq 10$. Some of these levels can be recognized in figure 3 and are listed in table 3.

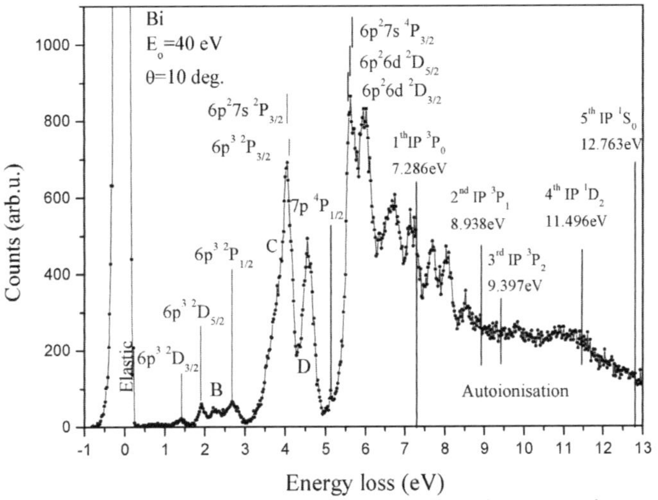

Figure 3. Electron energy-loss spectrum of bismuth at 40 eV impact energy and scattering angle of 10°. The ionisation limits (IP) are indicated by vertical lines.

The first detected bismuth lines in spectra of peculiar star 73 Dra at 472.2nm [7] correspond to the transitions between the $6p^3\ ^2D_{3/2}$ (1.416 eV) and $6p^27s\ ^4P_{1/2}$ state (ΔE = 2.6247 eV). Both these states can be seen in figure 3. The search for lines from the BiI spectra of chemically peculiar stars was hampered by line blending and the paucity of strong BiI lines. Wahlgren *et al* [18] tested the presence of two intense lines (λ = 206.23 nm and λ = 227.73 nm) of BiI in spectra χ Lipi star. In order to test the existence of observable BiI lines, synthetic spectra were calculated for the solar abundance and a 5.0 dex enhancement for the λ = 206.23 nm line using gf = 0.378 determined by Wiese and Martin [29] was used. A vary small difference in blended feature profile resulted from this large enhancement level, indicated that BiI lines in the χ Lipi would be noticeable. Nevertheless, none of the strongest BiI lines that were observed in absorption or arc spectra from laboratory measurements were located in the HR 7775 star.

The spectrum shown in figure 3 contains well-resolved features that correspond to low-lying excited $^2D_{3/2}$, $^2D_{1/2}$ and $^2P_{3/2}$ states (energy losses of 1.416 eV, 1.914 eV and 2.685 eV respectively) with the $6s^26p^3$ configuration. Due to the generally lower resolution in electron spectrometry the state $6s^26p^3\ ^2P_{3/2}$ is not visible. The presence of bismuth dimers could not be avoided. The three features at

2.20 eV, 3.97 eV and 4.52 eV correspond to the optically observed states of Bi_2. The excited dimer states are labelled as B, C, and D according to Huber and Herzberg [30].

Table 3. Electron levels observed in electron energy-loss spectra of bismuth at 40 eV ($\theta = 10^o$). Source abbreviation: C, Clearman [3]; A, Assous[4]; M, Mathews *et al* [6]; Moo, Moore [28].

Line No	Energy (eV)	Assignment	Source of data	This experiment (eV)
1	7.348		C	7.34
2	7.537	$6p^2(^3P_1)8s$ J=3/2	C	7.55
	7.539	$6p^2(^3P_1)8s$ $^4P_{3/2}$	A	
	7.540	J=3/2 or 5/2	M	
	7.540	$6p^2(^3P_1)8s$ J=3/2	Moo	
3	7.621	$6p^2(^3P_2)6d$	C	7.62
	7.664	$6p^2(^3P_2)6d$ $^4F_{3/2}$	A	
	7.579	-	M	
	7.624	$6p^2(^3P_2)6d$ J=1/2	Moo	
4	7.776	$6p^2(^3P_2)6d$	C	7.77
	7.777	$6p^2(^3P_2)6d$ $^4D_{3/2}$	A	
	7.779	$6p^2(^3P_2)6d$ J=3/2	Moo	
5	7.914	$6p^2(^3P_1)8s$ J=5/2	C	7.89
	7.917	Series B J=1/2 or 3/2	M	
	7.902	$6p^2(^3P_1)8s$ J=1/2	Moo	
6	7.966	-	C	7.98
	8.051	$6p^2(^3P_1)7d$	A	
	7.969	J=1/2 or 3/2	M	
	7.969	$6p^2(^3P_1)7d$ J=3/2	Moo	
7	8.135	-	M	8.16
	8.135	$6p^2(^3P_1)9s$ $^2P_{1/2}$	Moo	
8	8.318	$6p^2(^1D_2)7s$	C	8.27
	8.246	$6p^2(^3P_1)9p$ J=1/2	Mo	
9	8.434	$6p^2(^3P_2)7d$	C	8.43
	8.346	-	M	
	8.410	$6p^2(^3P_1)10s$ $^4P_{3/2}$	Moo	
10	8.634	Series B	M	8.61
	8.570	$6p^2(^3P_1)9s$ $^2P_{3/2}$	Mo	
11	8.875	Series B or D	M	8.88
	8.865	$6p^2(^3P_1)10s$ $^2P_{3/2}$	Moo	
12	9.094	$6p^2(^3P_2)10p$	C	9.09
	9.097	Series X	M	
	9.097	$6p^2(^3P_1)10d$ J=1/2	Moo	
13	9.264	Series Y	M	9.27
14	-	-	-	9.97
15	-	-	-	11.2

Above the first ionisation limit we identified fifteen autoionising states. These states are listed in table 3. Eleven states are identified between the first and the second ionisation limits, two states between the second and the third and two between the third and the fourth. The assignment of the different series is adopted from [6] and in accordance with the two-limit multichannel quantum defect theory. The two states at 9.97 and 11.2 eV, which have not been observed previously in either photoabsorption or in arc spectra, lie in the energy range between the third and fourth ionisation limits of bismuth atom. The feature at 11.2 eV is very broad (about 0.9 eV half width at full maximum).

4. Conclusion

In this work we have presented relative DCSs for the electron-impact excitation of the $6p^27s$ $^4P_{1/2}$ state in bismuth. To the best of our knowledge the differential cross sections at 60 eV are reported for the first time. The measured differential cross sections are very similar in shape to the cross sections obtained for the excitation of the 6p7s $^3P_{0,1}$ state in lead. In the autoionising region we have identified several energy levels up to the fifth ionisation limit. Not many of the BiI lines have been observed in spectra chemically peculiar stars with exception of the 472.2 nm line. In conclusion, our measurements show great similarity between bismuth and lead at least for electron-atom collisions processes which we studied.

5. Acknowledgment

This work has been partly supported by the MESTD of the Republic of Serbia under project #OI 171020 and motivated by COST Action CM1301 "Chemistry for Electron-Induced Nanofabrication (CELINA)".

6. References

[1] Zeeman P, Back E and Goudsmit S 1930 *Z Phys* **66** 1
[2] Mrozowski S 1942 *Phys. Rev.* **62** 526
[3] Clearman H E 1952 *J. Opt. Soc. Am.* **42** 373
[4] Assous A 1972 *J. Opt. Soc. Am.* **62** 544
[5] George S, Munsee J H and Verges J 1985 *J Opt. Soc. Am. B* **2** 1258
[6] Mathews C W, Ginter M L, Ginter D S and Brown C M 1989 *J. Opt. Soc. Am. B* **6** 1627
[7] Guthrie B N G 1972 *Ap. Sp. Sci.* **15** 214
[8] Jacobs J M and Dworetsky M M 1982 *Nature* **299** 535
[9] Colwey C R 1987 *The Observatory* **107** 188
[10] Wahlgren G M, Brage T, Gilliland R L, Johansson S G, Leckrone D S, Lindler J and Lindzen U 1994 *ApJ* **L67** 435
[11] Wahlgren G M , Lecrone D S, Brage T, Proffitt C R and Johansson S 1998 ASP Conf. Ser. 143 *Scientific Impact of the Goddard High Resolution* ed. J C Brant, T B Ake and C C Petersen (San Francisco: APS) 330
[12] Fink M, Ingram J C and Yates A C 1970 *J. Phys. B* 1970 **3** 536
[13] Neerja and Tripathi A N 2001 *Eur. Phys. J. D* **13** 5
[14] Williams W, Trajmar S and Bozinis D G 1975 *J. Phys. B: Atom. Molec. Phys.* **8** L96
[15] Lvov B V 1965 *Opt. Spectrosk.* **19** 282
[16] Bouzza S and Bauche J 1988 *Z Phys.* **10** 1
[17] Lecrone D S, Proffitt C R, Wahlgren G M, Johansson S G and Brage T 1999 *AJ* **117** 1454
[18] Wahlgren G M, Brage T, Brandt J C, Fleming J, Johansson S G, Lecrone D S, Proffitt C R, Reader J and Sansonetti C J 2001 *ApJ* **551** 520
[19] Dolk L, Litzen U and Wahlgren G M 2002 *A&A* **388** 692
[20] Kim Y-K and Stone P M 2001 *Phys. Rev. A* **64** 052707
[21] Kim Y-K and Stone P M 2007 *J. Phys. B: At. Mol. Opt. Phys.* **40** 1597
[22] Marinković B P, Filipović D M, Pejčev V and Šević D 2008 *Int. J. Mass. Spectrom.* **221** 76
[23] Predojević B, Pejčev V, Filipović D M, Šević D Tomčik B and Marinković B P *J. Phys. B: At. Mol. Opt. Phys.* **44** 055208
[24] Chutjian A 1979 *Rev. Sci. Instrum.* **50** 347
[25] Brinkman R T and Trajmar S 1981 *J. Phys. E: Sci. Instrum.* **14** 245
[26] Milisavljević S, Rabasović S M, Šević D, Pejčev V, Filipović M D, Lalita Sharma, Rajesh Srivastava, Stauffer A D and Marimković B P 2007, *Phys. Rev A* 022714
[27] Panajotović R, Pejčev V, Konstantinović M, Filipović D, Bočvarski V and Marinković B 1993 *J. Phys. B: At. Mol. Opt. Phys.* **26** 1005
[28] Moore C E 1958 *Nat. Bur. Stand. Cir* 467 Vol III 219
[29] Wiese W L and Martin G A 1980 *Transition Probabilities for Atom and Atomic Ions Part II Transition Probabilities* (Washington DC: US Gov. Printing Office)
[30] Huber K P and Herzberg G 1979 *Molecular Spectra and Molecular Structure IV. Constant of Diatomic Molecules* Van Nostrand-Reinhold, New York

Journal of Physics: Conference Series **565** (2014) 012020 doi:10.1088/1742-6596/565/1/012020

Studying the complex absorption profiles of Si IV in 21 HiBALQSO spectra

D Stathopoulos[1, 2], E Danezis[1], E Lyratzi[1, 2], A Antoniou[1], LČ Popović[3], D Tzimeas[1] and M S Dimitrijević[3, 4]

[1]University of Athens, Faculty of Physics, Department of Astrophysics, Astronomy and Mechanics Panepistimioupoli, Zographou 157 84, Athens, Greece
[2]Eugenides Foundation, 387 Sygrou Av., 17564, Athens, Greece
[3]Astronomical Observatory, Volgina 7, 11060 Belgrade 38, Serbia
[4]Laboratoire d' Etude du Rayonnement et de la Matiere en Astrophysique, Observatoire de Paris

E-mail:dstatho@phys.uoa.gr

Abstract. We investigate the physical conditions and kinematics of broad absorption line region clouds of Si IV in 21 HiBAL Quasars. We use the Danezis et al. method [1], [2], [3] in order to fit and analyze the broad absorption troughs of Si IV resonance lines in the UV region of the electromagnetic spectrum. We find that the BAL flow is not smooth but instead plasma clouds are formed in it. BAL troughs present multicomponent structure which indicates the existence of more than one absorbing cloud in the line of sight, where every absorbing cloud produces a Si IV doublet. We show that the blending of these doublets produces the apparent broad absorption troughs we observe. One of our main achievements is that we managed to decompose and deblend each complex absorption trough to the individual doublets that it consists of. Apart from that, we succeeded in deblending the resonance lines of every doublet. By achieving accurate fits to the BAL troughs we calculated some physical and kinematical parameters that describe the plasma clouds in the line of sight. These parameters are: the radial outflow velocities of the clouds, the random velocities of ions inside each plasma cloud, the apparent optical depth in the center of every absorption component, the FWHM and the equivalent width. As a final step we correlate these physical parameters in order to draw useful conclusions.

1. Introduction

Broad absorption line quasars (BALQSOs) are a subtype of QSOs, defined by the presence of broad and blue-shifted absorption features with typical velocities of 0.1c (e.g. [4]). The intrinsic fraction of BALQSOs from the total quasar population is estimated to be about 15-20% [5], [6] although this fraction rises to 17-23%, after correcting for optical/UV selection effects [7], [8]. High – ionization BAL quasars (HiBALQSOs) show absorption from C IV, N V, Si IV and Lyα (in order of decreasing typical strength (λ_{vac}> 1215 Å) and are classified by means of their UV C IV 1549 BAL profile.

Broad absorption line QSOs (BALQSOs) are a marvellous manifestation of outflows which are a fundamental characteristic of AGNs and are very commonly detected e.g. [9], [6], [10], [11], [12], [13], [14] and references therein. Observationally BAL quasars show troughs characterized by FWHM ~ 2000 – 20,000 km s^{-1} wide arising from resonance line absorption in gas with blueshifetd (outflowing) velocities up to 66,000 km s^{-1} [15]. The structure of the absorption troughs ranges from apparently very smooth (P-Cygni like) troughs to multiple troughs which set in at very near zero outflow velocity [16], [17] and/or complexes of many individual absorption lines [18]. Usually these absorption features are of the P-Cygni type, although in some cases the absorption features are detached from the emission line by as much as 30,000 km s^{-1}. The absorption in the majority of cases is at wavelengths shortward of line center, which indicates that the absorbing gas is flowing outward from the nucleus, and the high ionization level and high outflow velocities of the gas strongly suggest that these systems are closely associated with the nuclear regions (these lines are called intrinsic). However, there are cases that absorption is redshifted [19]. So, one of the difficulties in understanding the outflows (or inflows) revealed by absorption features in quasar spectra is their amazing diversity.

BALQSOs also present Narrow Absorption Lines (NALs) which have widths of several hundreds of km s^{-1}(FWHM < 200 – 300 km s^{-1}, [20]). A useful (but physically arbitrary) definition of NALs is that they are narrow enough not to blend important UV doublets, e.g., the CIV pair with separation ~ 500 km s^{-1}. Thus we require FWHM < 200 to 300 km s^{-1}. NALs with velocity shifts v < 5000 km s^{-1} from systemic are also called "associated" absorption lines (AALs) because of their plausible physical relationship to the AGN [10]. So narrow lines are separated into two subcategories: narrow lines far from the emission redshift ($z_a << z_e$) and narrow associated lines near the emission redshift ($z_a \approx z_e$). Narrow lines ($z_a << z_e$) which are due to gas clouds unrelated to the quasar that coincidentally fall along our line of sight to the quasar. On the other hand $z_a \approx z_e$ systems could be either intervening or intrinsic. This means that these systems are due to gaseous regions within the host galaxy of the quasar or within a galaxy cluster which contains the quasar. These absorbers may not be ejected from the quasar, but are still affected by the quasar luminosity and environment.

However there are absorption lines with intermediate widths (FWHM ~ 300 – 2000 km s^{-1}, [10]) which are called mini-BALs and are as common as BALs [21]. They appear at the same range of blueshifted velocities as the BALs, and they also clearly form in AGN outflows. Mini-BALs could represent a completely different type of absorber, or a particular line of sight through the same absorbers that we observe in other cases as BALs. Examples of high-velocity mini-BALs can be found in [22] and [23].

In this paper we present the physical parameters which are derived after fitting the doublets of Si IV absorption troughs in 21 HiBALQSOs. By deblending the doublets we show that the ratio of the doublet's components does not follow the theoretically expected value which indicates that the troughs present non-black saturation. Finally, using the values of the calculated parameters (outflow velocity, apparent optical depth, random velocities, FWHM, equivalent width) we try to describe statistically the behaviour of these parameters as a function of outflow velocity.

2. Data and method of analysis

The 21 QSOs were selected from the SDSS catalogue of broad line QSOs of Reichard et al. study [6]. We need to point out that in order to conclude to the final 21 HiBALQSOs, we used two criteria. Because this current work is a part of a wider study of HiBALQSO C IV and Si IV emission and absorption resonance lines, we demanded the redshift to be high enough (1.813 < z < 3.777) so that the Si IV doublet would lie within the wavelength range that the spectra cover. Another criterion we applied is that the BALQSOs, had blended C IV and Si IV doublets. From the HiBALQSOs that meet these two criteria we chose randomly 21. The spectra were taken from the Sloan Digital Sky Survey (SDSS) Data Release 7 and they cover the optical range 3800-9200 Å, at a resolution of 1800-2100. The SDSS imaging survey uses a wide-field multi-CCD camera [25]. In table 1, one can see in the first column the SDSS name of each BALQSO, in column 2 there is the modified Julian date, plate and fiber, and in the third column the redshift.

In this work we use the Danezis et al. method [1], [2], [3], which has been applied successfully to the spectra of hot emission stars, in order to deblend the observed complex features to their components and to calculate a number of physical parameters of the regions that create them. Here we study the absorption lines of Si IV λλ 1393.755, 1402.77 Å in the spectra of 21 HiBALQSOs.

Table 1. SDSS HiBALQSO Catalogue

#	SDSS Object	Redshift	MJD-Plate-Fiber
1	J023252.80-001351.2	2.025	51820-0407-158
2	J015024.44+004432.9	1.990	51793-0402-485
3	J031828.91-001523.2	1.990	51929-0413-170
4	J001502.26+001212.4	2.857	51795-0389-465
5	J001025.90+005447.6	2.845	51795-0389-332
6	J003551.98+005726.4	1.905	51793-0392-449
7	J015048.83+004126.2	3.703	51793-0402-505
8	J004041.39-005537.3	2.092	51794-0393-298
9	J004732.73+002111.3	2.879	51794-0393-588
10	J005419.99+002727.9	2.522	51876-0394-514
11	J010336.40-005508.7	2.442	51816-0396-297
12	J000103.85-104630.2	2.081	52143-650-133
13	J102517.58+003422.0	1.888	51941-0272-501
14	J004323.43-001552.4	2.806	51794-0393-181
15	J104109.86+001051.8	2.259	51913-0274-482
16	J110041.20+003631.9	2.017	51908-0277-437
17	J000056.89-010409.7	2.111	51791-0387-098
18	J001438.28-010750.1	1.813	51795-0389-211
19	J023908.99-002121.4	3.777	51821-0408-179
20	J104841.03+000042.8	2.022	51909-0276-310
21	J000913.77-095754.5	2.076	52141-651-519

3. Spectral fitting

The first step in the fitting process is the identification of spectral lines in the studied wavelength range in order to identify possible blends between the absorption troughs. Then we determine the continuum for which we use the power law indices derived by [11]. We observed that among the 21 HiBALQSOs there are cases that present multiple troughs or detached troughs or troughs consisting of multiple kinematic components (more than one doublet) which set in at very near zero outflow velocity and extend up ~ 25000 km s^{-1}.

In order to fit the Si IV absorption troughs we choose the most appropriate distribution among the following: Gauss, Lorentz, Voigt or the new Rotation [1] and Gauss-Rotation [2] distributions. In this study performing x^2 tests, we found that the best fit is achieved by using the Gauss distribution.

We observed two types of absorption troughs. In the first case the absorption trough is fitted using only one Si IV doublet, which is created by an individual absorbing region (cloud). However, in the second case we have troughs which cannot be simulated adequately using only one Si IV doublet but are fitted by using more than one doublet. In this situation the absorption trough is produced by more than one cloud. This means that the line function that describes the observed profile comes from the solution of the radiative transfer equation through more than one cloud [1], [2], [3]. As for the number of doublets we increase them until a standard F – test yields no further significant gain (95% confidence level) in the goodness of the fit, as measured by the reduced x^2. We must point out that we fit each doublet member using a Gaussian, i.e. a Gaussian for the red and a Gaussian for the blue component of a doublet.

During the fitting process we require that for a given trough the Si IV resonance lines should have almost the same width and the same outflow velocity. The width (FWHM) and central positions ($V_{outflow}$) of the two Gaussians have been fit simultaneously to reproduce the Si IV absorption troughs. As for the ratio of optical depths between the blue and the red component of a doublet, according to theory it is ~ 2:1 (see [26], [27]). However, because the doublets' members are blended, the ratio of their optical depths is not always 2 (see also [26]). So, by relaxing the constraint of optical depth between the doublet lines, we perform repeated fits and compare the results by x^2 test, in order to conclude to the best fit.

In the studied spectra we observed that the whole Si IV absorbing region comprises of one or more absorption doublets. This means that the whole Si IV absorbing region is created by more than one discrete cloud. If we want to fit the whole Si IV spectral region we need to solve the radiative transfer equation for a complex atmosphere that contains more than one discrete cloud. So, we need an interpolation polynomial which includes the line functions of every individual line ([1], [2], [3]). From this interpolation polynomial, in this current analysis we calculate the outflow velocity ($V_{outflow}$), the random velocities (V_{rand}), the apparent optical depth ($\tau_{apparent}$), the full width at half maximum (FWHM) and the equivalent width (EW) of each individual blue and red component of a doublet.

In figure 1, we give two examples of fitted spectra. In both cases the doublets' names are provided (A, B, C, D). Below the fit we present the residuals. Below the residuals the different components that compose the final profile are shown (with blue colour we indicate the 1393.755 Å line and with red colour the1402.77 Å line)

4. Results

In our study we identified up to five Si IV absorbing clouds. One HiBALQSO has four absorbing clouds, one HiBALQSO has three absorbing clouds, ten HiBALQSOs have two absorbing clouds and nine HiBALQSOs have one cloud. In this study we calculated for every one of the 21 HiBALQSOs the following parameters: the outflow velocities ($V_{outflow}$), the random velocities (V_{rand}), the apparent optical depth ($\tau_{apparent}$), the full width at half maximum (FWHM) and the equivalent width (EW) of each individual blue and red component of a doublet. We classify the calculated outflow velocities in five classes. The first one has the value of 24090 km s^{-1}, the second one has values between 13551 and 17208 km s^{-1}, the third one between 7744 and 10325 km s^{-1}, the forth one between 5377 and 6883 km s^{-1} and the fifth one between 2151 and 4732 km s^{-1}. Based on this classification we also classify the random velocities (V_{rand}), the FWHM, the apparent optical depth ($\tau_{apparent}$) and the equivalent width (EW). In table 2, we present the mean values and the errors of the calculated parameters of these five classes.

Table 2. Mean values and errors of the calculated parameters of the five classes.

Class	$<V_{outflow}>$ (km s^{-1})	$<V_{random}>$ (km s^{-1})	<FWHM> (km s^{-1})	$<\tau^{blue}_{app}>$	$<\tau^{red}_{app}>$	<EW> (Å)
1	24090 ± 2170	2280 ± 114	4190 ± 230	0.10 ± 0.02	0.09 ± 0.02	2.40 ± 0.05
2	14430 ± 520	1530 ± 78	3040 ± 180	0.20 ± 0.01	0.18 ± 0.01	2.30 ± 0.05
3	8640 ± 220	1460 ± 82	2820 ± 160	0.30 ± 0.05	0.27 ± 0.04	2.60 ± 0.05
4	6110 ± 260	580 ± 72	1190 ± 70	0.60 ± 0.04	0.50 ± 0.03	2.50 ± 0.05
5	3460 ± 110	410 ± 90	960 ± 50	1.20 ± 0.06	1.00 ± 0.05	3.10 ± 0.06

5. Discussion and conclusions

In the majority of cases the Si IV resonance lines where blended. By deblending them, we managed to calculate the apparent optical depth of the blue as well as the red component of the doublet. The ratio of optical depths τ_b/τ_r is not 2 as specified by atomic physics ($\tau_b/\tau_r = (g_b \cdot f_b \cdot \lambda_b)/(g_r \cdot f_r \cdot \lambda_r) \simeq 2$, where λ_b, λ_r are the transitions' wavelengths, f_b,f_r are the oscillator strengths and g_b, g_r are the statistical weights of the levels [27]). In fact this ratio ranges from ~ 1.1-1.2 which indicates that the broad absorption troughs present non-black saturation. Non-black saturation is an indication of partial coverage of the background source by the absorbing clouds. This phenomenon occurs when the absorbing clouds are smaller than the background source, allowing part of the light from the source to reach the observer unabsorbed. This effect also indicates that all absorption lines are intrinsic to the QSO. This picture is further justified by the width of the lines (FWHM > 300 km s^{-1}).

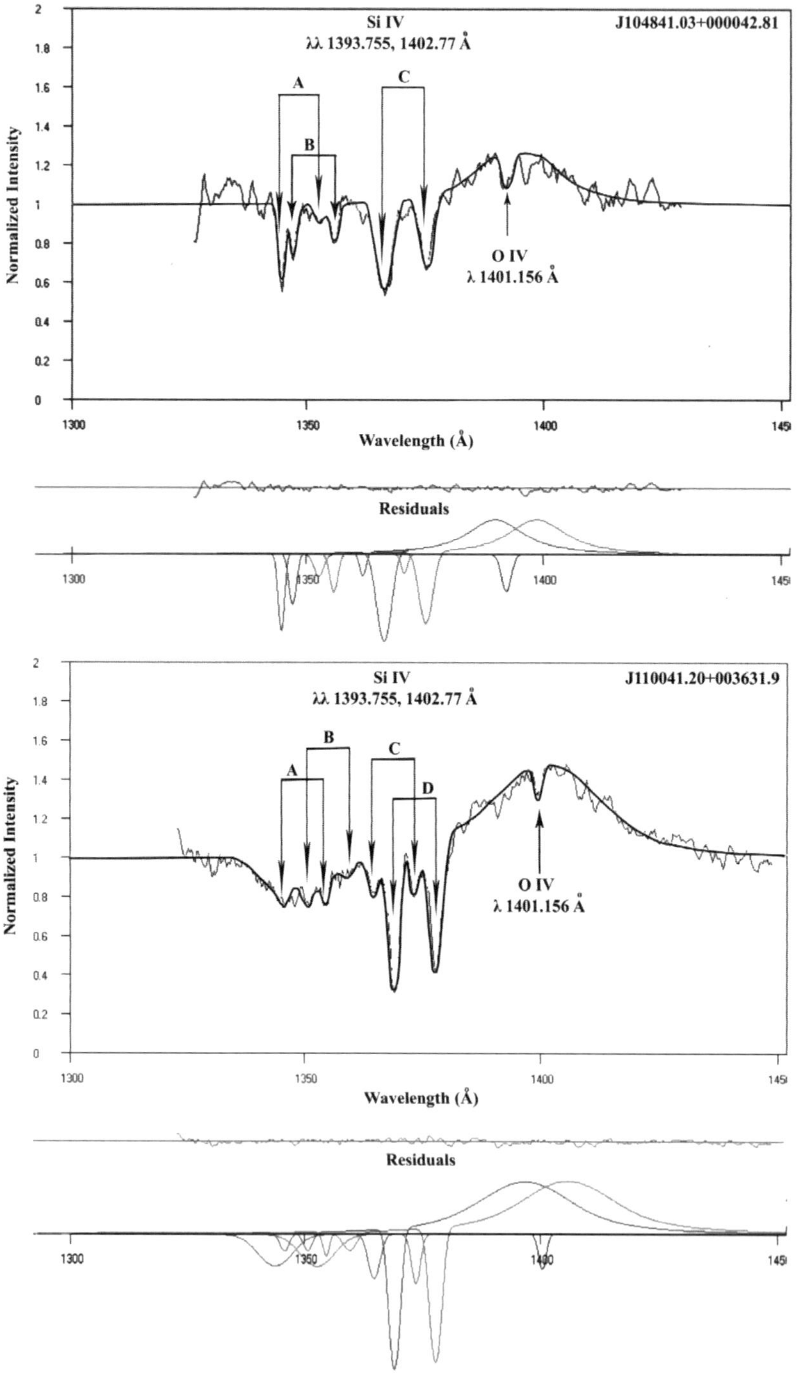

Figure 5. Two examples of fitted spectra. In both cases the doublets' names are provided (A, B, C, D). Below the fit the residuals are presented. Below the residuals the different components that compose the final profile are shown (with blue colour we indicate the 1393.755 Å line and with red colour the1402.77 Å line).

In figure 2a we can see that the apparent optical depth (of the blue component) of the lines decreases as the outflow velocity increases. This fact can be interpreted in two different ways. There is the possibility that, as the clouds accelerate outwards, its typical density will decrease which can explain why the optical depths decrease as the outflow velocity increases. According to this picture it seems that absorption is stronger at lower velocity. This picture is supported by figure 3a which shows that the equivalent width of resonance lines decreases with increasing outflow velocity. However, we cannot be sure that this is the real case because the decrease in the apparent optical depths can be due to a decrease in the covering fraction. So, in order to conclude we need to calculate the true optical depths as well as the covering fractions.

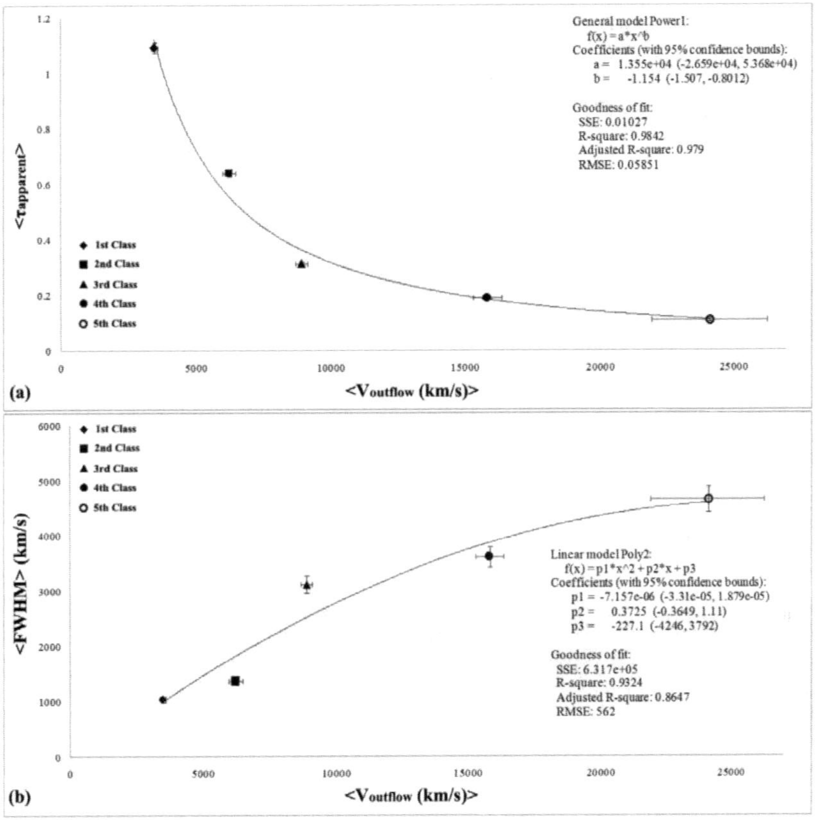

Figure 2. (a): Mean apparent optical depth (of the blue components) as a function of mean outflow velocity between the studied HiBALQSOs. (b): Mean FWHM (of the blue components) as a function of mean outflow velocity between the studied HiBALQSOs. In all graphs we give the equation of the curve, the coefficient of determination R^2, the adjusted R^2, the SSE (sum of squares due to error) and the RMSE (root mean square error).

The widths (FWHM) of the Si IV resonance lines range from 960-4190 km s^{-1}, a range of values which exceeds the thermal width by many orders of magnitude for silicon in a 10^4-10^5 K gas. Therefore, the broad line environment encompasses a range of non-thermal velocities along our line of sight. These velocities are possibly due to bulk or turbulent motions involved with the absorbing flow. Indeed, outflow lines tend to have profiles that are broad and smooth compared to thermal widths ([28], [29], [30], [31], see [32] for a possible explanation of how turbulent motions broaden the lines).In figure 2b, we present the behaviour of FWHM as a function of outflow velocities where we can see that the FWHM increases with increasing outflow velocity. In order to explain the increase of FWHM with increasing outflow velocity we can assume that temperature rises to extreme values and so the medium would be extremely hot that would dissolve the clouds. Apart from that, such a medium would be unable to absorb any Si IV photon. On the other hand we can assume that the turbulent velocities are not constant along the line of sight but instead are increasing as the clouds accelerate outwards.

Figure 3b indicates that lines with high FWHM (higher FWHM corresponds to higher outflow velocity according to figure 2b) have low values of optical depths. This means that absorption lines at high outflow velocities tend to be broader but shallower. Whether this is an evolutionary effect of clouds or a geometrical effect (due to changes in the covering factor) or a combination needs to be checked in order to reach to conclusions using time variability of independent HiBALQSOs and calculations of the covering fractions.

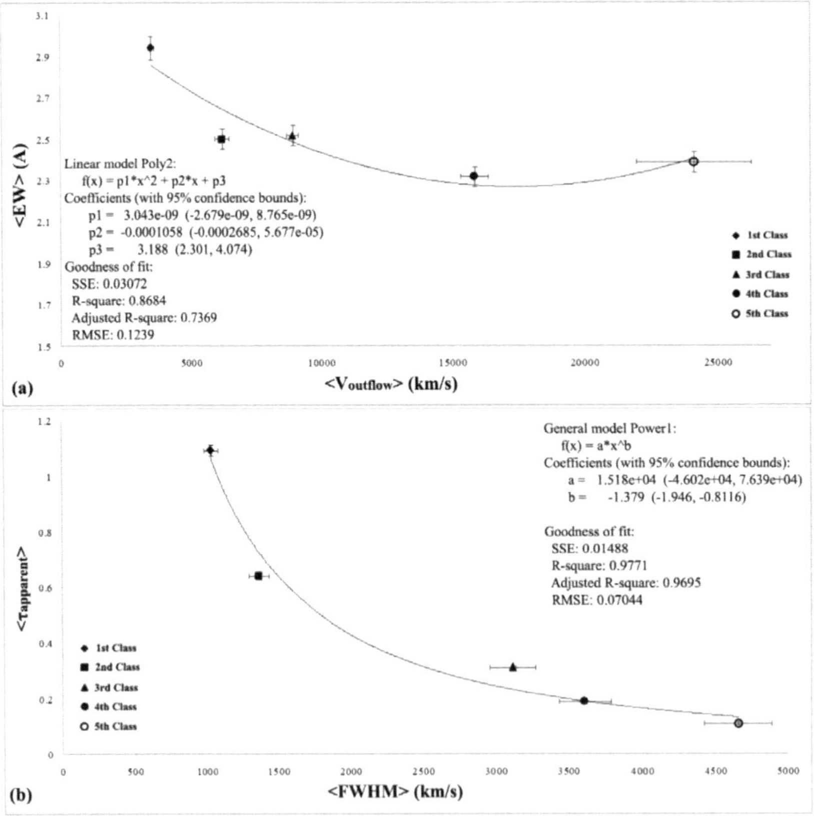

Figure 3. (a) Mean equivalent width as a function of mean outflow velocity between the studied HiBALQSOs. (b) Mean apparent optical depth as a function of mean FWHM. In all graphs we give the equation of the curve, the coefficient of determination R^2, the adjusted R^2, the SSE (sum of squares due to error) and the RMSE (root mean square due to error).

Acknowledgements

This research project is in progress at the University of Athens, Department of Astrophysics, Astronomy and Mechanics. We are thankful to the Special Account for Research Grants for the financial support. We also thank for their financial support the Ted and Erica Spyropoulos Foundation.

This work was also supported by the Ministry of Science and Technological Development of Serbia through the projects "Influence of coalitional processes on astrophysical plasma line shapes" and "Astrophysical spectroscopy of extragalactic objects". We also acknowledge the use of the data from SDSS DR7survey.

References

[1] Danezis E et al 2003 *Ap&SS* **284** 1119-42
[2] Danezis E et al 2007 *PASJ* **59** 827-834
[3] Lyratzi E, Danezis E and Popović L Č 2007 *PASJ* **59** 357-371
[4] Weymann R J, Morris S L, Foltz C B and Hewett P C 1991 *ApJ* **373** 23-53
[5] Hewett P C and Foltz C B 2003 *AJ* **125** 1784-94
[6] Reichard T A et al 2003 *AJ* **126** 2594-2607
[7] Knigge C, Scaringi S, Goad M R and Cottis C E 2008 *MNRAS* **386** 1426-35
[8] Gibson R R et al 2009 *ApJ* **692** 758-777
[9] Crenshaw D M et al 1999 *ApJ* **516** 750-68
[10] Hamann F and Sabra W A2004 *ASPC* **311** 203-12
[11] Trump J R et al 2006 *ApJ* **165** 1-18
[12] Nestor D, Hamann F and Rodriguez H P 2008 *MNRAS* **386** 2055-64
[13] Dunn J P et al 2008 *AJ* **136** 1201-20
[14] Ganguly R and Brotherton M S 2008 *ApJ* **672** 102-07
[15] Foltz C et al *PASP* 1983 **95** 341-46
[16] Korista K T, Voit G M and Morris S L 1993 *ApJS* **88** 357-81
[17] North M, Knigee K and Goad M 2006 *MNRAS* **365** 1057-66
[18] Turnshek D A 1984 *ApJ* **280** 51-65
[19] Hall P Bet al 2013 *MNRAS* **434** 222-56
[20] Misawa T et al 2005 *ApJ* **629** 115-30
[21] Rodriguez H P, Hamann F and Hall P 2011*MNRAS* **411** 247-59
[22] Jannuzi B T et al 1996 *ApJ* **470** 11
[23] Hamann F, Barlow T A and Junkkarinen V 1997 ApJ **478** 87-93
[24] Hamann F et al 1997 *ASPC* **128** 19-24
[25] Gunn J E et al 1998 *AJ* **116** 3040-81
[26] Borguet B C J et al 2013 *ApJ* **762** 49
[27] Arav N et al 2001 *ApJ*, **546**, 140-49
[28] Hamann F and Ferland G 1999 *ARA&A* **37** 487-531
[29] Srianand R and Petitjean P 2000 *A&A* **357** 414-28
[30] Ganguly R et al 2006 *ApJ* **645** 868-89
[31] Schaye J, Carswell R F and Kim T S 2007 *MNRAS* **379** 1169-94
[32] Antoniou A et al 2014 *ASR* 54 1308-18

Dynamic Characteristics of Excited Atomic Systems

N N Bezuglov[1], M S Dimitrijevic [3], A N Klyucharev[1,2], A A Mihajlov[4]
[1]Department of Physics, St Petersburg State University, 198504, St Petersburg, Russia
[2]University ITMO, St. Petersburg, 197101, Russia
[3]Astronomical Observatory, Volgina 7, Belgrad, Serbia
[4] Institute of Physics, University of Belgrade, P.O. Box 57, 11001, Belgrad, Serbia

E-mail: anklyuch@gmail.com

Abstract. The dynamics of excited atom interactions with other atoms, which often lead to associative ionization, is largely governed by stochastic diffusion of the valence electron through Rydberg states prior to the ionization. Such processes are associated with random changes of the energy state of the highly excited electron, and they are likely to influence the nuclear dynamics, especially at subthermal collision energies. Possibilities of manipulation of the chaotic dynamics of Rydberg states require a detailed exploration. For an electron in a given Rydberg state moving in a microwave field, which can be generated via interaction with another atom or molecule, there exists critical field strength, above which motion of the electron in the energy space is chaotic. Recently a way to block the dynamic chaos regime was shown, if a given Rydberg state is located somewhat above the middle between the two other states with the orbital quantum number differing by one, whereby level shifts can be controlled by employing Stark/Zeeman shifts in external DC electric/magnetic fields. The stochastic effects in collisions involving Rydberg particles, in which the initial and final reaction channels are connected via intermediate highly excited collision complexes with multiple crossings of energy levels, can be treated using the dynamic chaos approach (Chirikov criterion, Standard and Keppler mapping of time evolution of the Rydberg electron, solution of the Fokker-Plank- and Langevin-type of equations, etc.). Such approach to obtaining dynamics characteristics is a natural choice, since the treatment of Rydberg electron dynamics as a kind of diffusion process allowing one to bypass the multi-level-crossing problem, which can hardly be solved by conventional quantum chemistry methods.

1. Introduction

As demonstrated by the results of research on the subject of Rydberg atoms (both theoretical and experimental), when the principal quantum number value *n* of a selectively excited atom increases, an optical electron in the terms of the quasi-classical approximation of quantum mechanics begins to show its trajectory instability (see, e.g., [1]). In addition to that the evolution of quasi-molecular collision complexes during the time of one collision leads to multiple quasi-crossings of the initial term with neighboring terms and therefore to migrating excitation across the grid of crossing energy terms with the distance between them $\Delta W \approx n^{-3}$ [2]. Within the framework of today's ideas of the determinancy and chaos in the physics of atomic collisions this allowed speaking about the manifestation of the effect of the deterministic (having a cause) dynamic chaos in chemi-ionization processes in case of thermal and subthermal collisions involving Rydberg atoms (RA) with values of $n > 5$ [3]. In recent research on Rydberg atoms cooled to temperature of $3 \cdot 10^{-4}$ K [4], [5], where many-particle long-range dipole-dipole interaction was considered to be the main ionization channel, marked instability of the plasma of cold atoms has been observed. It is highly probable that manifestations of deterministic chaos in the ensemble of frozen RA are also observed. On the one hand this circumstance limits the selectivity of processes in magneto-optical traps, operation of computer logic based on dipole-dipole interaction involving Rydberg atoms. On the other hand, however, the possibility arises to control to a certain extent the parameters of excited media.

Poincaré in the late 19th century [6] was the first to suggest the mathematical basis of the theory of chaos in the world around in his works that demonstrated the existence of "integrable" and "nonintegrable" systems of equations. As integrable the systems were understood with smooth response to a weak disturbance when the possibility of analytical solution remains, e.g. Newton and Lagrange trajectories. In the most general case the system of motion equations cannot be considered to be absolutely integrable, as it is impossible to exclude absolutely the probability of a strong disturbance leading to the failure of the solution stability.

The reason for this instability, which eventually leads to global chaos, is related to the appearance of dynamic resonances. Within the framework of KAM theory (Kolmogorov, Arnold, Moser) [7] this leads to Hamilton inderterminancy of solutions with respect to the operator's own solutions. In other words, it leads to stochasticity effects of both trajectory calculations within the framework of quasi-classics and the approximate nature of solutions of Shrödinger equations. Physical literature often gives physical arguments in addition to mathematical ones in order to interpret the nature of chaos in specific phenomena.

A recent survey [8] has summarized to a certain extent the research on the subject of Rydberg condensed matter. This research started in 1990 with the publication of the results of experiments that analyzed the mass composition of heavy particles in a thermionic converter of thermal energy into electrical energy in alkali-metal vapors [9]. Later it was suggested to include ultracold Rydberg plasma into the category of Rydberg matter [8].

In all above examples, we deal with many-particle systems in a field of a long-range potential.

2. Rydberg atom.

The use of the name "Rydberg atoms" in literature began in the late 19th century after Rydberg's pioneer work on spectroscopy of highly excited atoms was published.

Below the parameters of an excited hydrogen atom within the framework of Bohr-Sommerfeld quasi-classical model are given depending on the values of the principal quantum number.

1.	Radius of the outer (optical) electron orbit	$r \approx n^2$
2.	Excited atom lifetime	$\tau \approx n^3$
3.	Binding energy of an optical electron (energy threshold for RA photo-ionization)	$E \approx n^{-2}$
4.	Distance between adjacent levels	$\Delta W \approx n^{-3}$
5.	Average speed of an optical electron rotation in orbit	$\upsilon \approx n^{-1}$
6.	Polarizability	$\alpha \approx n^7$
7.	Dipole moment	$\lvert d \rvert \approx n^2$
8.	Energy of interaction with an electric field of E intensity	$W_E \approx n^2 E$
9.	Energy of interaction with a magnetic field of H induction	$W_B \approx \alpha^2 n^4 B^2$
10.	RA interaction with a black body radiation – experimental volume walls at temperature T.	$W_T \approx \alpha^3 T^2$

Due to the latter circumstance, for $n < 40$ most chemi-ionization experiments usually record charged particles via associative ionization channel with molecular ions formation. On the other hand, there is a known diagnostics of RA concentration based on their ionization by wall radiation.

The active phase of experimental and theoretical research on RA began with Fermi's work, who suggested considering a hydrogen RA as a system consisting of an ionic core and an optical electron loosely-bound to it.

Further development of the theory of ionization processes involving RA was driven by chemi-ionization studies and the emergence of a theoretical model of a dipole resonance mechanism [11], within which in general experimental and theoretical results are currently compared (see Fig. 1). In its turn, the analysis of the first results of chemi-ionization of hydrogen-like atoms of alkali metals led its authors [2] to the idea of collisional diffusion ionization of RA as a manifestation of the dynamic chaos effect during a single collision RA + A.

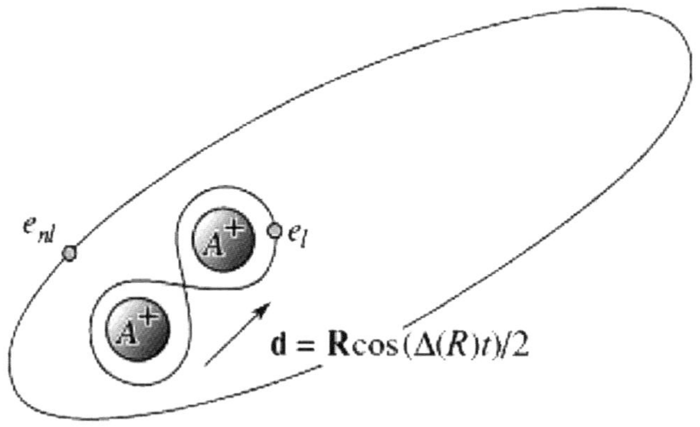

Figure 1 A microwave electric field generated in a collisional Rydberg quasi-molecule (PA + A). d is the dipole moment of subsystem ($A^+ - A$) with the interparticle distance R, ΔR is the exchange splitting of molecular ion term A_2^+. RA geometry is according to Fermi.

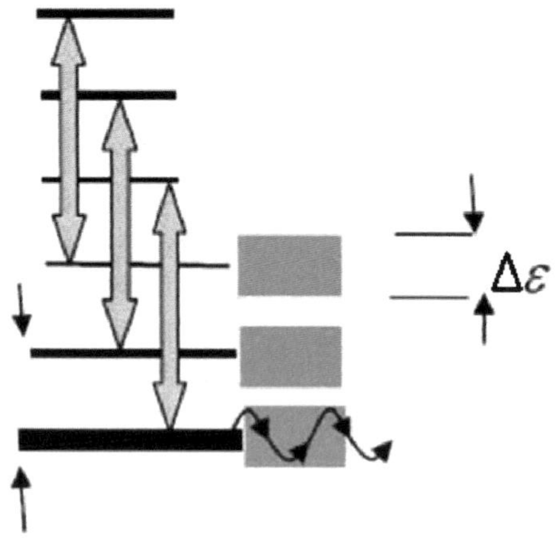

Figure 2 Nonlinear dynamic resonance. Within the framework of quantum mechanics, they correspond to the case of m-photon resonances; $\Delta\varepsilon$ is the oscillation of RE energy that illustrates the initial stage of global chaos.

3. Stochastic dynamics of a hydrogen atom in an external microwave field

Within the framework of quantum mechanics, the condition of multi-photon resonance when total energy of photons coincides with the energy distance between the atomic terms corresponds to the effective mechanism of mixing two fixed atomic states. In classical mechanics, the emergence of an additional stationary member in the potential of a disturbing field – the effect of a dynamic nonlinear resonance – corresponds to the concept of resonance. The condition of resonance overlapping, known in literature as Chirikov criterion [12] (see Fig. 2), leads to realization of K-system. It should be noted that the calculation of the effects of dynamic chaos in real situations is possible only when the numerical algorithms used are stable over large time spans. Moreover, standard numerical integration methods are unacceptable, since an error of such calculations grows exponentially over time. Paper [13] proposed a method which makes it possible to calculate correctly the conditions of dynamic chaos in atomic systems in steady-state and alternating electric and magnetic fields. Figure 3 illustrates a Rydberg electron (RE) reaching ionization limit in the conditions of global dynamic chaos. Distinctly irregular movement of RE in the energy space of bound states is evident. The lower limit of stochasticity is due to large distances between levels for $n < n_{min}$ It is the deterministic motion area of RE.

For the development of diffusion ionization of a Rydberg electron (RE), the finite time is needed that depends on the chaotic diffusion coefficient value, which is proportional to the values of the principal quantum number of degree three. This implies that the diffusion ionization of RA during one atom-atom collision is primarily realized within the range of slow collisions.

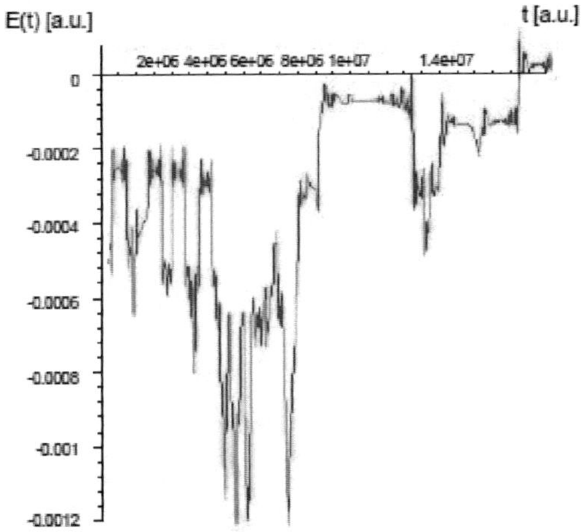

Figure 3 RA stochastic ionization development.

4. Specific features of RE stochastic diffusion under the conditions of double Stark resonance or Föster resonance

Large values of dipole moments for transitions between Rydberg states of an excited atom give rise to optical nonlinear effects in relatively weak electric fields. One of the manifestations of these effects include the case of Föster resonance [14], that occurs when level ℓ of a series is strictly in the middle between the two levels of (ℓ -1) or (ℓ +1) series. This configuration corresponds to a two-photon resonance for the transition $\{\ell+1,n\}\rightarrow\{\ell n\}\rightarrow\{(\ell+1)(n-1)\}$ for {s,p} and {p,d} series of hydrogen-like alkali atoms for the values of electrical fields of the order of 1 V cm^{-1}, but is absent in a hydrogen atom (see Figure 4). Incidentally, hence the term "double Stark resonance" originated.

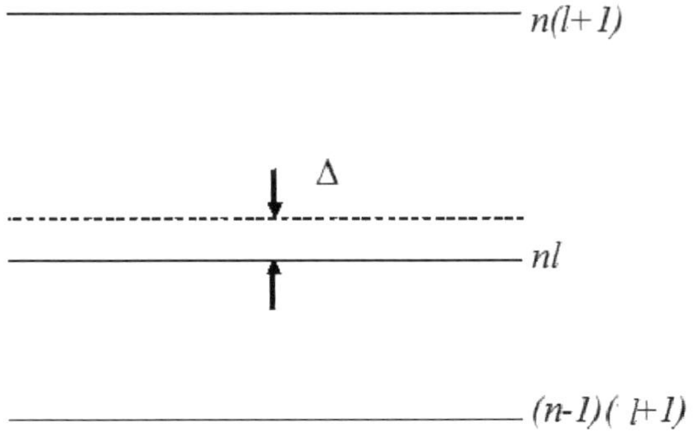

Figure 4 The scheme of terms of a highly excited atom, corresponding to the emergence of double Stark resonance (Förster resonance) for s-p series $\{\ell+1,n\}\rightarrow\{\ell,n\}\rightarrow\{\ell+1,n-1\}$.

Literature considers the effect of double Stark resonance as an effective method for manipulating atoms in laser fields. This method has good prospects in solving practical problems of quantum information science. This is primarily due to the possibility of blocking dipole transitions that are directly related to the dynamics of global chaos, because the scheme of terms corresponding to the case of Föster resonance is completely analogous to the case of a three-dimensional quantum oscillator, in which only "short" transitions to adjacent levels are allowed. Thus, for RA the role of "long" transitions with large variations of ΔW is small, while near transitions are blocked. This situation will negatively affect the processes of stochastic diffusion development.

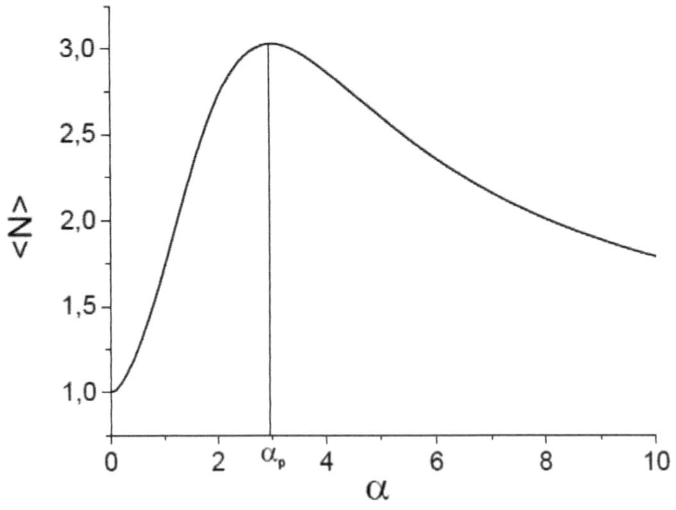

Figure 5 Time required for RA diffusion ionization as a function of Sommerfeld parameter. <N> is the number of cycles of RE motion along the trajectory required for a RE to go into a continuum.

Figure 5 from [15,16,17] shows the dependence of the diffuse ionization time required for RE to achieve ionization limit in the conditions of global chaos for different values of Sommerfeld parameter α, which characterizes the difference value between the spectrum of a hydrogen-like atom and a hydrogen atom ($\alpha \equiv 0$). Maximum dependence at $\alpha = 2{,}81$ corresponds to the case of two-photon Stark resonance for s-p series at a fixed value of the initial effective quantum number $n_{ef} = 30$ and the microwave electric field intensity that is several times higher than its threshold value F_c ($F_c \approx n^{5}\omega^{-1/3}$), where ω has the value of the order of the RE rotation speed in the Keplerian orbit, $\omega \approx n^{-5}$, i.e. it is assumed that the effect of a field on a RE is of resonant nature.

It can be seen that the maximum dependence, which corresponds to the maximum suppressing of the stochastic diffusion effect, corresponds to the value $\alpha = 2{,}81$, i.e. to the case of Föster resonance for s-p series of alkali atoms. The features of radiation kinetics caused by Föster resonance in the system of RA terms are directly related to the probabilities of spontaneous transitions between them, and therefore to the lifetimes of highly excited atoms. Figure 6 shows the total probability of radiative transitions in RA for a fixed n = 30. Apparently, in the vicinity of the two-photon Stark resonance ($\alpha = 2{,}81$) the probability of radiative transitions decreases by more than an order of magnitude, and hence the lifetime of RA increases.

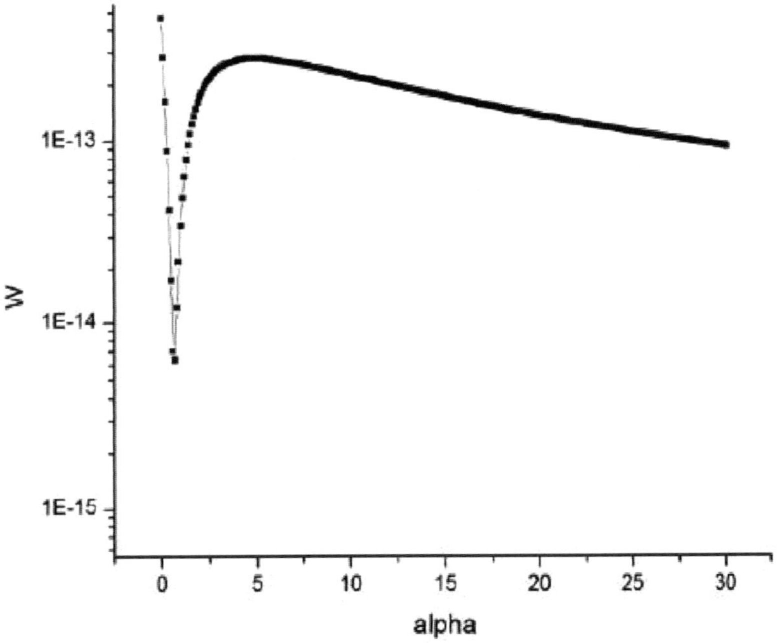

Figure 6 The total probability of spontaneous transitions from the fixed state nS (n=30) depending on the Sommerfeld parameter value. Position of the minimum on the curve corresponds to the manifestation of the Förster resonance effect (atomic units).

5. Conclusion

The study of the emission spectra of celestial bodies forms the basis of today's astrophysics. For the typical parameters of the ionized atmospheres of planets and cooling stars (dwarfs) the manifestations of dynamic chaos can be controlled by external electric and magnetic fields. Thus, astrophysical literature has not yet explained the cause of an anomaly in the infrared spectrum of "white dwarfs" for the RA values of n of the order of 10. It should be noted that this range of n values corresponds to the maximum values of chemi-ionization processes in thermal and subthermal atomic collisions. In terrestrial conditions, RA chaotic diffusion processes in a single collision should influence the stability effectiveness of operation of computer logistics devices.

The material given in the report suggests that the processes of collisional dynamics of Rydberg atoms are beginning to be considered within the framework of a dynamic chaos model of the parameters of an excited atom. Contrary to a traditional (deterministic) approach, chaotic diffusion develops during a single atom-atom collision. To understand these processes and their nature, as well as to ascertain their specific mechanisms is crucial for the models of laser-guided physicochemical reactions and parameters of excited atoms ensembles.

Acknowledgments
The authors are thankful to the Ministry of Education, Science and Technological Development of the Republic of Serbia for support of this work within projects 176002 and III4402.

References

[1] Bezuglov N N, Golubkov G V, Klyucharev A N *Ionization of excited atoms in thermal collisions in book The Atmosphere and Ionosphere* 2013 (Springer-Verlage)
[2] Devdariani A Z , Klyucharev A N, Penkin N P, Sebyakin Yu N 1988 *Opt. Spectrosc.* **93** 661
[3] Klyucharev A N, Bezuglov N N, Mihajlov A A, Ignatovic Lj M 2010 *Journal of Physics Conference Series* **257** 012027
[4] Tanner Paul J, Han Jianing, Shuman E S, Gallagher T F 2008 *Phys. Rev. Lett.* 043002
[5] Park Hyunwook, Shuman E S, Gallagher T F 2011 *Phys. Rev. A* **84** 052708
[6] Poincaré H. 1892 *Les Méthodes Nouvelles de la Mécanique Céleste. Tome 1.* (Paris: Gauthier-Villars)
[7] Gutswiller Martinc 1990 *Chaos in classical and Quantum Mechanics.* (New York: Springer-Verlage)
[8] Manykin M I, Zelener B B, Zelener B V 2010 *Pisma Zh. Eksp. Teor. Fiz.* **92** 696
[9] Aman C C, Petterson J B C, Holmid L 1990 *Chem. Phys.* **147** 189
[10] Fermi E. 1934 *Nuovo Cimento.* **11** 157
[11] Smirnov V M, Mihajlov A A 1971 *Opt. Spectrosc.* **30** 525
[12] Chirikov B V 1979 *Physics Reports* **52** 263.
[13] Efimov D K, Bezuglov N N, Klyucharev A N et al. 2014 *Opt. Spectrosc.* **111** 8
[14] Beterov I M, Riabtsev I I *AIP Conference Proceeding* 1995 Ser. **329** 472
[15] Zakharov M Yu, Bezuglov N N, Klyucharev A N et al. 2011 *Khim. Fiz.* **30** 2
[16] Mihajlov A A, Sreckovic V A, Ignjatovic Lj M, Klyucharev A N *Journal of Cluster Science* 2012 **23** 47
[17] Bezuglov N N, Klyucharev A N, Mihajlov A A, Sreckovic V A 2014 *Advances in Space Research* **54**, 014 1159

Helium-rich white dwarf atmospheres: the non-symmetric ion-atom absorption processes

V A Srećković[1], A A Mihajlov[1], Lj M Ignjatović[1], M S Dimitrijević[2,3,4]

[1] University of Belgrade,Institute of Physics, P. O. Box 57, 11001 Belgrade, Serbia
[2]Astronomical Observatory, Volgina 7, 11160 Belgrade 74, Serbia
[3]IHIS-Technoexperts, Bežanijska 23, 11080 Zemun, Serbia
[4]Observatoire de Paris, 92195 Meudon Cedex, France

Abstract.
Here, the processes of photo-association and absorption charge-exchange in He+H^+ collisions together with the process of molecular ion HeH^+ photo-dissociation are considered as factors of influence on the opacity of the atmospheres of helium-rich white dwarfs in the far UV region. It is shown that they should be taken into account even in the cases of the atmospheres of white dwarfs with H:He $=10^{-5}$. Also, it is established that in the cases of white dwarfs with H:He $\gtrsim 10^{-4}$, especially when H:He $\approx 10^{-3}$, these processes have to be included *ab initio* in the corresponding models of their atmospheres.

1. Introduction

This paper is a continuation of our work about some atomic collision processes in white dwarfs atmospheres [1] and transport properties in it [2]. Recently it has been shown in [3, 4, 5, 6], that in order to consider the contribution of the absorbtion processes connected with binary ion-atom systems to the opacity of the solar photosphere it is not enough to take into account only the processes of absorbtion charge exchange in (H + H^+)-collisions and the molecular ion H_2^+ photo-dissociation. It has been established that in the very important far UV and EUV spectral regions they have to be considered together with the processes of the absorbtion charge exchange and photo-association in non-symmetric (H + X^+)-collisions and molecular ion HX^+ photo-dissociation, where X is one of metal atoms. Namely, it has been proved that only in such case the total efficiency of the ion-atom absorbtion processes in the mentioned spectral regions approaches the efficiency of the relevant concurrent processes in the whole solar photosphere.

Results from these papers suggest that it is useful to consider again the situation of ion-atom absorbtion processes in the atmospheres of helium-rich white dwarfs. Let us remind that in the previous papers (see e.g. [7]), dedicated to some DB white dwarf atmospheres (where "D" stands for "degenerate" or "dwarf" and B-stand for spectral type classification where helium lines dominate), the processes of absorbtion charge exchange in (He + He^+)-collisions:

$$\varepsilon_\lambda + \mathrm{He}^+ + \mathrm{He} \longrightarrow \mathrm{He} + \mathrm{He}^+ \quad (1)$$

and the processes of molecular ion He_2^+ photo-dissociation were studied

$$\varepsilon_\lambda + \mathrm{He}_2^+ \longrightarrow \mathrm{He} + \mathrm{He}^+. \quad (2)$$

Here ε_λ is the energy of a photon with wavelength λ, He = He($1s^2$), He^+ = $He^+(1s)$ and He_2^+ = $He_2^+(1^2\Sigma_u^+)$. By a direct comparison of their efficiencies with the main concurrent process of inverse "bremssthrallung" in (free electron + He)-collisions, i.e.

$$\varepsilon_\lambda + e + \mathrm{He} \longrightarrow e' + \mathrm{He}, \tag{3}$$

the significance of these symmetric ion-atom absorption processes for the atmospheres of the considered DB white dwarfs was established in [7].

It was established that the processes given by Eqs. (1) and (2) significantly influence the opacity of the considered DB white dwarf atmospheres, with an effective temperature $T_{eff} \geq 12000$ K. But, the same comparison demonstrated also that the dominant role in those atmospheres generally still belongs to the concurrent absorbtion process given by Eq. (3), while the processes (1) and (2) can be treated as dominant (with respect to this concurrent process) only in some layers of that atmospheres, and only within the part 50 nm$< \lambda <$ 250 nm of the far UV and EUV region.

Hence it follows that in the case of helium-rich white dwarf atmospheres it would certainly be useful to include into consideration some new ion-atom absorbtion processes, which is principally allowed in accordance with the composition of such atmospheres.

Here we will examine the significance of non-symmetric ion-atom absorption processes with participation of the hydrogen component. We mean the processes of molecular ion HeH^+ photo-dissociation

$$\varepsilon_\lambda + \mathrm{HeH}^+ \longrightarrow \mathrm{He}^+ + \mathrm{H} \tag{4}$$

and the processes of absorption charge exchange and photo-association in the (He + H^+)-collisions, namely

$$\varepsilon_\lambda + \mathrm{He} + \mathrm{H}^+ \longrightarrow \mathrm{He}^+ + \mathrm{H}, \tag{5}$$

$$\varepsilon_\lambda + \mathrm{He} + \mathrm{H}^+ \longrightarrow (\mathrm{HeH}^+)^*, \tag{6}$$

where H = H($1s$), He= He($1s^2$), He^+ = $He^+(1s)$, and HeH^+ and $(HeH^+)^*$ denote the molecular ion in the ground and the first excited electronic states which are adiabatically correlated with the states of the systems He + H^+ and He^+ + H respectively at the infinite internuclear distance. Already in [3] it was noted that these processes, whose significance was practically neglected for the solar photosphere, could be rather important in the case of helium-rich white dwarf atmospheres.

The main aim was to study when the non-symmetric absorbtion processes given by Eqs. (4) - (6) can significantly influence the opacity of helium-rich white dwarf atmospheres in UV and EUV spectral region, and to show that the processes (4) - (6) deserve to be included *ab initio* in the corresponding white dwarf atmosphere models.

Therefore the relevant spectral characteristics of the processes given by Eqs. (4) - (6) are determined here for the atmospheres of different helium-rich white dwarfs with T_{eff} = 12000 K and 14000 K, log g = 8 and 7, and for the values of the ratio H:He from 10^{-5} to 10^{-3}. The necessary expressions for these spectral characteristics are given in Section 2. Then, with their help, in the Section 3 the values are calculated of the parameters characterizing the relative efficiency of the non-symmetric processes given by Eqs. (4) - (6) with respect to the efficiency of all ion-atom processes, as well as with respect to the electron-atom process given by Eq. (3), which comprise the main direct results of this work.

Here, beside the electron-atom processes given by Eq. (3) the process of photoionization of hydrogen atoms, is also considered, namely

$$\varepsilon_\lambda + \mathrm{H} \longrightarrow e + \mathrm{H}^+, \tag{7}$$

where H=H(1*s*). It was treated as a concurrent process, potentially necessary in the region $\lambda < \lambda_H$ where $\lambda_H \approx 91.1$ nm corresponds to the threshold of atom H photoionization. That is why in this work the process given by Eq. (7) was again included into the consideration and carefully examined.

One can see that in this work we take into account only such non-symmetric ion-atom absorption processes where, apart from the dominant helium component of the considered atmospheres, only the hydrogen component participates, although they contain also a lot of metal components [8]. This is due to the fact that the existing atmosphere models do not provide the necessary data (about the relevant metals' abundances) which would be needed for the present calculations.

2. The relevant spectral characteristics

2.1. The symmetric ion-atom, electron-atom and hydrogen photo-ionization processes

We have to introduce the spectral coefficients $\kappa_{sim}(\lambda; \log\tau)$, $\kappa_{e-He}(\lambda; \log\tau)$ and $\kappa_{phi}(\lambda; \log\tau)$ which characterize the efficiencies of the symmetric ion-atom absorption processes given by Eqs. (1) and (2) together, the electron-atom processes Eq. (3) and hydrogen photoionization process Eq. (7) respectively. As in [7] we can take these coefficients in the known form:

$$\kappa_{sim}(\lambda; \log\tau) = K_{sim}(\lambda, T) \cdot N_{He} \cdot N_{He^+}, \tag{8}$$

$$\kappa_{e-He}(\lambda; \log\tau) = K_{e-He}(\lambda, T) \cdot N_{He} \cdot N_e, \tag{9}$$

$$\kappa_{phi}(\lambda; \log\tau) = \sigma_{phi}(\lambda; H) \cdot N_H, \tag{10}$$

where N_{He^+}, N_e and N_H are the local densities of ions He^+, free electrons and atoms H, $K_{sim}(\lambda, T)$ and $K_{e-He}(\lambda, T)$ - adequately defined spectral rate coefficients, and $\sigma_{phi}(\lambda; H)$ - spectral cross-section for atom H photoionization. The absorption coefficient $K_{sim}(\lambda, T)$ is determined in the way which is described in detail in [7], and $K_{e-He}(\lambda, T)$ - by means of the data from [9], and $\sigma_{phi}(\lambda; H)$ is taken from [10].

Here we take into account the fact that all the values of the rate coefficients $K_{sim}(\lambda, T)$ and $K_{e-He}(\lambda, T)$ which are needed for our calculations have already been determined in [7], and that the photo-ionization cross-section $\sigma_{phi}(\lambda; H)$ is given by a known analytical expression. Therefore in the further text we will simply treat these characteristics and, consequently, the spectral absorption coefficients $\kappa_{sim}(\lambda, \log\tau)$, $\kappa_{e-He}(\lambda, \log\tau)$ and $\kappa_{phi}(\lambda; \log\tau)$ as known quantities.

2.2. The non-symmetric ion-atom processes

As the relevant characteristics of the processes given by Eqs. (4), (5) and (6) we will use the corresponding spectral absorption coefficients. They are defined as functions of $\log\tau$, where τ is Rosseland optical depth of the part of the examined atmosphere above the considered layer for the wavelength λ. They are denoted here as $\kappa^{(bf)}_{nsim}(\lambda; \log\tau)$, $\kappa^{(ff)}_{nsim}(\lambda; \log\tau)$ and $\kappa^{(fb)}_{nsim}(\lambda; \log\tau)$, in accordance with the fact that the mentioned processes can be treated as bound-free, free-free and free-bound respectively. These coefficients are determined here within the corresponding atmosphere models, by means of the local temperature and the densities of He atoms and H^+ ions, and used in a similar form, namely

$$\kappa^{(bf,ff,fb)}_{nsim}(\lambda; \log\tau) = K^{(bf,ff,fb)}_{nsim}(\lambda; T) N_{He} N_{H^+}, \tag{11}$$

where $T \equiv T(\log\tau)$, $N_{He} \equiv N_{He}(\log\tau)$ and $N_{H^+} \equiv N_{H^+}(\log\tau)$. It is understood that the photo-dissociation rate coefficient $K^{(bf)}_{nsim}(\lambda; T)$ is given by the known relations

$$K^{(bf)}_{nsim;X}(\lambda; T) = \sigma^{(phd)}_{HeH^+}(\lambda, T) \cdot \chi^{-1}(T; HeH^+), \tag{12}$$

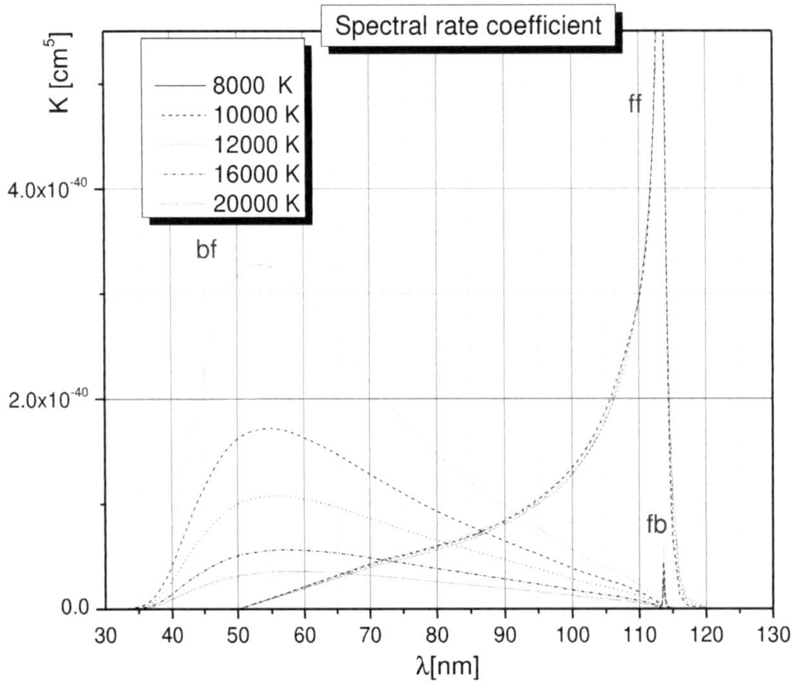

Figure 1. The plots of the bound-free (bf), free-free (ff) and free-bound (fb) spectral rate coefficients $K^{(bf,ff,fb)}_{HeH^+}(\lambda;T)$ for the molecular ion HeH$^+$.

$$\chi(T;HeH^+) = \left[\frac{N_{He}\cdot N_{H^+}}{N_{HeH^+}}\right], \tag{13}$$

where $\sigma^{(phd)}_{HeH^+}(\lambda,T)$ is the mean thermal cross-section for the molecular ion HeH$^+$ photo-dissociation, N_{HeH^+} denotes the local density of these molecular ions, and $\chi(T;HeH^+)$ is determined under condition of local thermodynamical equilibrium (LTE) with given T, N_{He} and N_{H^+}. Let us note that these expressions contain no correction factors which take into account the influence of stimulated emission, since within the actual range of ε_λ/kT ratio for the considered cases the changes due to these factors will be of the order of magnitude of 10^{-3}%.

The total efficiency of the non-symmetric processes given by Eqs. (4), (5) and (6) is characterized here by the spectral absorption coefficient $\kappa_{nsim}(\lambda;\log\tau)$ given by the relations

$$\kappa_{nsim}(\lambda;\log\tau) = K_{nsim}(\lambda;T)N_{He}N_{H^+}, \tag{14}$$

$$K_{nsim}(\lambda;T) = K^{(bf)}_{nsim}(\lambda;T) + K^{(ff)}_{nsim}(\lambda;T) + K^{(fb)}_{nsim}(\lambda;T), \tag{15}$$

where the rate coefficients $K^{(bf,ff,fb)}_{nsim}(\lambda,T)$ are determined by means of the necessary characteristics of the considered molecular ion, in a way which was described in detail in [3, 5, 4].

The adiabatic potential curves of the molecular ion in the ground state (HeH$^+$) and the first excited electronic state ((HeH$^+$)*), as well as the corresponding dipole matrix element are determined by fitting the corresponding data from [11] and [12] (see [3]).

The behavior of the rate coefficients $K^{(bf,ff,fb)}_{nsim}(\lambda,T)$ is illustrated here by Fig. 1.

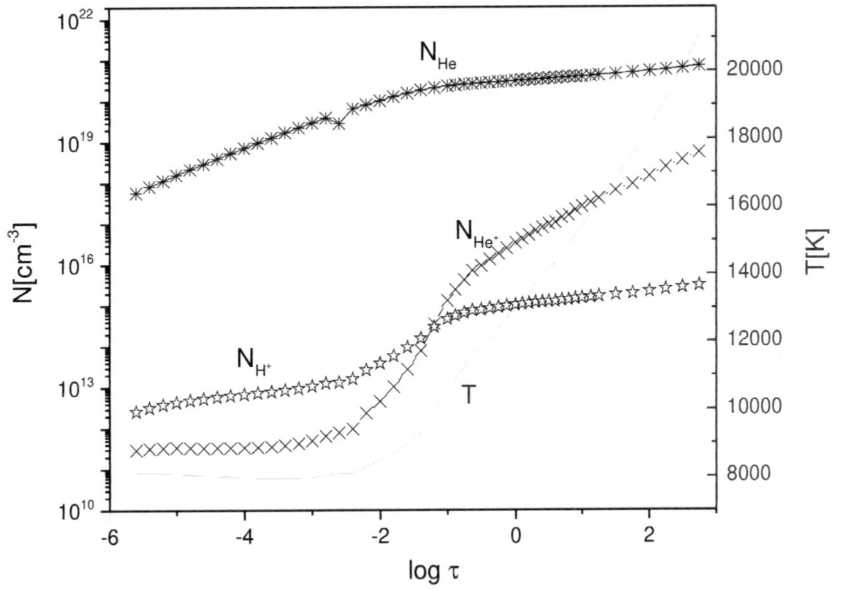

Figure 2. The local densities N_{He}, N_{He^+} and N_{H^+} and the temperature T as functions of $\log\tau$, where τ is Rosseland optical depth, according to the model of the DB white dwarf atmosphere from [13] for: T_{eff}=12000 K, $\log g = 8$ and H:He=10^{-5}.

3. Results and Discussion

3.1. DB white dwarfs

As it is well known, the above defined spectral absorption coefficients depend on the wavelength, local temperature, and local particle densities, based on the corresponding models of helium-rich white dwarf atmospheres characterized by certain values of T_{eff}, $\log g$ and the ratio of the hydrogen and helium species (H:He).

Here we will start from DB white dwarf atmospheres with T_{eff} = 12000 K and 14000 K, $\log g = 8$ and 7 and H:He = 10^{-5}. For the calculations we will use the equilibrium models which are presented in [13]. Although newer atmosphere models for helium-rich white dwarfs now exist (see e.g. the review article of [14]), only the models from [13] contain in a tabular form all the relevant data which are needed for our calculations.

The behavior of the densities of free electrons and ions He^+ and H^+ in the atmosphere of a DB white dwarf with T_{eff} = 12000 K and log g = 8 is illustrated by Fig. 2, which shows that the processes given by Eqs. (4) - (6) could be of interest already in the case H:He = 10^{-5}. Namely, this figure suggests that for $T_{eff} \lesssim$ 14000 K, the ion H^+ density is even larger than that of He^+ in significant parts of DB white dwarfs' atmospheres ($\log\tau < -1$).

In accordance with the aim of this work we have to estimate first the relative efficiency of the non-symmetric processes given by Eqs. (4) - (6) with respect to the total efficiency of all the above mentioned ion-atom absorption processes, which is characterized by the spectral

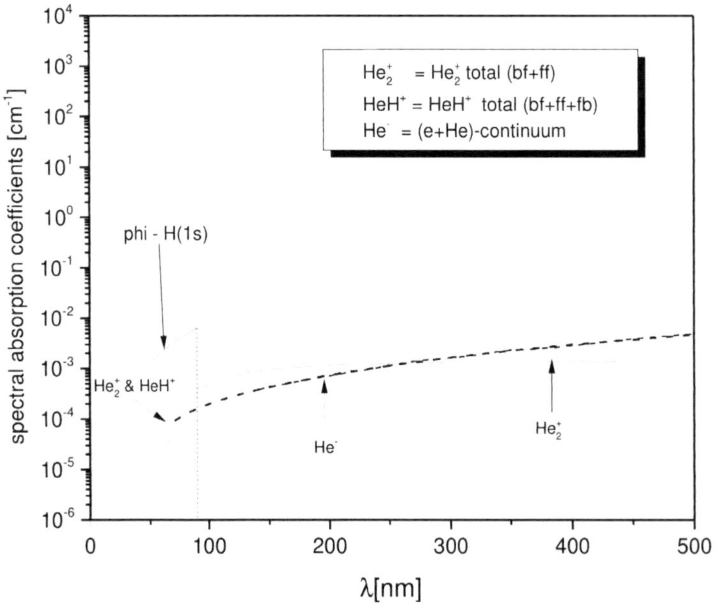

Figure 3. The plots of all considered absorbtion processes for $\log\tau = 0$ in the case of a DB white dwarf with T_{eff}=12000 K, log $g = 8$ and H:He=10^{-5}, from [19].

absorption coefficient

$$\kappa_{ia}(\lambda; \log\tau) = \kappa_{nsim}(\lambda; \log\tau) + \kappa_{sim}(\lambda; \log\tau). \tag{16}$$

The results shows that the non-symmetric processes (4) - (6) are dominant within a significant part of the considered atmosphere ($-5.6 \leq \log\tau \lesssim -0.75$), which corresponds to the part in Fig. 2 where $N_{H^+} > N_{He^+}$.

In order to establish how the inclusion of the non-symmetric processes (4) - (6) into the consideration influences the relative efficiency of the ion-atom absorption processes with respect to the efficiency of the concurrent electron-atom process (3), we calculated their ratio.

The inclusion of the ion-atom non-symmetric absorption processes causes a very significant increase in the relative efficiency of the ion-atom absorption processes in just that region $\log\tau < 0.75$, i.e. where the symmetric processes can be practically neglected.

Then, we established the fact that the relative efficiency of the ion-atom absorption processes with respect to the efficiency of the concurrent is also similar in the cases of DB white dwarf atmospheres with the same value of H:He, but with $T_{eff} = 14000$ K and $\log g = 8$, and $T_{eff} = 12000$ K and $\log g = 7$. Based on the above mentioned, it can be concluded that the non-symmetric processes (4) - (6) have a visible significance for the atmospheres of the considered DB white dwarfs with H:He= 10^{-5} and should be included in their models.

It is necessary to draw attention to the fact that this conclusion refers to a spectral region $\lambda > \lambda_H$ where the hydrogen photo-ionization process (7) is impossible. In order to estimate the partial efficiencies of the mentioned processes in the case of the considered DB white dwarf

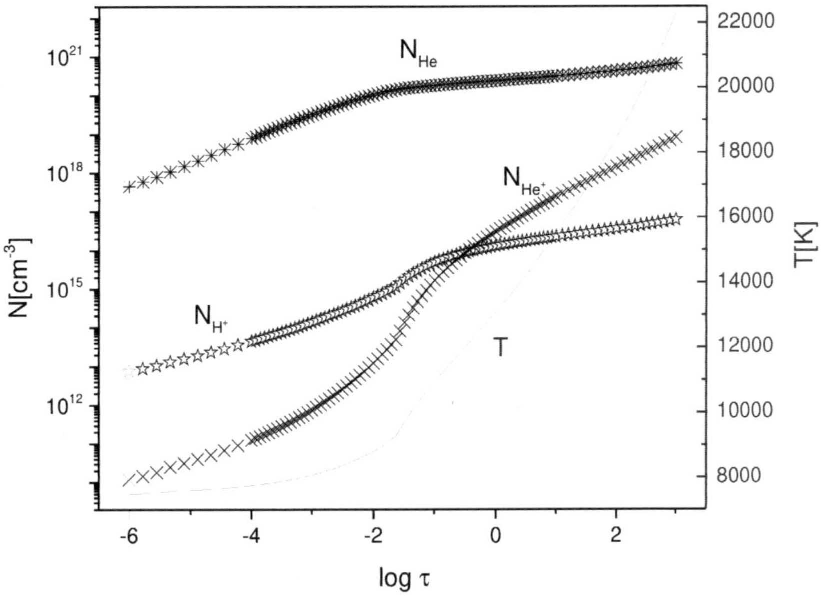

Figure 4. The local densities N_{He}, N_{He^+} and N_{H^+}, and the temperature T as functions of $\log\tau$, where τ is Rosseland optical depth, according to the developed model of helium-rich white dwarf atmosphere for: T_{eff}=12000 K, log $g = 8$ and H:He=10^{-4}.

in the whole region $\lambda > \lambda_{He}$ the corresponding plots of all discussed absorption processes for $\log\tau = 0$ are presented in Fig. 3. From figure one can see that in the region $\lambda_{He} < \lambda < \lambda_H$ the process (7) alone gives the dominant contribution to the opacity of the considered atmosphere. Here it was established that in the considered case this dominance exists for any $\log\tau < 0$.

However, the main results of the research of DB white dwarf atmospheres is the establishment of the fact that the inclusion of the non-symmetric processes causes an increase of the total efficiency of ion-atom absorption processes in the region $0.75 < \log\tau < 2$, where the symmetric processes (1) and (2) are dominant, which is not negligible but rather reaches several percent.

3.2. Other cases of the helium rich white dwarfs.

The mentioned result is important for our further research since it leads to an expectation that the significance of the considered non-symmetric ion-atom absorption processes could be much greater in the cases of the atmospheres of some other helium-rich white dwarfs. We mean the atmospheres with the same or similar T_{eff} and $\log g$, but with the values of the ratio H:He which are larger by one or even two orders of magnitude.

In this context let us note that in [15] some DC white dwarfs with $T_{eff} \approx 12500$ K, $\log g = 8$ and H:He $= 2 \cdot 10^{-4}$ are described. Then, in [16] some weakly magnetic DZ white dwarfs with $\log g = 8$, $T_{eff} \approx 7000$ K and H:He $\approx 10^{-3}$ are discussed. We will remind also that in [17] some DZ white dwarfs with $\log g \approx 8$, $T_{eff} > 12000$ K and H:He $= 10^{-4}$ and 10^{-3} are mentioned, as well as a number of other DZ white dwarfs with $\log g \approx 8$, the values of T_{eff} from about 6500 K to about 10000 K, and the values of H:He from about $10^{-3.2}$ to about $10^{-4.4}$. From the above

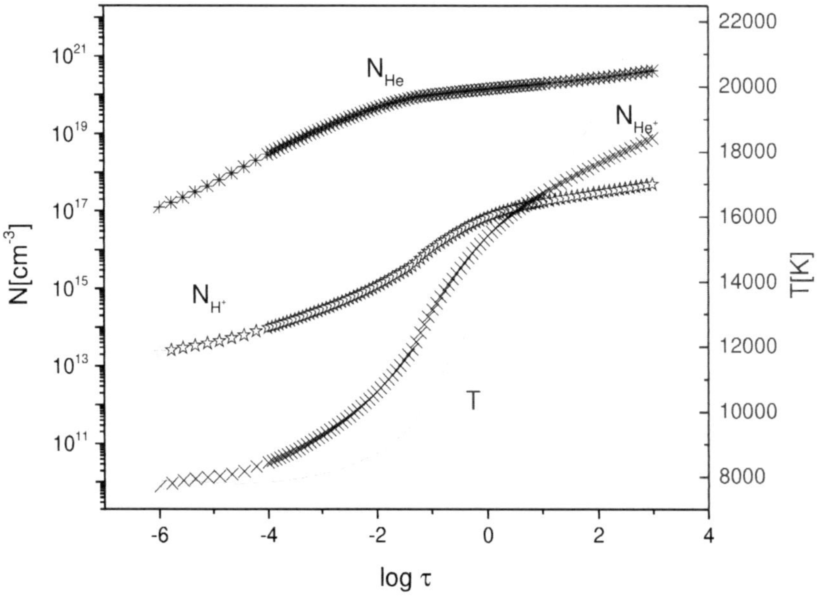

Figure 5. Same as in Fig. 4, but for H:He=10^{-3}.

mentioned result it follows that the contribution of the non-symmetric processes given by Eqs. (4) - (6) to the opacity of such atmospheres should be very significant. Namely, although the ion H^+ density cannot increase proportionally to the ratio H:He, an increase of this ratio of 10 or 100 times, has to cause an increase of N_{H^+} of at least several times. So, the mentioned increase of several percent in region $-1 < \log \tau \leq 2$ in the case of DB white dwarf atmospheres has to become an increase of several tens of percent in the cases of the helium-rich white dwarfs with $10^{-4} \lesssim$ H:He $\lesssim 10^{-3}$.

In order to check our expectations we performed calculations simulating the behavior of T, N_{He}, N_{H^+} and other particle densities, in helium-rich white dwarf atmospheres with $T_{eff} = 12000$ K, $\log g = 8$ and H:He $> 10^{-5}$. All the needed calculations have been performed on the basis of the models taken from [18]. In Figs. 4 and 5 the corresponding densities N_{H^+}, N_{He^+} and N_e are shown as functions of $\log \tau$ for H:He $= 10^{-4}$ and 10^{-3} respectively. One can see, by comparing these figures and Fig. 2, that the considered increase of the ratio H:He indeed causes a very significant increase of N_{H^+}.

From these calculations one can see that for H:He= $\gtrsim 10^{-4}$ the inclusion of the non-symmetric processes given by Eqs. (4) - (6) causes the ion-atom absorption processes to become absolutely dominant with respect to the electron-atom process (3) in the greatest part of this region, namely for 51 nm $\leq \lambda \lesssim 110$ nm, while for $\lambda > 110$ nm the efficiency of ion-atom processes stays close to the efficiency of the process given by Eq. (3).

In order to obtain the complete picture of the discussed absorption processes in the helium-rich white dwarf atmospheres in the cases H:He= 10^{-4} and 10^{-3} it is necessary again to include into the consideration the hydrogen photo-ionization process given by Eq. (7). The significance of partial absorption processes in such atmospheres within the whole region $\lambda > \lambda_{He}$ is illustrated

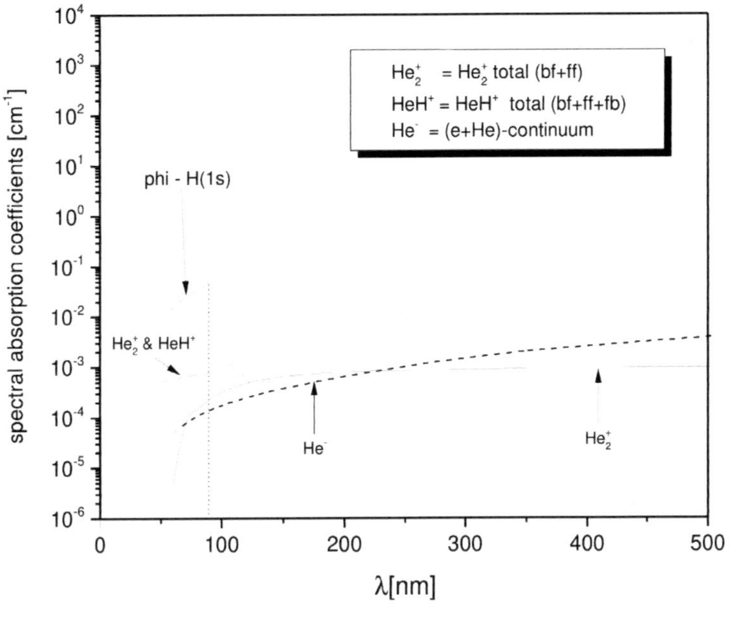

Figure 6. The plots of all considered absorbtion processes for $\log \tau = 0$ in the case of a helium-rich white dwarf with T_{eff}=12000 K, $\log g = 8$ and H:He=10^{-4}, from [19].

in Fig's 6 and 7 where the corresponding plots of these processes are presented for $\log \tau = 0$. One can see that, as in the case of DB white dwarf, the process (7) in the region $\lambda_{He} < \lambda < \lambda_H$ gives the dominant contribution to the opacity of the considered atmospheres. Then, it has been established that in these cases this dominance also holds for any $\log \tau < 0$.

In accordance with our considerations, it is necessary to remind that the considered ion-atom absorption processes (symmetric and non-symmetric) can naturally be significant in the helium-rich white dwarf atmospheres with $T_{eff} <$ 20000K, since at the higher temperatures electron-ion absorption processes completely dominate with respect to the considered ion-atom and electron-atom processes. Therefore, it is necessary here to take into account the papers [20] and [21], where the data about numerous helium-rich white dwarfs are presented. Namely, from these one can see that the values H:He which correspond to the helium-rich white dwarfs with $T_{eff} <$ 20000K are mainly situated between 10^{-5} and 10^{-4}, while the values H:He $> 10^{-4}$, especially H:He close to 10^{-3}, have to be treated as certain extremes. But our results are supported by the following facts: from the results obtained it follows that the non-symmetric ion-atom processes cannot be neglected already for the case of atmospheres of DB white dwarfs with H:He=10^{-5}, and from these results it follows that the effect of inclusion of the non-symmetric processes in the consideration fully manifests already for H:He $= 10^{-4}$, i.e. far from the extremum (H:He=10^{-3}, and remains practically the same further on. This makes important not just a neighborhood of the extremum value H:He=10^{-3}, but rather the whole region of H:He$> 10^{-5}$.

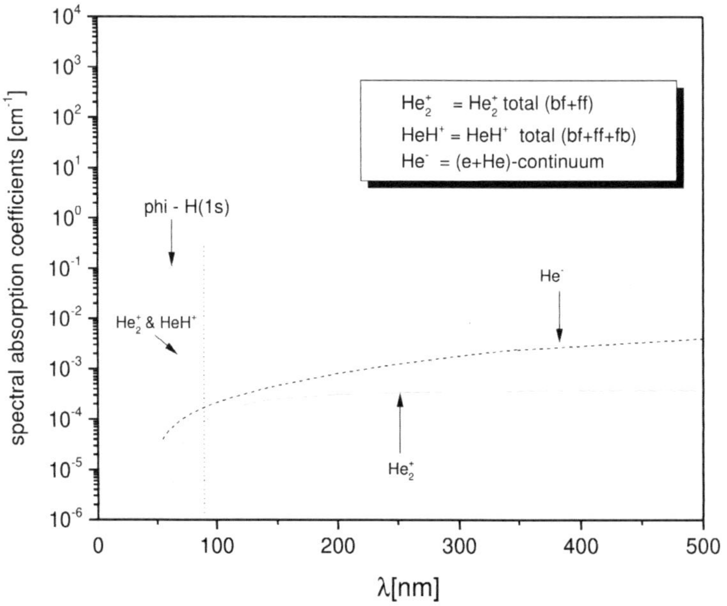

Figure 7. Same as in Fig. 6, but for H:He=10^{-3}.

4. Conclusions

From the presented material it follows that the investigated non-symmetric ion-atom absorption processes have to be treated as one of the important channels of influence on the opacity of the atmospheres of helium-rich white dwarfs in the far UV and EUV region. So, it has been shown that even in the case of DB white dwarfs with H:He = 10^{-5} such processes should be included in the models of their atmospheres. However, the main result of this research is the establishment of the fact that in the cases of helium-rich white dwarfs with H:He > 10^{-5}, and particularly with H:He =10^{-4}, these non-symmetric ion-atom absorption processes have to be included *ab initio* in the models of the corresponding atmospheres, since in the greater part of the considered far UV and EUV region they could be completely dominant with respect to the referent electron-atom and symmetric ion-atom absorbtion processes.

Besides, it has been shown that in all the considered cases (H:He = 10^{-5}, 10^{-4} and 10^{-3}) the hydrogen photo-ionization processes Eq. (7) yield a dominant contribution to the opacity of the corresponding atmospheres in the region $\lambda_{He} < \lambda < \lambda_H$.

As a task for further investigations in this area is future study of the ion-atom non-symmetric absorption processes with participation of some metal components of the considered atmospheres.

Acknowledgments

The authors are thankful to the Ministry of Education, Science and Technological Development of the Republic of Serbia for the support of this work within the projects 176002 and III4402.

References

[1] Srećković V A, Mihajlov A A, Ignjatović L M, & Dimitrijević M S 2013, A&A, 552, A33
[2] Srećković V A, Ignjatović Lj M, Mihajlov A A, & Dimitrijević M S 2010, MNRAS, 406, 590
[3] Mihajlov A A, Ignjatović Lj M, Srećković V A, Dimitrijević M S, & Metropoulos A 2013, MNRAS, 431, 589
[4] Ignjatović L M, Mihajlov A A, Srećković V A, & Dimitrijević M S 2014, MNRAS, 441, 1504
[5] Srećković V A, Mihajlov A A, Ignjatović L M, & Dimitrijević M S 2014, Adv. Space Res., 54, 1264
[6] Mihajlov, A A, Srećković V A, Ignjatović Lj M, Dimitrijević M S,& Metropoulos A 2012, Journal of Physics Conference Series, 397, 012054
[7] Ignjatović Lj M, Mihajlov A A, Sakan N M, Dimitrijević M S, & Metropoulos A 2009, MNRAS, 396, 2201
[8] Bues I 1970, A&A, 7, 91
[9] Somerville W B 1965, ApJ, 141, 811
[10] Bethe H A, & Salpeter E E 1957, Quantum Mechanics of One- and Two-Electron Atoms, New York: Academic Press, 1957,
[11] Green T A, Michels H H, Browne J C, & Madsen M M 1974b, J. Chem. Phys., 61, 5186
[12] Green T A, Browne J C, Michels H H, & Madsen M M 1974a, J. Chem. Phys., 61, 5198
[13] Koester D 1980, A&AS, 39, 401
[14] Koester D 2010, Mem. Soc. Astron. Italiana, 81, 921
[15] Wegner G, & Koester D 1985, ApJ, 288, 746
[16] Dufour P, Bergeron P, Schmidt G D, Liebert J, Harris H C, Knapp G R, Anderson S F, & Schneider D P 2006, ApJ, 651, 1112
[17] Dufour P, Bergeron P, Liebert J, Harris H C, Knapp G R, Anderson S F, Hall P B, Strauss M A, Collinge M J, & Edwards M C 2007, ApJ, 663, 1291
[18] Koester D. 2013, private comunication
[19] Ignjatović L M, Mihajlov A A, Srećković V A, & Dimitrijević M S 2014, MNRAS, 439, 2342
[20] Bergeron P, Wesemael F, Dufour P, et al. 2011, ApJ, 737, 28
[21] Voss B, Koester D, Napiwotzki R, Christlieb N, & Reimers D 2007, A&A, 470, 1079

AUTHOR INDEX